AF770202

Earthquake Geology

Earthquake Geology

Editor:

Nelson Bolívar, Ph.D

ArclerPress

www.arclerpress.com

Earthquake Geology

Editor: Nelson Bolívar, Ph.D

© 2017 Arcler Press LLC
708 3rd Avenue, 6th Floor
New York
NY 10017
United States of America
www.arclerpress.com

ISBN: 978-1-68094-484-6

Library of Congress Control Number: 2016951347

This book contains information obtained from highly regarded resources. Reprinted material sources are indicated. Copyright for individual articles remains with the authors as indicated. A Wide variety of references are listed. Reasonable efforts have been made to publish reliable data and views articulated in the chapters are those of the individual contributors, and not necessarily those of the editors or publishers. Editors or publishers are not responsible for the accuracy of the information in the published chapters or consequences of their use. The publisher believes no responsibility for any damage or grievance to the persons or property arising out of the use of any materials, instructions, methods or thoughts in the book. The editors and the publisher have attempted to trace the copyright holders of all material reproduced in this publication and apologize to copyright holders if permission has not been obtained. If any copyright holder has not been acknowledged, please write to us so we may rectify.

Notice: Registered trademark of products or corporate names are used only for explanation and identification without intent of infringement.

Arcler Press LLC publishes wide variety of books and eBooks. For more information about Arcler Press and its products, visit our website at www.arclerpress.com

About the Editor

Nelson Bolívar, Ph.D

Nelson Bolivar has a PhD. in physics from the University of Lorraine in France finished in 2014. His expertise is in quantum systems and condensed matter. His interest includes spintronic devices and correspondences between general relativity and condensed matter. He is currently an associate professor at the Central University of Venezuela

Contents

List of Contributors

Dimitrios Konstantinidis
Department of Civil and Environmental Engineering, University of California, Berkeley, CA, USA

Nicos Makris
Department of Civil Engineering, University of Patras, 26500 Patras, Greece

Giulio Riga
Geologist, Independent Researcher, Lamezia Terme, Italy

Paolo Balocchi
Geologist, Independent Researcher, Modena, Italy

Paul Sava
Center for Wave Phenomena, Colorado School of Mines

S. Nielsen
Durham University, Earth Sciences, Durham, UK

E. Spagnuolo
EPFL, Lausanne, Switzerland

M. Violay
Department of Geology, University of Otago, Otago, New Zealand

S. Smith
School of Earth, Atmospheric and Environmental Sciences, Manchester University, Manchester, UK

G. Di Toro
Dipartimento di Geoscienze Address, Universita degli Studi ` di Padova Division, Padova, Italy

A. Bistacchi
Department of Earth and Environmental Sciences, Universita degli Studi di Milano Bicocca, Milan, Italy

Alexander Ivanchin
Institute of Monitoring of Climatic and Ecological Systems (IMCES SB RAS), Tomsk, Russia

Liangquan Ge,
College of Applied Nuclear Technology and Automation Engineering, Chengdu University of Technology, Chengdu, China

Jiankun Zhao,
College of Applied Nuclear Technology and Automation Engineering, Chengdu University of Technology, Chengdu, China

Yaoyao Luo
College of Applied Nuclear Technology and Automation Engineering, Chengdu University of Technology, Chengdu, China

Giulio Riga
Geologist, Independent Researcher, Lamezia Terme, Italy

Paolo Balocchi
Geologist, Independent Researcher, Modena, Italy

Youjin Su
Seismological Bureau of Yunnan Province, Kunming, China

Hui Hu
Yunnan Observatory, Chinese Academy of Sciences, Kunming, China

Zhe Yin
Department of Information Management, Beijing University, Beijing, China
Mathematics Department, Yanbian University, Yanji, China
Baris Baykant Alagoz
Department of Electric-Electronics Engineering, Inonu University, Malatya, Turkey

Serkan Alagoz
Department of Physics, Inonu University, Malatya, Turkey

Ryoya Ikuta
Graduate School of Science and Technology, Shizuoka University

Yuta Mitsui
Graduate School of Science and Technology, Shizuoka University

Yuri Kurokawa
Graduate School of Science and Technology, Shizuoka University

Masataka Ando
Graduate School of Science and Technology, Shizuoka University

Sheetal P. Karia
Applied Physics Department, S. V. National Institute of Technology, Surat, India

Kamlesh N. Pathak
Applied Physics Department, S. V. National Institute of Technology, Surat, India

Kunvar S. Yadav
Applied Physics Department, S. V. National Institute of Technology, Surat, India

Nikita P. Chaudhary
Applied Physics Department, S. V. National Institute of Technology, Surat, India

Nilesh C. Patel. Ranjan Jana
Applied Physics Department, S. V. National Institute of Technology, Surat, India

Telman Aliev
Institute of Control Systems of the Azerbaijan National Academy of Sciences, Baku, Azerbaijan

Qambar Quluyev
Institute of Control Systems of the Azerbaijan National Academy of Sciences, Baku, Azerbaijan

Farhad Pashayev
Institute of Control Systems of the Azerbaijan National Academy of Sciences, Baku, Azerbaijan

Ulkar Sattarova
Institute of Control Systems of the Azerbaijan National Academy of Sciences, Baku, Azerbaijan

Narmin Rzayeva
Institute of Control Systems of the Azerbaijan National Academy of Sciences, Baku, Azerbaijan

Igor Petrovich Dobrovolsky
Institute of Physics of the Earth, Russian Academy of Sciences, Moscow, Russia

Salam Al-Hashmi
Earthquake Monitoring Center, Sultan Qaboos University, Muscat, Oman

Adrian Rawlins
University of Newcastle upon Tyne, Newcastle upon Tyne, UK

Frank Vernon
Institute of Geophysics and Planetary Physics, University of California, San Diego, USA

Baurov Yury Alexeevich
Closed Joint Stock Company Research Institute of Cosmic Physics, Korolev, Russia

Baurov Alexey Yur'evich
Closed Joint Stock Company Research Institute of Cosmic Physics, Korolev, Russia

Baurov Alexandr Yur'evich (Jr.)
Closed Joint Stock Company Research Institute of Cosmic Physics, Korolev, Russia

Spitalnaya Alexandra Alfredovna
Closed Joint Stock Company Research Institute of Cosmic Physics, Korolev, Russia

Abramyan Ara Arshavirovich
Closed Joint Stock Company Research Institute of Cosmic Physics, Korolev, Russia

Solodovnikov Vladimir Alexandrovich
Closed Joint Stock Company Research Institute of Cosmic Physics, Korolev, Russia

Carlos A. Vargas
Department of Geophysics, National University of Colombia, Bogotá, Colombia;

Emanuel D. Kästle
Department of Geophysics, National University of Colombia, Bogotá, Colombia;

Youjin Su
Seismological Bureau of Yunnan Province, Kunming, China

Hong Fu
Seismological Bureau of Yunnan Province, Kunming, China

Hui Hu
National Astronomical Observatories/Yunnan Observatory, Chinese Academy of Sciences, Kunming, China

Robert Shcherbakov
Department of Earth SciencesUniversity of Western Ontario

Donald L. Turcotte
Department of GeologyUniversity of California

John B. Rundle
Department of PhysicsUniversity of California

Kristy F. Tiampo
Department of Earth SciencesUniversity of Western Ontario

James R. Holliday
Department of PhysicsUniversity of California

Preface

An earthquake is always an unexpected phenomenon. Modern science is not able to predict the time or the place or the earthquake strength. The problem of locating the focus of a starting earthquake has not even been set due to the poor level of understanding the processes preceding its start. At present the main earthquake hypothesis is the "explosive" relaxation of the high elastic stresses accumulated in the lithosphere. Understanding a fault's slip behavior, as well as its length and connectivity, is important for constraining the magnitude range and frequency of earthquakes that a particular fault is likely to produce.

This book, Earthquake Geology, presents contributions from researchers of different countries in the world that point out the study of seismoinduced phenomena associated with recent and historical earthquakes. First chapter aims to estimate the response of freestanding full-scale equipment to 2% in 50 years hazard level motions, and the results are used to generate ready-to-use fragility curves and second chapter emphasizes on earthquake forecast with the seismic sequence hierarchization method. Third chapter encompasses the micro-earthquake monitoring with sparsely sampled data. In fourth chapter, we analyze in detail the features of the experimental weakening curves and provide a general fit which is purely empirical, with the synthesis of a large number of experiments and their result in terms of frictional breakdown energy Gf. Fifth chapter presents how to locate the focus of a starting earthquake and sixth chapter presents research on earthquake radon anomalies. Seventh chapter highlights on seismic sequence structure and earthquakes triggering patterns and eighth chapter emphasizes on predicting earthquakes with microsequences and reversed phase repetitive patterns. Application of commensurability in earthquake prediction is discussed in ninth chapter and tenth chapter gives out the co-planarity and symmetry theory of earthquake occurrence. Eleventh chapter presents a numerical investigation of earthquake shielding with seismic crystals and twelfth chapter presents an evaluation of strain accumulation in global subduction zones from seismicity data. Thirteenth chapter reveals on modification in atmospheric refractivity and GPS based TEC as earthquake precursors, and fourteenth chapter proposes a seismic-acoustic system for monitoring the earthquake origin process. The aim of fifteenth chapter is to calculate hydrodynamical phenomena: Earth's tidal and precursory variations in level of liquid in wells (boreholes) using identical systems of equations and to clarify data on distribution of hydrodynamical precursors on the Earth's surface. The objective of sixteenth chapter is to provide a wavelet transform method to detect P and S-phases in three component seismic data. In seventeenth chapter, basic models and standard mechanisms of earthquakes are briefly considered, results of processing of information on the earthquakes in the context of global spatial anisotropy caused by the existence of the vector Ag, are presented, and an analysis of them is given. Eighteenth chapter concentrates more on the actual relationship between earthquakes and solar activity and treats the effects causing the correlation only in the aspect of geomagnetic field strength variations. Nineteenth chapter presents a study on correlation of tidal forces with global great earthquakes, and an analysis and verification of forecasting the locations of future large earthquakes is given in last chapter. Forecasts of the locations of future major earthquakes play an important role in earthquake preparedness and determining earthquake insurance costs. Many such forecasts have been carried out with examples in this chapter.

Editor
Nelson Bolívar, Ph.D

INTRODUCTION

Seismic waves are produced when some form of energy stored in Earth's crust is suddenly released, usually when masses of rock straining against one another suddenly fracture and "slip." Earthquakes occur most often along geologic faults, narrow zones where rock masses move in relation to one another.

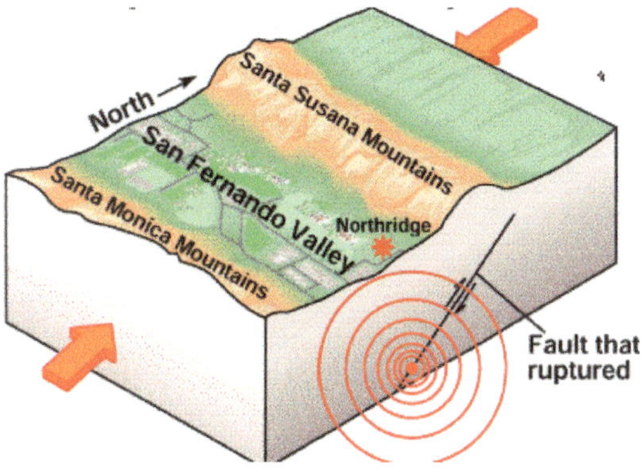

What are Earthquakes?

An earthquake is a tremor in the Earth's crust, caused by movements below its surface. These events can vary widely in intensity, from seismic activity that is barely detectable using sophisticated devices, to devastating temblors that can level cities and trigger tsunamis and sometimes even volcanic activity. The study of these tremors is known as seismology, a word derived from a Greek word meaning "to shake."

The Earth's outer layer, or crust, is composed of two sections: the lithosphere, a Greek word meaning "rocky sphere," and the athenosphere, a thick layer of liquid that rests on top of the upper mantle. The liquid rock of the upper mantle keeps the crust in constant motion, with the edges of continental plates being pulled slowly apart or together as they float on the athenosphere. The movement of these plates is what triggers earthquakes. In addition to plate boundaries, temblors also occur along faults, cracks in the lithosphere caused by the stresses created as tectonic plates move.

There are a number of different types of faults, but most can be divided into three categories: strike slip faults, thrust faults, and normal faults. A strike slip fault occurs in an area where two plates are sliding past each other, while a thrust fault happens when plates are being pushed together. A normal fault is the result of plates being pulled apart. The largest normal faults in the world are along the deep sea ocean ridges of the Pacific and Atlantic, where plates are pulling apart, crashing into the continental plates and causing thrust faults. Earthquakes along each fault have different characteristics that help seismologists to identify them.

The roots of a quake lie in stresses placed on the lithosphere as it drifts on the surface of the Earth. Pressure builds up along a fault line, which finally fails, often deep below the crust of the Earth, in an area called the focus. The corresponding spot on the surface of the planet is called the epicenter, and usually the greatest concentration of damage occurs here. When the fault fails, it triggers seismic waves, very low frequency sound waves that come in several shapes, and which can cause the earth to ripple, heave, buckle, or tear apart. The waves can continue for hours after the earthquake was triggered, and aftershocks, further smaller tremors, can continue for months and possibly even years later.

The intensity of an earthquake is called its magnitude. Various scales were proposed to measure this factor until 1935, when the Richter scale was developed. Under this scale, each order of magnitude is 10 times more intensive than the last. A quake that measure a 2 on the Richter scale is 10 times more intense than a 1, while a 3 is 100 times greater. Most quakes around the world are below a 4.5, the magnitude at which it can start to damage buildings, and every year there is at least one greater than an 8, with the largest modern earthquake ever recorded occurring in Chile in 1960; it measured a 9.5.

Earthquake Geology

Earthquake, any sudden shaking of the ground caused by the passage of seismic waves through Earth's rocks. Seismic waves are produced when some form of energy stored in Earth's crust is suddenly released, usually when masses of rock straining against one another suddenly fracture and "slip." Earthquakes occur most often along geologic faults, narrow zones where rock masses move in relation to one another. The major fault lines of the world are located at the fringes of the huge tectonic plates that make up Earth's crust. Little was understood about earthquakes until the emergence of seismology at the beginning of the 20th century. Seismology, which involves the scientific study of all aspects of earthquakes, has yielded answers to such long-standing questions as why and how earthquakes occur.

Global seismic centres in 1975–99:
earthquakes of magnitude 5.5 and greater

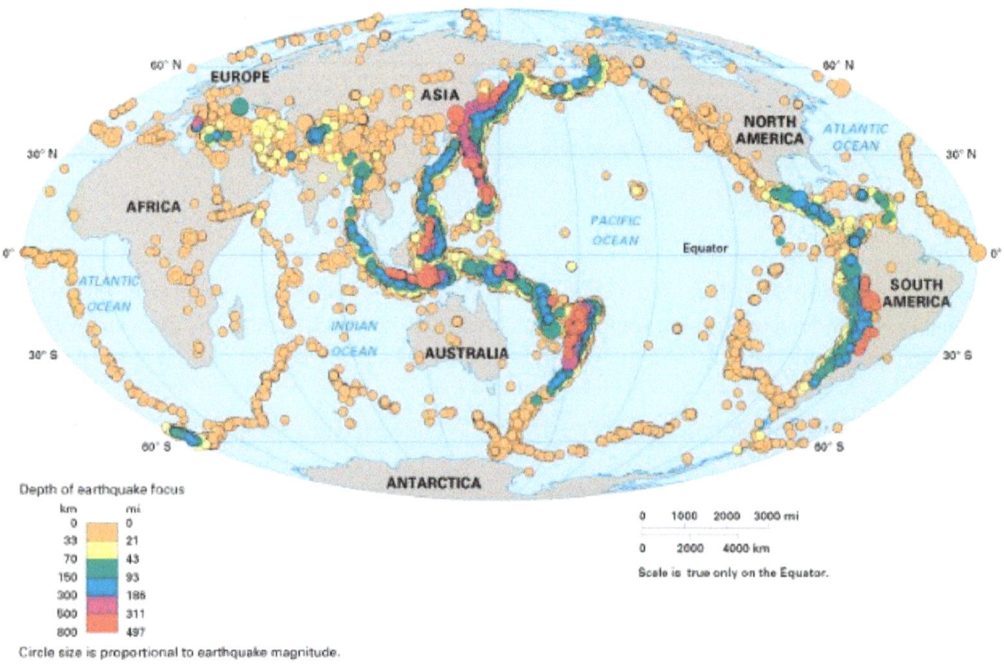

About 50,000 earthquakes large enough to be noticed without the aid of instruments occur annually over the entire Earth. Of these, approximately 100 are of sufficient size to produce substantial damage if their centers are near areas of habitation. Very great earthquakes occur on average about once per year. Over the centuries they have been responsible for millions of deaths and an incalculable amount of damage to property.

The Nature of Earthquakes

Causes of earthquakes

Earth's major earthquakes occur mainly in belts coinciding with the margins of tectonic plates. This has long been apparent from early catalogs of felt earthquakes and is even more readily discernible in modern seismicity maps, which show instrumentally determined epicentres. The most important earthquake belt is the Circum-Pacific Belt, which affects many populated coastal regions around the Pacific Ocean—for example, those of New Zealand, New Guinea, Japan, the Aleutian Islands, Alaska, and the western coasts of North and South America. It is estimated that 80 percent of the energy presently released in earthquakes comes from those whose epicentres are in this belt. The seismic activity is by no means uniform throughout the belt, and there are a number of branches at various points. Because at many places the Circum-Pacific Belt is associated with volcanic activity, it has been popularly dubbed the "Pacific Ring of Fire."

Natural forces

Earthquakes are caused by the sudden release of energy within some limited region of the rocks of the Earth. The energy can be released by elastic strain, gravity, chemical reactions, or even the motion of massive bodies. Of all these the release of elastic strain is the most important cause, because this form of energy is the only kind that can be stored in sufficient quantity in the Earth to produce major disturbances. Earthquakes associated with this type of energy release are called tectonic earthquakes.

Tectonics

Tectonic earthquakes are explained by the so-called elastic rebound theory, formulated by the American geologist Harry Fielding Reid after the San Andreas Fault ruptured in 1906, generating the great San Francisco earthquake. According to the theory, a tectonic earthquake occurs when strains in rock masses have accumulated to a point where the resulting stresses exceed the strength of the rocks, and sudden fracturing results. The fractures propagate rapidly through the rock, usually tending in the same direction and sometimes extending many kilometers along a local zone of weakness. In 1906, for instance, the San Andreas Fault slipped along a plane 430 km (270 miles) long. Along this line the ground was displaced horizontally as much as 6 meters (20 feet).

Earthquakes have different properties depending on the type of fault slip that causes them (as shown in the figure). The usual fault model has a "strike" (that is, the direction from north taken by a horizontal line in the fault plane) and a "dip" (the angle from the horizontal shown by the steepest slope in the fault). The lower wall of an inclined fault is called the footwall. Lying over the footwall is the hanging wall. When rock masses slip past each other parallel to the strike, the movement is known as strike-slip faulting. Movement parallel to the dip is called dip-slip faulting. Strike-slip faults are right lateral or left lateral, depending on whether the block on the opposite side of the fault from an observer has moved to the right or left. In dip-slip faults, if the hanging-wall block moves downward relative to the footwall block, it is called "normal" faulting; the opposite motion, with the hanging wall moving upward relative to the footwall, produces reverse or thrust faulting.

Can an Earthquake Affect the Earth's Rotation?

Earthquakes in 2010 and 2011 -- in Chile and in northeast Japan -- shortened the length of an Earth day fractionally by changing the distribution of the planet's mass, in much the same way that a figure skater speeds up a spin by drawing her arms in closer to her body. Both quakes caused the Earth to rotate more quickly, trimming a total of 3.06 microseconds off our 24-hour day.

The 8.9 magnitude event in Japan and the 8.8 magnitude temblor in Chile were both thrust earthquakes. This type of event creates inward motion, which can shorten days. Other types of earthquakes -- such as a horizontal strike-slip quake in which two plates slide horizontally past one another -- have no effect on Earth's rotation.

Hold on a microsecond:

- A microsecond is a millionth of a second.
- One Earth day is about 86,400 seconds long. In a year, its length varies by about 1,000 microseconds, depending on seasonal variations such as the shift in the jet stream.
- The measurement of a day is accurate only to about 20 millionths of a second. So the effect of an earthquake can be estimated, but not accurately measured.

What Is Earth Pressure?

Earth pressure refers to the natural movement of the earth's soil. The movements will often apply pressure on existing structures. It is sometimes referred to as lateral earth pressure, which is measured by the amount of tension that occurs against basement or retaining walls. There are three types of forces, including at rest, passive or active earth pressure.

There is the common saying that a house creaks due to the fact that it is "settling," which is a direct reference to earth pressure. Natural movements and changes in the soil cause the foundation to move. Most of these changes and movements are slight. Despite the fact that most cases are slight enough to go undetected, there are instances where the amount of pressure is strong enough to result in visible cracks, leaning, crumbling or even complete relocation.

The effects of earth pressure may take time to accumulate before they become visible. Since the earth's soil is constantly evolving and changing at different rates during certain periods of time, many years may pass before any effects are seen at all. Some types of soils or landscapes are more vulnerable to change, meaning that structures built in those areas are more likely to fail or sustain damage sooner.

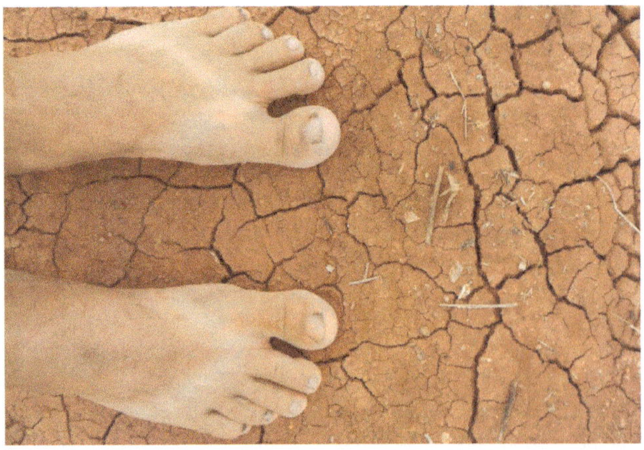

There are a few theories that attempt to explain the type and amount of active earth pressure. One of those is the Rankine Theory. It assumes that the soil and horizontal structures do not experience pressure. Rather there is only pressure in vertical walls. Failures occur within a certain parameter and manifest as a wedge in the structure as a result of a force that is parallel to the wall.

Another theory, which is called the Coulomb Theory, states that there is active pressure between a structure and the earth's soil. The pressure does not just occur within vertical structures, but horizontal ones as well. Pressure forces and friction from the soil occur in other areas that are not parallel to the structure, according to the theory.

Both of these theories involve mathematical calculations that can be used to determine the amount of soil pressure. The calculations are used to predict both active and passive earth pressure. With passive pressure, the wall or structure moves towards the soil. In active cases, structures are pushed away from the soil. In contrast, at rest cases indicate no movement, either due to insignificant amounts of pressure or no changes in the earth's composition.

CHAPTER 1

EXPERIMENTAL AND ANALYTICAL STUDIES ON THE RESPONSE OF 1/4-SCALE MODELS OF FREESTANDING LABORATORY EQUIPMENT SUBJECTED TO STRONG EARTHQUAKE SHAKING

Dimitrios Konstantinidis and Nicos Makris

Keywords: Laboratory equipment, Nonstructural components, Scaling, Rocking, Sliding, Fragility curves

INTRODUCTION

Freestanding laboratory equipment on various floor levels of research laboratories, hospitals and other critical facilities may slide, rock, or even overturn during an earthquake. Rocking response is very sensitive to the geometry and mass distribution of the rocking object and the nature of the ground motion. Minor variations in the input can result in overturning (Yim et al. 1980; Makris and Roussos 2000; Makris and Konstantinidis 2003a). Therefore, rocking is in principle an undesirable response for the equipment since it is often the cause of mechanical damage or total loss in the event of overturning. Yet, even if overturning does not occur, the high acceleration spikes that develop during impact of the rocking equipment are a major concern, since they can result in damage of valuable scientific research material stored in the equipment by directly disturbing it or indirectly by causing failure in the electronic components of the equipment that maintain the controlled environment the material is stored in. Sliding is the most favorable mode of response. Nonetheless, excessive sliding displacements may block a path or doorway that services evacuation or result in impact with walls or neighboring equipment. In practice, excessive sliding is prevented by restraining the equipment. Although this may succeed in reducing sliding displacements, it substantially ampli- fies accelerations (Konstantinidis and Makris 2005b). The problem of equipment sliding has been studied in the past at various scales by Shao and Tung (1999), Lopez Garcia and Soong (2003a,b), and Hutchinson and Chaudhuri (2006). In an earlier paper (Konstantinidis and Makris 2009), the authors presented experimental and analytical studies on the seismic vulnerability of freestanding laboratory equipment located on various floor levels of a research laboratory building located at the University of California, Berkeley, campus—herein referred to as the UC Science Building. The equipment of interest included low-temperature refrigerators, freezers, incubators, and other heavy equipment. The study investigated the response of equipment to moderately strong motions (50 and 10% probability of being exceeded in 50 years) which resulted in Peak Ground Displacements (PGD) or Peak Floor Displacements (PFD) that could be accommodated by the shake table at the Pacific Earthquake Engineering Research (PEER) Center, University of California, Berkeley. Shake table tests showed that there was no incidence of overturning due to excessive uplift. Uplift rotations ranged from very low, for two of the three specimens, to moderate, for the third, never exceeding 50% of the stockiness, α (the angle between a vertical line and the line that passes through the pivoting point and the center of mass of the equipment). For motions in this hazard level, the equipment tested exhibited excessive sliding displacements, reaching up to 60 cm. The results of the tests were used to develop a dimensionless Engineering Demand Parameter (EDP) (a parameter that quantifies the response of the equipment), as a function of the Intensity Measure (IM)

Citation: Dimitrios Konstantinidis, Nicos Makris, Experimental and analytical studies on the response of 1/4-scale models of freestanding laboratory equipment subjected to strong earthquake shaking, DOI 10.1007/s10518-010-9192-8.

(a parameter of the excitation that corresponds to a certain seismic hazard level). Ready-to-use fraglity curves, which give the probability that the EDP will exceed a specific limit c, were generated. The results of the shake table tests to moderate (POE of 50% in 50 years and 10% in 50 years) hazard motions were used by Konstantinidis and Makris (2009) to validate the commercially available dynamic simulation software Working Model 2D (2000). However, experimental validation of Working Model for full-scale equipment to motions with large hazard level (POE of 2% in 50 years) could not be performed because the ground and floor displacements which resulted from these motions could not be accomodated by the shake table at the UC Berkeley PEER Center, which has a horizontal displacement capacity of ±15 cm. To address this challenge, shake table tests on quarter-scale wooden block models of the full-scale equipment prototypes were conducted. The reduction in length by a factor of 4 corresponds to a reduction in time by a factor of $\sqrt{4} = 2$. This can be shown by dimensional analysis for both sliding (Makris and Black 2003, 2004) and rocking motions (Makris and Konstantinidis 2003b), where accelerations are preserved in the model and the protoype, thereby tacitly accepting Froude similitude. The friction coeffient between the wooden block models and the shake table was considerably larger than the full-scale equipment to floor friction coefficient, thus resulting in rocking and overturning response rather than sliding. Simulations in Working Model using large friction coefficients validated the software's capability to capture the overturning behavior. The experimental validation of Working Model for sliding full-scale equipment and rocking wooden-block models together with a numerical validation for pure sliding and for pure rocking blocks provided confidence in its ability to estimate the response of equipment. In this study, Working Model is used to estimate the response of freestanding full-scale equipment to 2% in 50 years hazard level motions, and the results are used to generate ready-to-use fragility curves.

FRICTION AND SHAKE TABLE TESTS OF WOODEN BLOCK MODELS

Three quarter-scale wooden blocks were made to model the three pieces of prototype equipment that were used for shake table tests presented in the earlier work published by the authors (2009). Figure 1 is a photograph of the three wooden block models resting on the shake table. The FORMA incubator prototype is also shown on the shake table, while the other two

Figure. 1 The three 1/4-scale wooden block models together with the FORMA incubator prototype resting on the shake table at the UC Berkeley PEER Center. The other full-scale prototype equipment can be seen in the background

Table 1 Geometric characteristics of the full-scale prototype equipment and the 1/4-scale wooden block models

Equipment	Height [cm]	Width [cm]	α [rad]	p [rad/s]
FORMA: prototype	228.6	62.2	0.27	2.49
FORMA: model	57.2	15.6	0.27	4.98
KELVINATOR: prototype	213.4	66.0	0.30	2.57
KELVINATOR: model	53.4	16.5	0.27	5.14
ASP: prototype	180.3	58.4	0.31	2.79
ASP: model	40.1	14.6	0.27	5.58

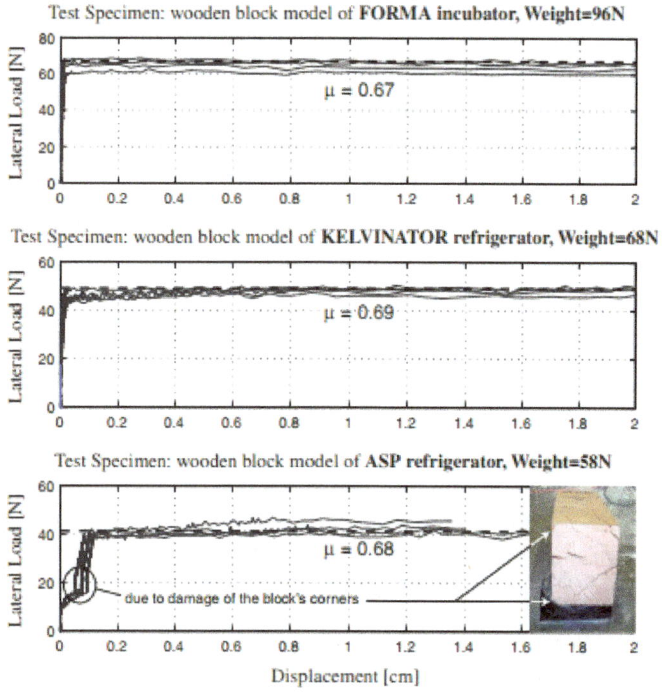

Figure. 2 Recorded load-displacement plots for the wooden block models obtained from quasi-static pull tests. The wood-concrete interface exhibits a nearly perfect rigid-plastic behavior

equipment prototypes (KELVINATOR and ASP refrigerators) can be seen in the background. Table 1 presents the geometrical characteristics of the full-scale prototype equipment and the quarter-scale wooden block models. Figure 2 shows results obtained from quasi-static pull tests on the wooden blocks. The behavior of the contact interface between the wooden blocks and the concrete surface of the shake table atop which the blocks rested is nearly rigid-plastic without exhibiting any difference between the static and kinetic values of the friction coefficient. The average value of the friction coefficient from the quasi-static pull tests on the three wooden blocks is about $\mu_s = \mu_k = 0.68$

Table 2 Results of shake table tests on the wooden block models subjected to compressed 2% in 50 years Loma Prieta earthquake motions. Listed are also the outcomes predicted by Working Model simulations on full-scale prototypes subjected to the corresponding uncompressed motions

Equipment prototype	Compressed earthquake motion	PTA [g]	Shake table experiment outcome	Working model simulation outcome ($\mu = 0.68$)
FORMA	Los Gatos PC, ground	0.54	Overturn	No overturn[b]
	Los Gatos PC, ground	0.60	Overturn	Overturn
	Los Gatos PC FP, ground	0.57	Overturn	Overturn
	Corralitos, ground	0.84	Overturn	No overturn[b]
	Corralitos, ground	0.83	Overturn	Overturn
	Gilroy Hist. Bldg., ground	0.59	Overturn	Overturn
	Gilroy Hist. Bldg., ground	0.58	Overturn	No overturn[a]
	Gilroy Hist. Bldg., ground	0.59	Overturn	No overturn[b]
	Gilroy Hist. Bldg., 6th floor	0.96	Overturn	Overturn
KELVINATOR	Los Gatos PC, ground	0.54	Overturn	Overturn
	Los Gatos PC, ground	0.60	Overturn	Overturn
	Los Gatos PC FP, ground	0.57	Overturn	Overturn
	Corralitos, ground	0.84	No overturn	No overturn
	Corralitos, ground	0.83	No overturn	No overturn
	Gilroy Hist. Bldg., ground	0.59	Overturn	Overturn
	Gilroy Hist. Bldg., ground	0.58	Overturn	Overturn
	Gilroy Hist. Bldg., ground	0.59	Overturn	Overturn
	Gilroy Hist. Bldg., 6th floor	0.96	Overturn	Overturn
ASP	Los Gatos PC, ground	0.54	Overturn	No overturn
	Los Gatos PC, ground	0.60	Overturn	Overturn
	Los Gatos PC FP, ground	0.57	Overturn	No overturn
	Corralitos, ground	0.84	Overturn	No overturn
	Corralitos, ground	0.83	Overturn	No overturn
	Gilroy Hist. Bldg., ground	0.59	Overturn	Overturn
	Gilroy Hist. Bldg., ground	0.58	Overturn	Overturn
	Gilroy Hist. Bldg., ground	0.59	Overturn	Overturn
	Gilroy Hist. Bldg., 6th floor	0.96	Overturn	Overturn

Table 2 lists the compressed records that were used as input motions for the shake table tests on the wooden models together with the outcome of each experiment. The ground motions used are from a seismic hazard study for the UC Science site performed by Somerville (2001). The simulated floor motion used was obtained from a study by Lee and Mosalam (Comerio 2005; Lee and Mosalam 2005), who conducted dynamic analysis of a sophisticated structural model of the building. Since in most cases the wooden blocks overturned during the shake table tests, it was decided not to instrument the blocks but only to record whether the block survived the motion or overturned. More than one experiment were conducted for each motion in an effort to examine the repeatability of the outcomes; the outcomes were indeed repeatable in all but few cases.

VALIDATION OF THE WORKING MODEL SOFTWARE

Since this study was part of a comprehensive study that employed the PEER-proposed Performance-Based Earthquake Engineering (PBEE) practical methodology, it was decided to utilize for the analysis a commercially available software; one that the practicing engineer could easily use to predict the response of laboratory equipment and other building contents alike. The shake table experiments presented in Konstantinidis and Makris (2009) indicated that the primary mode of response is sliding. However, rocking and possible overturning may happen for interfaces with larger coefficients of friction, μ, and for equipment with more slender configurations (smaller α). Therefore, the software of choice had to be able to capture both sliding and rocking response. Working Model 2D (2000) is a software that combines robust numerical techniques with sophisticated editing capabilities. Its main attraction is its capability to compute the motion of mechanically interacting rigid bodies under a variety of constraints and the action of time-varying forces. One of the most challenging tasks in the dynamic simulation of rigid bodies is the treatment of the contact interfaces. In Working Model, the satisfaction of all imposed constraints at the contact interfaces is enforced simultaneously during the numerical integration. In the tangential direction, the contact interface of adjacent bodies is modeled by static and kinetic Coulomb friction (Working Model 2000). Regardless of whether there is sliding or not, the rigid body that models the equipment while engaging in rocking motion can impact the rigid body that models the ground. During an integration step, two colliding bodies may overlap by a small amount. In Working Model, collisions are detected by finding intersections between the geometries of bodies. Since the bodies are assumed rigid, for any two points on the body $\mathfrak{B}$, $\|x_1 - x_2\| = \|X_1 - X_2\|$ for all time, where x is the one-to-one mapping $x = \chi(X, t)$, and X is the position in some reference configuration. This implies that the position and orientation of the edges of a rigid body are known for any time by tracking a master node. When intersection between edges is detected, Working Model computes forces sufficient to "repel" the bodies. Working Model employs an impulse-based collision model in which the coefficient of restitution is used (Working Model 2000). The numerical integration of the equations of motion in conjunction with the satisfaction of the constraint conditions (friction and restitution), is done using a robust Kutta-Merson method (5th order Runge-Kutta). Integration error as well as model assembly and collision overlap tolerances can be set to achieve the desired precision. With the available variable-timestep Kutta-Merson scheme, near collision, the timestep is reduced appropriately to restrict the overlap between bodies from exceeding the specified overlap tolerance. For all the simulations presented in this report, the overlap error tolerance was set to 10^{-5} cm.

Validation for pure sliding

We consider a rigid block resting on a base that is subjected to a horizontal excitation $\ddot{u}_g$. The interface coefficient of friction is μ. An in-depth study of the response of a sliding mass on a moving base has been presented by Younis and Tadjbakhsh (1984). The solution to a constant acceleration pulse with amplitude a_p and duration $T_p = 2\pi/\omega_p$ presented by Newmark (1965) is

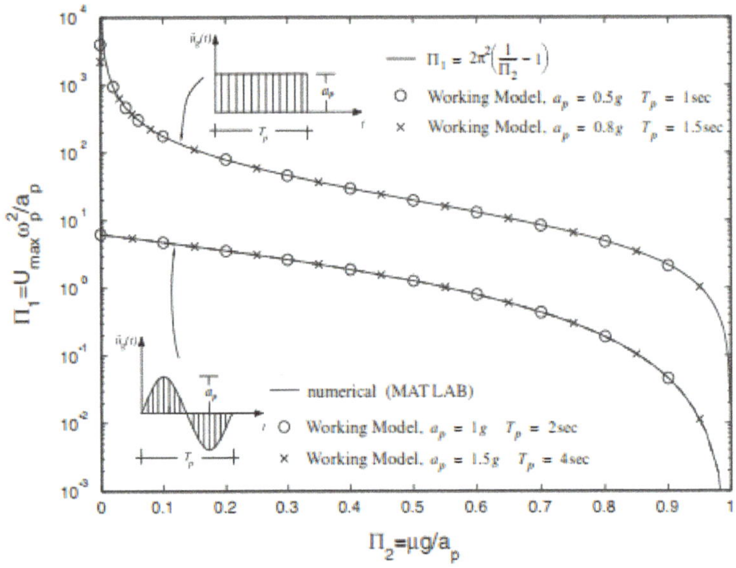

Figure. 3 Dimensionless displacement Π_1 as a function of the dimensionless strength Π_2 for rigid-plastic system subjected to rectangular and Type-A acceleration pulses with amplitude a_p and duration $T_p = 2\pi/\omega_p$

$$\frac{U_{max}\omega_p^2}{a_p} = 2\pi^2\left(\frac{a_p}{\mu g} - 1\right) \qquad (1)$$

where Umax is the maximum relative displacement of the block. Figure 3 plots with a solid line this dimensionless maximum relative displacement as a function of the dimensionless strength μ_g/ap. The ◦ and × points plot the results obtained with the software Working Model where various combinations of the values of a_p, T_p, and μ have been used. Working Model captures with high fidelity the closed-form solution given by Eq. (1), which is plotted with a solid line. Figure 3 also plots with a solid line the solution Umaxω2 p/a_p due to a Type-A (forwarddisplacement) pulse with acceleration amplitude a_p and duration $T_p = 2\pi/\omega_p$ as a function of μ_g/ap. The solution is obtained by numerically integrating in MATLAB (2002) the equation of motion associated with the Bouc-Wen model (Makris and Black 2003, 2004; Konstantinidis and Makris 2005a,b). The Bouc-Wen parameter values (Wen 1975, 1976), $\beta = \gamma = 0.5$, $n = 20$ and $u_y = 10^{-5}$ cm, were used to model the rigid-plastic behavior. Figure 3 shows that the Working Model solution is in excellent agreement with the solution obtained with MATLAB. More evidence of the accuracy of the solutions obtained with Working Model is offered in Fig. 4 which plots the sliding response of a rigid mass subjected to three strong earthquakes. Note that despite the large variability in the peak values of the sliding displacements, the solutions obtained with Working Model are in excellent agreement with the MATLAB numerical solution.

Validation for Pure Rocking

In this section, we present a comparison of results obtained with Working Model to results obtained by numerically integrating the equation of motion for pure-rocking

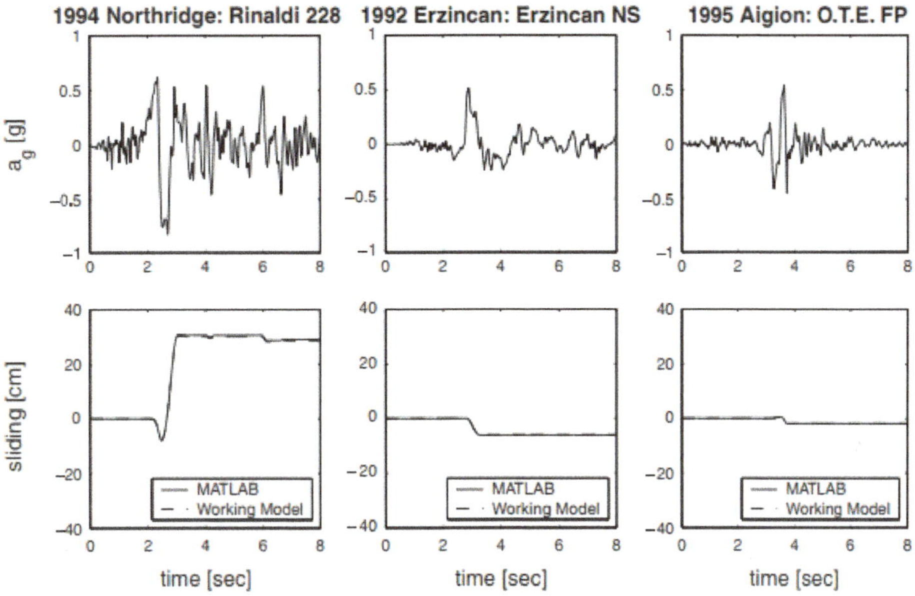

Figure. 4 Sliding displacement of a freestanding rigid mass (μ = 0.3) subjected to three strong earthquakes. The response obtained with the software Working Model is in excellent agreement with the numerical solution obtained with MATLAB.

(Makris and Roussos 2000; Makris and Konstantinidis 2003a), which yields time histories of the rotation and angular velocity of the rocking block. Standard ODE solvers available in MATLAB (2002) and a custom routine that detects impacts and imposes a reduction in angular velocity (to take into account energy lost upon impact) are used. Figure 5 plots the normalized rotation, θ/α, and angular velocity, θ/˙ p, histories of a rigid block with frequency parameter p = √3g/4R = 1.25 rad/sec (where R is the distance from the pivot point to the center of mass of the block) and stockiness α = 0.16 rad when subjected to two strong ground motions. The solid line is the numerical solution obtained with MATLAB. The block survives the Rinaldi 228 motion (left) recorded during the 1994 Northridge, California, earthquake yet topples when subjected to the TCU052NS motion (right) recorded during the 1999 Chi-Chi, Taiwan, earthquake. The dashed line is the prediction of the software Working Model where the coefficient of friction has been set to a high value in order to avoid slipping. The solutions obtained with the two numerical codes are in excellent agreement. Figure 6 plots the minimum overturning acceleration spectrum of the same block when subjected to a one-sine acceleration pulse (Type-A pulse) with acceleration amplitude a_p and duration $T_p = 2\pi/\omega_p$. Notice the multivaluedness of $a_p/\alpha g$ that implies that a block can survive a pulse with acceleration amplitude larger than the minimum acceleration pulse necessary to overturn it. This interesting fact was first exposed by Zhang and Makris (2001). The shaded area in Fig. 6 is the unsafe region (i.e., where overturning occurs), while the blank area is the safe region (i.e., where overturning does not occur). Note that the software Working Model successfully predicts this multivaluedness of the response where a safe region appears above the minimum overturning acceleration line. The overall performance of Working Model is very good, with only minor degradation for shorter-period pulses ($\omega_p/p > 10$, or $T_p < 0.5$ s) with large acceleration amplitudes ($a_p/\alpha g > 15$, or

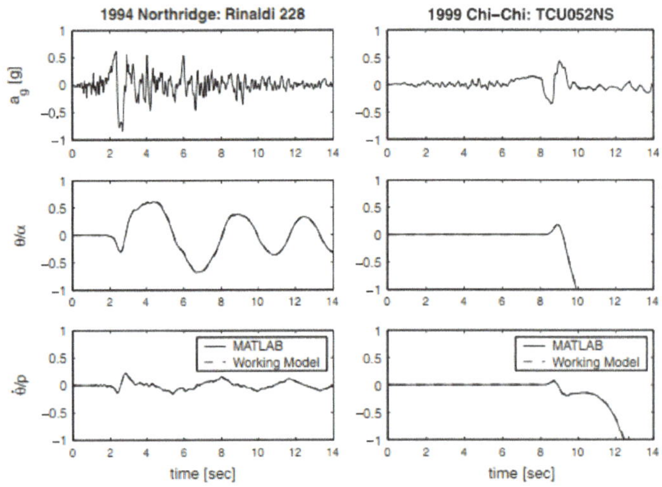

Figure. 5 Rotation and angular velocity of a rigid block with frequency parameter p = 1.25 rad/s and stockiness α = 0.16 rad subjected to two strong ground motions.

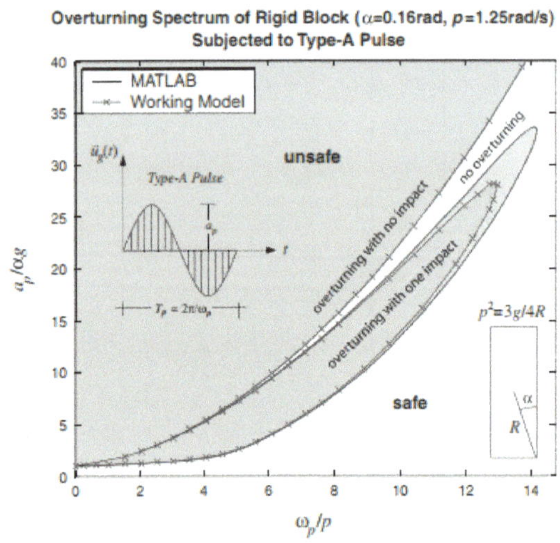

Figure. 6 Overturning acceleration spectrum of a rigid block with frequency parameter p = 1.25 rad/s and stockiness α = 0.16 rad subjected to a one-sine (Type-A) pulse

$a_p > 2.4g$). More evidence on the fidelity of Working Model is offered in Konstantinidis and Makris (2005b).

COMPARISON OF RESULTS FROM WORKING MODEL SIMULATIONS AND SHAKE TABLE TESTS ON QUARTER-SCALE WOODEN MODELS

The results of the shake table experiments that were carried on the quarter-scale wooden blocks (models) are of great value because they can validate the fidelity of results of numerical simulation studies conducted on the full-scale equipment (prototypes). Since the friction coefficients of the model-base interfaces were different from the friction coefficients of the prototype-base interfaces, two sets of simulation studies were conducted. First, the response of the full-scale prototype equipment was computed with Working Model by using a coefficient of friction at the sliding interface equal to $\mu_s = \mu_k = 0.68$, the value obtained from the slow pull tests on the wooden blocks. The results of this analysis demonstrated that Working Model can capture well the experimentally observed overturning behavior of the wooden blocks. Then, once the fidelity of Working Model had been validated, the response of the full-scale prototype equipment was computed with the friction coefficients of the equipment- floor sliding interfaces, presented in Konstantinidis and Makris (2009) and repeated here on the left column of Table 3. The bottom plot on the left column of Fig. 7 shows the time-compressed (by a factor of 2) acceleration history of the motion recorded at the Los Gatos Presentation Center during the 1989 Loma Prieta, California, earthquake. The plot above the acceleration history plots the resulting displacement history. The time-compressed motion was used as shake table input to test the quarter-scale wooden block model of the FORMA incubator. The final outcome of the experiment was that the wooden model overturned. The right-column plots of Fig. 7 show the response computed with Working Model for the full-scale FORMA incubator prototype subjected to the uncompressed motion. Note that the base displacement history of the uncompressed motion is 4 times larger in amplitude than the table displacement history of the compressed motion. The heavy solid lines of the top two windows on the right column of Fig. 7 plot the simulated sliding and uplift responses of the full-scale equipment with a coefficient of friction $\mu = 0.68$, the value obtained from the quasi-static pull tests on the wooden blocks. While the simulated response predicts that the incubator does not overturn (as the experiment on the scaled model showed), the negative rotation θ of the equipment is so large that it in fact exceeds the stockiness value α at around t = 7.5 s. Luckily, at the same time, the negative table acceleration which opposes the overturning of the equipment (positive θ is clockwise) marginally saves it from toppling. This behavior whereby the equipment prototype marginally survives, while its corresponding wooden model overturns, is observed for a few of the motions. As shown in Table 2, when subjected to the Corralitos motion recorded during the 1989 Loma Prieta, California, earthquake, the wooden block overturns. The full-scale FORMA incubator, on the other hand, survives the motion. Figure 8 (right) which plots the response computed with Working Model, shows that around t = 5.5 s the rotation θ exceeds α, and the equipment hangs on the verge of overturning; yet the restoring base acceleration spares the equipment. When the experiment was repeated on the shake table with the same motion but a slightly different Peak Table Acceleration (0.83g instead of 0.84g, due to imprecision of the shake table actuators), the wooden model overturned again. Interestingly, when the Working Model simulation was ran with PT A = 0.83g (Peak Table Acceleration), the result was overturning. Except for these

Table 3 Kinematic characteristics of the 2% in 50 years Loma Prieta motions used in this study. Together with the maximum computed equipment sliding displacement, Umax, and interface friction coefficient, μ, the motion PTA and $\omega_p = 2\pi/T_p$ produce the intensity measure, IM, and engineering demand parameter, EDP

FORMA	Los Gatos PC, ground	2.10, −75	0.54	45.6	3.17	0.77
	Los Gatos PC, ground	2.10, −75	0.60	47.1	3.58	0.72
$\mu = 0.23$ (slow pull)	Los Gatos PC, ground	2.10, −75	0.57	40.3	3.39	0.64
	Corralitos, ground	0.60, −61	0.84	19.9	5.45	2.65
	Corralitos, ground	0.60, −61	0.83	20.6	5.38	2.78
$\mu = 0.13$ (best fit)	Gilroy Hist. Bldg., ground	1.50, −84	0.59	35.3	3.57	1.06
	Gilroy Hist. Bldg., ground	1.50, −84	0.58	35.9	3.43	1.12
	Gilroy Hist. Bldg., ground	1.50, −84	0.59	37.6	3.55	1.14
	Gilroy Hist. Bldg., 6th floor	1.60, −120	0.96	69.6	6.40	1.14
KELVINATOR	Los Gatos PC, ground	2.10, −75	0.54	31.7	2.19	0.53
	Los Gatos PC, ground	2.10, −75	0.60	34.1	2.50	0.52
$\mu = 0.28$ (slow pull)	Los Gatos PC, ground	2.10, −75	0.57	32.8	2.36	0.52
	Corralitos, ground	0.60, −61	0.84	18.7	3.94	2.49
	Corralitos, ground	0.60, −61	0.83	18.8	3.88	2.53
$\mu = 0.17$ (best fit)	Gilroy Hist. Bldg., ground	1.50, −84	0.59	35.3	2.50	1.06
	Gilroy Hist. Bldg., ground	1.50, −84	0.58	38.2	2.39	1.19
	Gilroy Hist. Bldg., ground	1.50, −84	0.59	39.4	2.48	1.19
	Gilroy Hist. Bldg., 6th floor	1.60, −120	0.96	70.5	4.66	1.15
ASP	Los Gatos PC, ground	2.10, −75	0.54	28.8	1.71	0.49
	Los Gatos PC, ground	2.10, −75	0.60	30.6	1.98	0.47
$\mu = 0.31$ (slow pull)	Los Gatos PC, ground	2.10, −75	0.57	28.8	1.86	0.48
	Corralitos, ground	0.60, −61	0.84	15.9	3.20	2.12
	Corralitos, ground	0.60, −61	0.83	16.8	3.15	2.26
$\mu = 0.20$ (best fit)	Gilroy Hist. Bldg., ground	1.50, −84	0.59	33.8	1.97	1.02
	Gilroy Hist. Bldg., ground	1.50, −84	0.58	34.3	1.88	1.07
	Gilroy Hist. Bldg., ground	1.50, −84	0.59	35.1	1.96	1.06
	Gilroy Hist. Bldg., 6th floor	1.60, −120	0.96	64.0	3.81	1.05

marginal cases, Working Model managed to capture well the overturning of the wooden models. Working Model also predicted correctly the two cases when the wooden model does not overturn. Figure 9 shows the experimental outcome (left) of the wooden block model of the KELVINATOR refrigerator subjected to the compressed Corralitos record of the 1989 Loma Prieta, California, earthquake and the computed response of the full-scale prototype with μ = 0.68 to the uncompressed record (right). The response computed with Working Model verifies that the block survives the motion. Although they are presented in Table 2 for completeness, the Working Model simulations for the ASP refrigerator cannot be compared to the shake table experimental outcomes of

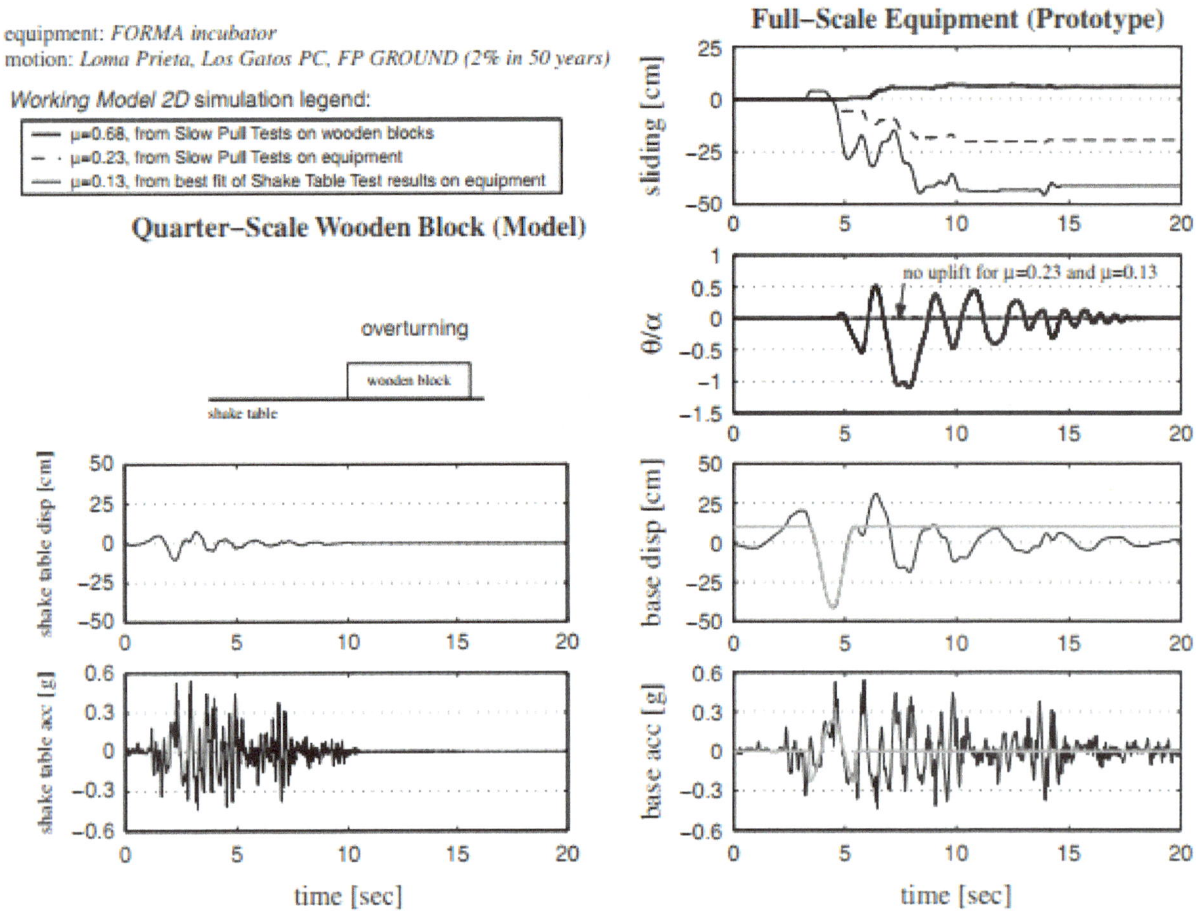

Figure. 7 Response of the FORMA incubator to the Loma Prieta, Los Gatos PC, FP Ground (2% in 50 years) motion. Left: Outcome of the shake table experiment on the quarter-scale wooden block model together with the table acceleration and displacement of the compressed (by a factor of 2) motion. Right: Computed response by Working Model simulations on the full-scale equipment prototype subjected to the uncompressed motion

the wooden model because the wooden model of the ASP refrigerator suffered damage that reduced its seismic stability. Figure 2 (bottom) shows a photograph of the damage on the corner of the wooden model and how this damage is also evident in the load-displacement curve. The reduced base (and therefore stability) due to the damage of the block's corners explains why, despite the large stockiness of the ASP refrigerator, the wooden model toppled in all cases, while the Working Model simulation predicted that the prototype topples in only 5 out of the 9 cases. More details on the experimental results on the wooden models and numerical simulation of full-scale equipment to 2% in 50 years are offered in the report by Konstantinidis and Makris (2005b). The comparison between experimental and computed results demonstrate Working Model's ability to capture the experimentally observed behavior and confirm the validity of the analytical models. This provides us with confidence that the program can compute the seismic response of laboratory equipment for a wider range of friction coefficients and hazard-level base motions.

SLIDING RESPONSE DUE TO 2% IN 50 YEARS HAZARD MOTIONS

When the values of μ from the quasi-static pull tests were used, the computed sliding and rocking time histories presented in the report by Konstantinidis and Makris(2005b) showed a response that was contrary to the primarily sliding response that was experimentally observed for the 10 and 50% in 50 years motions. In particular, the numerical simulations predicted

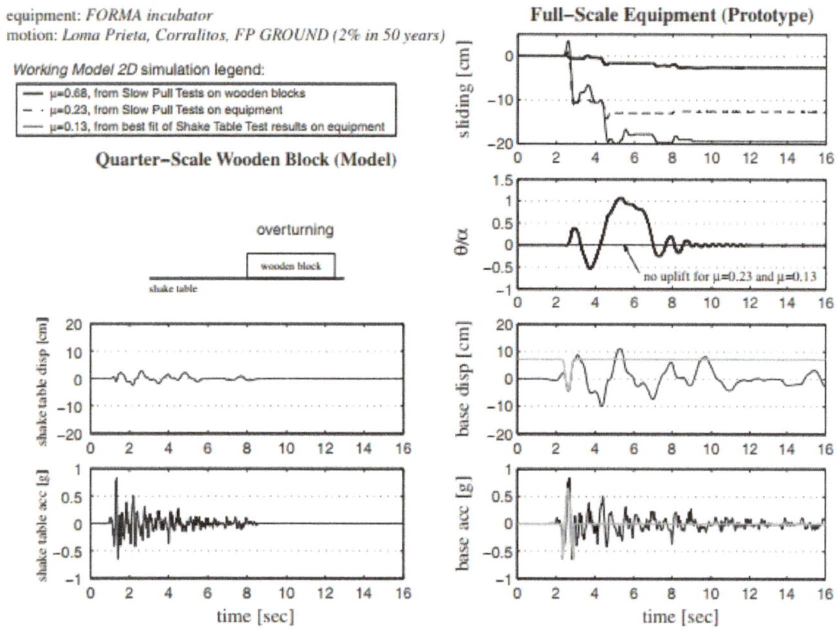

Figure. 8 Response of the FORMA incubator to the Loma Prieta, Corralitos, FP Ground (2% in 50 years) motion

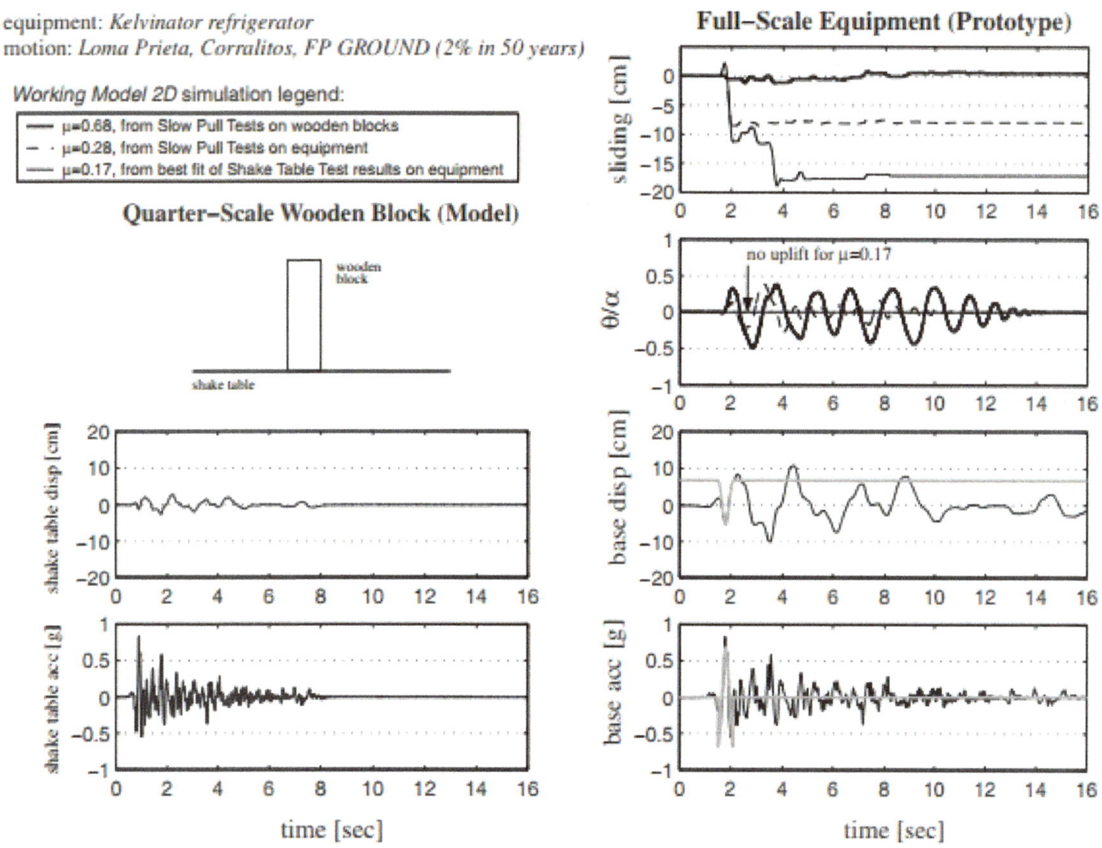

Figure. 9 Response of the KELVINATOR refrigerator to the Loma Prieta, Corralitos, FP Ground (2% in 50 years) motion

that the KELVINATOR and ASP refrigerators experience large rotations—and even overturning in a few occasions. In light of the earlier observations where the KELVINATOR and ASP refrigerators in fact experienced very small rotations, it was concluded that the friction coefficients from the slow pull test were too large, and consequently they caused the equipment to engage in rocking during the numerical simulations. A considerably improved agreement between experimental and numerical results was observed when μ was reduced in the numerical simulations. As explained in Konstantinidis and Makris (2009), part of the reason for the necessary reduction is possibly pressure-dependence ofμthat was not captured during the quasi-static pull tests. This would be similar to the pressure-dependent behavior of a teflon-steel interface observed by other investigators (Mokha et al. 1988; Constantinou et al. 1993). Moreover, a reduction in μ was necessary because the Working Model blockto-base interface is rigid-plastic (Coulomb), while the actual behavior is elastoplastic, where the elasticity originates from the flexibility of the equipment legs. Recent studies (Makris and Black 2004; Makris and Psychogios 2006) have shown that for the same value of μ, the sliding displacement increases with increasing yield displacement. Consequently, given that in all three pieces of equipment the behavior at the sliding interface is elastoplastic (finite yield displacement), when a rigid-plastic model is adopted to capture the behavior, a smaller value of μ is needed. The reduced values of μ were chosen so as to best-match the computed response to the experimentally observed one. For the present study, Working Model simulations for 2% in 50 years motions with the best-fitted friction coefficient values result in the predominant sliding response that is anticipated (plotted with solid black lines in Figs. 7, 8, 9). For this reason, the analysis presented in this paper uses the reduced (or best-fit) friction coefficients listed in Table III of Konstantinidis and Makris (2009), which is also repeated herein on the left column of Table 3. Table 3 lists the 2% in 50 years input motions used with the Working Model simulations and the computed peak sliding displacement, Umax , of the laboratory equipment.

REGRESSION ANALYSIS AND FRAGILITY CURVES

Since we are primarily concerned with sliding and not rocking, the parameters that govern the response of a sliding body become those that describe (a) the mechanical characteristics of the equipment-floor interface and (b) the kinematic characteristics of the base motion. As demonstrated in Konstantinidis and Makris (2009), the parameter that best characterizes the sliding resistance of the contact interface is the kinetic coefficient of friction μ. The impulsive character of near-fault ground motions can be adequately described by physically realizable trigonometric pulses. The input parameters of the model have an unambiguous physical meaning. The minimum number of parameters is two, which are either the acceleration amplitude, a_p, and duration, T_p, or the velocity amplitude, v_p, and duration, T_p (Makris 1997; Makris and Chang 2000; Mavroeidis and Papageorgiou 2003). The third column of Table 3 lists the defining parameters of Type-B trigonometric pulses that approximate the predominant pulses of each of the 2% in 50 years motions used in this study. A Type-B pulse, which results in a forward-and-back ground displacement, is defined by u_B g (t) = v_p 1 − cos(ωpt) /ω_p, 0 ≤ t ≤ T_p (Makris 1997). The velocity and acceleration histories are obtained by differentiating the expression for ground displacement, $u^B{}_g$ (t). The acceleration and displacement histories of such pulses are plotted with heavy grey lines on the bottom two windows of Figs. 7, 8 and 9 (right) together with the corresponding earthquake records (black lines).

Intensity Measure and Engineering Demand Parameter

The estimation of the sliding response is cast in a probabilistic framework. The PEER-proposed PBEE methodology (Porter 2003) suggests the identification of an Intensity Measure (IM) and Engineering Demand Parameter (EDP). The choices for the IM and EDP emerge from dimensional analysis in conjunction with previously published results on the response of a sliding block (Newmark 1965; Barenblatt 1996; Konstantinidis and Makris 2005a,b). For a pulse-type excitation, the maximum sliding displacement Umax can be expressed as a function of independent variables a_p, ω_p and μ_g

$$U_{max} = f(a_p, \omega_p, \mu g) \tag{2}$$

where $\omega_p = 2\pi/T_p$ is the circular frequency of the pulse, or if an earthquake motion is approximated by a pulse, the circular frequency of the pulse that approximates the predominant pulse of the earthquake. The dependent variable Umax and independent variables a_p, ω_p and μ_g, involve only two dimensions, those of length, L, and time, T . The quantities of interest have dimensions

$$[U_{max}] = L, \quad [a_p] = LT^{-2}, \quad [\omega_p] = T^{-1}, \quad [\mu g] = LT^{-2}$$

(3)

Buckingham's Pi-Theorem states that a dimensionally homogeneous equation with a total of k variables and r reference dimensions, can be reduced to a relationship among k − r independent dimensionless -products (Barenblatt 1996). Accordingly, in this case there are 4 − 2 = 2 dimensionless products. Two obvious choices for -products are

$$\Pi_1 = \frac{U_{max}\omega_p^2}{a_p}, \quad \Pi_2 = \frac{\mu g}{a_p} \tag{4}$$

and the two are related by a function

$$\Pi_1 = \phi(\Pi_2) \tag{5}$$

For a rectangular acceleration pulse, this function is (Newmark 1965)

$$\Pi_1 = 2\pi^2\left(\frac{1}{\Pi_2} - 1\right) \tag{6}$$

For the case of trigonometric pulses, such as a Type-B pulse, the response is again described by Eq. (5) (Makris and Black 2003; Konstantinidis and Makris 2005b), and the form of the function φ is obtained numerically. Figure 3 plots with solid lines the response due to a rectangular acceleration pulse excitation and to a one-sine acceleration pulse excitation (i.e., a Type-A pulse, defined by $u_g^A(t) = v_p\left[\omega_p t - \sin(\omega_p t)\right]/(2\omega_p), 0 \le t \le T_p$).

The dimensionless displacement ϕ 1 = Umaxω2 p/a_p is plotted on a logarithmic scale in order to illustrate the relative strengths of the two types of pulses. The closed-form solution due to the rectangular pulse is given by Eq. (6), while the solution due to the one-sine pulse is obtained numerically using standard ODE solvers available in MATLAB (2002). The choices for IM and EDP used in this study fall naturally from E_qs. (4) and (6), and are

$$\mathrm{IM} = \frac{\mathrm{PTA}}{\mu g} - 1 \tag{7}$$

And

$$\Delta = \frac{U_{max}\omega_p^2}{\text{PTA}} \tag{8}$$

The chosen EDP will henceforth be designated. The last two columns of Table 3 list the IM and EDP. Note from Eqs. (4), (6) and (7) that the IM chosen is exactly the quantity in parenthesis in Eq. (6) except that the IM uses for simplicity PTA instead of ap. The two values are expected to be close.

Figure 10 (bottom) plots the Engineering Demand Parameter, = $U_{max}\omega_p^2$/PTA, as a function of the Intensity Measure, IM = PTA/μg−1, obtained from Working Model simulations on the three pieces of equipment subjected to the 2% in 50 years hazard level motions listed in Table 3. It is obvious that the data exhibits considerable scatter, which suggests that Δ has to be treated as a random variable. When a random variable, Δ expresses a quantity that is only positive (δ > 0), it is common to assume that the variable is lognormally distributed. In this study we hypothesize that the EDP is lognormally distributed, Δ and we test this hypothesis against the experimental results. The reason behind the name lognormal is that the lognormally distributed variable is related to a normally distributed variable X by X = ln Δ. Note that attains only positive values, δ > 0, while the corresponding X variable is unrestricted, −∞ < x < ∞. It is assumed that Δ takes the form

$$\Delta = \zeta\left(\frac{\text{PTA}}{\mu g} - 1\right)^{\eta} Z = \zeta \text{IM}^{\eta} Z \tag{9}$$

where ζ and η are regression parameters and Z is a lognormally distributed random variable. Another important feature of the scatter in Fig. 10 (bottom) is that it increases with increasing IM (funneling effect), i.e., the variance is non-constant. However, the logarithmic transformation

$$X = \ln \Delta = \ln \zeta + \eta \ln \text{IM} + \ln Z \tag{10}$$

stabilizes the variance (homoscedasticity), as seen on the top left plot of Fig. 10, and therefore ordinary least-squares can be applied to estimate the parameters ζ and η. We obtain

$$\Delta = 0.370\left(\frac{\text{PTA}}{\mu g} - 1\right)^{0.953} Z = 0.370\text{IM}^{0.953} Z \tag{11}$$

The lognormally distributed variable has the probability density function

$$F_{\Delta}(\delta) = \int_0^{\delta} \frac{1}{\sqrt{2\pi}\sigma_X \delta'} e^{-\frac{1}{2}\left(\frac{\ln\delta' - m_X}{\sigma_X}\right)^2} d\delta' \tag{12}$$

where the two defining parameters mX and σX of the distribution are in fact the mean and standard deviation of the corresponding normally distributed variable X (Crow et al. 1988). The cumulative distribution function of Δ is given by

$$F_{\Delta}(\delta) = \int_0^{\delta} \frac{1}{\sqrt{2\pi}\sigma_X \delta'} e^{-\frac{1}{2}\left(\frac{\ln\delta' - m_X}{\sigma_X}\right)^2} d\delta' \tag{13}$$

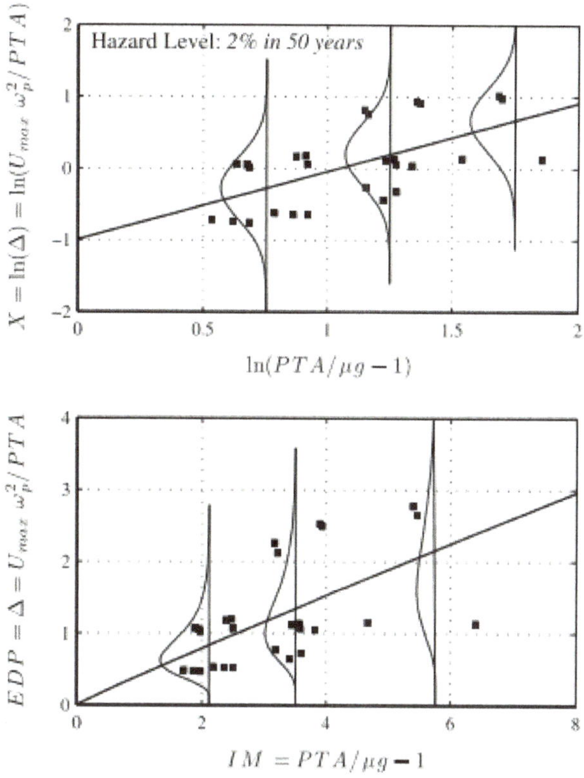

Figure. 10 Top: The normal random variable X = ln Δ and the linear fit which gives its mean m_X. Bottom: The associated lognormal random variable, together with the power-law fit for its mean m. σ_X is constant, while σ increases with IM

Using the substitution $u = (\ln\delta' - m_X)/\sigma_X$, we obtain

$$F_{\Delta}(\delta) = \int_{-\infty}^{(\ln\delta - m_X)/\sigma_X} \frac{1}{\sqrt{2\pi}} e^{-\frac{u^2}{2}} du = \Phi\left(\frac{\ln\delta - m_X}{\sigma_X}\right) \tag{14}$$

where Φ is the cumulative distribution function of a standard normal variable (i.e., mean of zero and standard deviation of one). Note from Eq. (10) that the mean of X is m_X (IM) = ln(ζ IMη). Also, the standard deviation of X can be estimated from

$$\sigma_X = \sigma_{\ln Z} = \sqrt{\frac{1}{n-1}\sum_{i=1}^{n}\left(\ln\frac{\delta_i}{\zeta \text{IM}_i^{\eta}}\right)^2} \tag{15}$$

where n is the sample size. Using the results of the 27 Working Model simulations on the full-scale equipment (Table 3), we obtain σX = 0.445. The mean m (IM) and standard deviation σ (IM) of Δ can be shown to be related to the mean m_X (IM) and standard deviation σ_X of the normal variable X through

$$m_{\Delta}(\text{IM}) = e^{m_X(\text{IM}) + \frac{\sigma_X^2}{2}} = \zeta \text{IM}^{\eta} e^{\frac{\sigma_X^2}{2}} \tag{16}$$

and

$$\sigma_\Delta(\mathrm{IM}) = e^{m_X(\mathrm{IM})+\frac{\sigma_X^2}{2}}\sqrt{e^{\sigma_X^2}-1} = \zeta \mathrm{IM}^\eta e^{\frac{\sigma_X^2}{2}}\sqrt{e^{\sigma_X^2}-1} \tag{17}$$

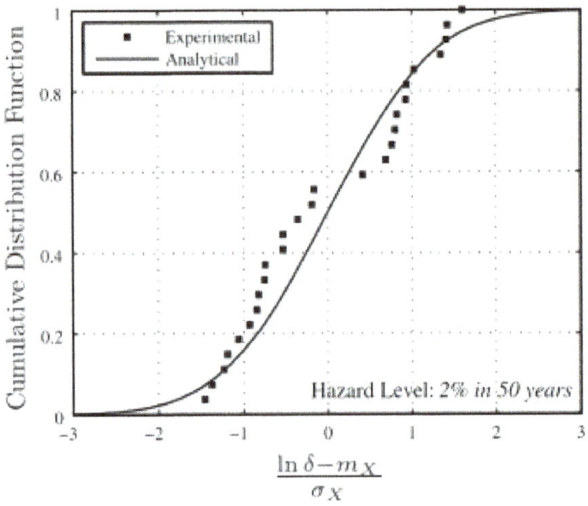

Figure. 11 The empirical cumulative distribution function, F* $_\Delta$, and the theoretical cumulative distribution function F $_\Delta$ of its hypothesized lognormal distribution

Figure 10 (top) shows how X is distributed with a constant standard deviation at three different IM values, while Fig. 10 (bottom) shows how $_\Delta$ is distributed with an increasing σ (IM). In order to not reject the hypothesis that $_\Delta$ has a lognormal distribution, the KolmogorovSmirnov goodness-of-fit test is performed (Scheaffer and McClave 1995). Figure 11 plots the empirical cumulative distribution function, F* $_\Delta$ Φ = i/n. Together is plotted the cumulative distribution function of the hypothesized lognormal distribution, F $_\Delta$ = ((ln δ − m_X)/σ_X). The maximum distance between the distribution functions presented in Fig. 11 is

$$S = \max_\delta \left|F^*_\Delta(\delta) - F_\Delta(\delta)\right| = 0.161 \tag{18}$$

Tables for the Kolmogorov-Smirnov test (Daniel 1990) provide S_{cr} for n = 27 and various significance levels, a,

$$S_{cr}(a, n=27) = \begin{cases} 0.305, & a = 0.01 \\ 0.284, & a = 0.02 \\ 0.254, & a = 0.05 \\ 0.229, & a = 0.10 \\ 0.200, & a = 0.20 \end{cases}$$

Since S < S_{cr}, the hypothesis that $_\Delta$ is lognormally distributed is not rejected.

Fragility curves Fragility is viewed as a conditional probability of failure. For the problem at hand, where we want to characterize the seismic response of heavy laboratory equipment in a probabilistic framework, we define fragility as the probability P_f that the EDP, $\Delta = U_{max}\omega_p^2/\mathrm{PTA}$, for a piece of equipment will exceed a certain threshold (capacity), c, given the IM. For the lognormally distributed random variable, $_\Delta$

$$P_f \equiv P(\Delta > c|\mathrm{IM}) = 1 - P(\Delta < c|\mathrm{IM}) = 1 - F_\Delta(c) \tag{19}$$

Substituting Eq. (14) and m_X (IM) = ln(ζ IM$^\eta$) into Eq. (19) gives

$$P_f = 1 - \Phi\left(\frac{\ln c - m_X(\mathrm{IM})}{\sigma_X}\right) = 1 - \Phi\left(\frac{1}{\sigma_X}\ln\frac{c}{\zeta \mathrm{IM}^\eta}\right) \tag{20}$$

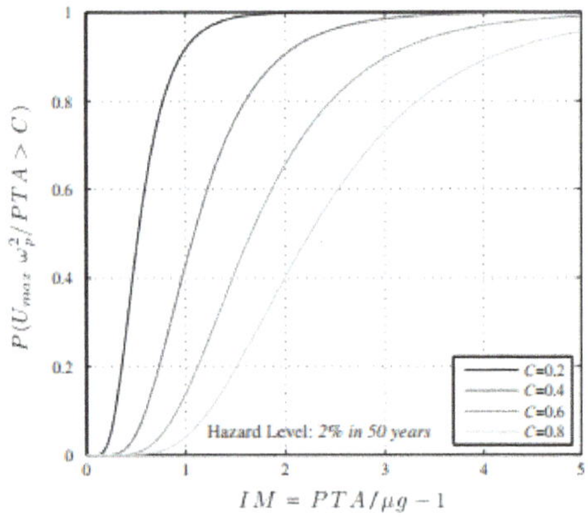

Figure. 12 Fragility curves for four different c's

where Φ is the standard normal cumulative distribution function, and

$$\sigma_X = 0.445, \quad \zeta = 0.370, \quad \eta = 0.953 \tag{21}$$

Once the σ_X is estimated from Eq. (15) and the regression parameters ζ and η are obtained from ordinary least-squares fit on Eq. (10), the fragility curves can be generated using Eq. (20) for different capacities c. Figure 12 shows fragility curves for four values of capacity c (0.2, 0.4, 0.6, and 0.8). Typically, fragility curves are plotted against the EDP. In this study, however, they are plotted against the IM, as this makes their use more direct. Konstantinidis and Makris (2009) offer an example of how to use such fragility curves.

CONCLUSIONS

This paper investigates the response of freestanding laboratory equipment subjected to strong earthquake shaking (2% in 50 years hazard level) via model testing. The shake table displacement capacity could not accommodate the ground and floor motions with 2% probability of being exceeded in 50 years. Instead, tests were conducted on quarter-scale wooden block models of the equipment so that the table displacements were reduced by a factor of four. The large friction coefficient of the wooden block-to-base interface caused the wooden blocks to rock and overturn rather than slide. The results from the shake table tests on the wooden models were compared to results from numerical simulations produced by Working Model in order to evaluate the ability of the software to reliably capture the overturning potential of equipment. Working Model's accuracy in predicting overturning and estimating sliding displacements gave us confidence on its overall ability to compute the response of a piece of equipment for different friction coefficients. Numerical simulations for the full-scale equipment under 2% in 50 years motions resulted in a response that is dominated by sliding, with maximum sliding displacements up to 70 cm. A physically motivated Intensity Measure, IM, and the associated Engineering Demand

Parameter, EDP, for the 2% in 50 years hazard level were identified with the help of dimensional analysis. Relationships for the mean and standard deviation of the EDP in terms of the IM were developed. Finally, the paper presents ready-to-use fragility curves, which give the probability that the EDP will exceed a specified threshold c for a given IM.

REFERENCES

1. Barenblatt GI (1996) Scaling, self-similarity, and intermediate asymptotics. Cambridge University Press, Cambridge
2. Comerio MC (ed) (2005) PEER testbed study on a laboratory building: exercising seismic performance assessment.
3. Report No. PEER 2005-12, Pacific Earthquake Engineering Research Center. University of California, Berkeley
4. Constantinou MC, Tsopelas P, Kim YS, Okamoto S (1993) NCEER-Taisei Corporation research program on sliding seismic isolation systems for bridges: experimental and analytical study of a friction pendulum system (FPS). Technical Report NCEER-93-0020, National Center for Earthquake Engineering Research. State University of New York at Buffalo, New York
5. Crow EL, Shimizu K (eds) (1988) Lognormal distributions. Marcel Dekker, New York
6. Daniel WW (1990) Applied nonparametric statistics, 2nd edn. PWS-KENT, Boston
7. Hutchinson TC, Chaudhuri SR (2006) Bench-shelf system dynamic characteristics and their effects on equipment and contents. Earthq Eng Struct Dyn 35(13):1631–1651
8. Konstantinidis D, Makris N (2005a) Seismic response analysis of multidrum classical columns. Earthq Eng Struct Dyn 34(10):1243–1270
9. Konstantinidis D, Makris N (2005b) Experimental and analytical studies on the seismic response of freestanding and anchored laboratory equipment. Report No. PEER 2005-07, Pacific Earthquake Engineering
10. Research Center. University of California, Berkeley
11. Konstantinidis D, Makris N (2009) Experimental and analytical studies on the response of freestanding laboratory equipment to earthquake shaking. Earthq Eng Struct Dyn 38(6):827–848
12. Lee TH, Mosalam KM (2005) Seismic demand sensitivity of reinforced concrete shear-wall building using
13. FOSM method. Earthq Eng Struct Dyn 34(14):1719–1736
14. Lopez Garcia D, Soong TT (2003a) Sliding fragility of block-type non-structural components. Part 1: unrestrained components. Earthq Eng Struct Dyn 32(1):111–129
15. Lopez Garcia D, Soong TT (2003b) Sliding fragility of block-type non-structural components. Part 2: restrained components. Earthq Eng Struct Dyn 32(1):131–149
16. Makris N (1997) Rigidity-plasticity-viscosity: can electrorheological dampers protect base-isolated structures from near-source ground motions? Earthq Eng Struct Dyn 26(5):571–591
17. Makris N, Black CJ (2003) Dimensional analysis of inelastic structures subjected to near-fault ground motions.
18. Report No. EERC 2003-05, Earthquake Engineering Research Center. University of California, Berkeley
19. Makris N, Black CJ (2004) Dimensional analysis of rigid-plastic and elastoplastic structures under pulse-type excitations. J Eng Mech (ASCE) 130(9):1006–1018
20. Makris N, Chang SP (2000) Response of damped oscillators to cycloidal pulses. J Eng Mech (ASCE) 126(2):123–131
21. Makris N, Konstantinidis D (2003a) The rocking spectrum and the limitations of practical design methodologies. Earthq Eng Struct Dyn 32(2):265–289
22. Makris N, Konstantinidis D (2003b) The rocking spectrum, existing design guidelines, and a scale invariance. In: Proceedings of the fédération internationale du Béton symposium, Athens, 6–9 May 2003
23. Makris N, Psychogios T (2006) Dimensional response analysis of yielding structures with first-mode dominated response. Earthq Eng Struct Dyn 35(10):1203–1224
24. Makris N, Roussos Y (2000) Rocking response of rigid blocks under near-source ground motions. Géotechnique 50(3):243–262
25. MATLAB (2002) High-performance language software for technical computing. The MathWorks, Inc., Natick 123
26. Bull Earthquake Eng (2010) 8:1457–1477 1477
27. Mavroeidis GP, Papageorgiou AS (2003) A mathematical representation of near-fault ground motions. Bull Seismol Soc Am 93(3):1099–1131
28. Mokha A, Constantinou MC, Reinhorn AM (1988) Teflon bearings in aseismic base isolation: experimental studies and mathematical modeling. Technical Report NCEER-88-0038, National Center for Earthquake Engineering Research. State University of New York at Buffalo, New York
29. Newmark NM (1965) Effects of earthquakes on dams and embankments. Fifth rankine lecture. Géotechnique 15:139–160
30. Porter KA (2003) An overview of PEER's performance-based earthquake engineering methodology. In: Proceedings of the ninth international conference on applications of statistics and probability in civil engineering (ICASP9), San Francisco 6–9 July 2003
31. Scheaffer RL, McClave JT (1995) Probability and statistics for engineers, 4th edn. Duxbury Press, Belmont
32. Shao Y, Tung CC (1999) Seismic response of unanchored bodies. Earthq Spectra 15(3):523–536
33. Somerville PG (2001) Ground motion time histories for the UC Lab building. URS Corporation, Pasadena
34. Wen YK (1975) Approximate method for nonlinear random vibration. J Eng Mech Div (ASCE) 101(EM4): 389–401

35. Wen YK (1976) Method for random vibration of hysteretic systems. J Eng Mech Div (ASCE) 102(EM2):249– 263
36. Working Model (2000) User's manual. MSC. Software Corporation, San Mateo
37. Yim CS, Chopra AK, Penzien J (1980) Rocking response of rigid blocks to earthquakes. Earthq Eng Struct Dyn 8(6):565–587
38. Younis C, Tadjbakhsh G (1984) Response of sliding rigid structure to base excitation. J Eng Mech (ASCE) 110(3):417–432
39. Zhang J, Makris N (2001) Rocking response of free-standing blocks under cycloidal pulses. J Eng Mech (ASCE) 127(5):473–483

CHAPTER 2 EARTHQUAKE FORECAST WITH THE SEISMIC SEQUENCE HIERARCHIZATION METHOD

Giulio Riga and Paolo Balocchi

Keywords: Branching Structure, Earthquake, Microsequence, Hierarchization, Foreshock

INTRODUCTION

In some cases the main shock is triggered by one or more lower magnitude early events, called foreshocks, which in turn can trigger additional seismic events. Even a strong earthquake is followed in time and space by groupings of earthquakes with a lower magnitude, called aftershocks, which generally are triggered by the same main event.

These aspects related to the aftershocks behavior have been incorporated in the triggering models, where the probability of occurrence of an earthquake in a given time and place is associated with the past history of the earthquakes occurred in the surrounding areas. The most famous model is the short term ETAS (Epidemic Type Aftershock Sequence) [1] [2] , where each event, regardless of its magnitude, increases the probability of triggering seismic events in the surrounding areas. In this system, a sequence of aftershocks may be seen as a "trigger cascade" [3] that extends geographical scope of the main initial shock. Conversely, according to our opinion, this approach highlights how the seismic sequence is a seqeunce of energy release and accumulation phases that develop before and after the strongest shocks [4] .

A long term challenge consists in determining which earthquakes are connected, directly or indirectly, and in identifying the earthquakes sequences based on causal connection [5] . In the past, various declustering algorithms were proposed to subdivide earthquakes into main shock and aftershocks [6] - [11] . Some of these methods are arbitrary rules characterized by various parameters which jeopardise their objectivity. To solve these issues, a series of methods that seek to evaluate the correlations between the earthquakes has been developed [6] - [11] .

We believe that the study and identification of branched structures, without resorting to a physical modeling, can provide valuable information on the evolution of the seismic sequence and on the process of strong earthquake preparation as well as on triggering relationships between earthquakes.

An unquestionably crucial aspect in the analysis of the magnitude values' historical series is the identification of the graph patterns that may help to explain some of the dynamics issues related to the preparation of an earthquake, especially those anticipating an energy release cycle and allow us to identify the critical stages.

For example, the branching structure is a pattern that not only allows us to evaluate graphically the extent of the preparation process of an earthquake but also contributes to outline the past and future evolution of an area's seismicity.

In the initial developmental stages, the branching structures develop slowly, while in the final phases before the energy release stage, the decrease of the magnitude values is very rapid.

These are identified by considering that after an energy shock, the magnitude values gradually decrease up to the trigger point by following a hierarchization process.

Normally, each branching structure is connected with one or more medium-strong earthquakes concerning a certain area. In some cases, after the branching structure's trigger point formation, various orders foreshocks with increasing magnitude may form, which play a crucial role in the preparation phase of an earthquake.

In this case, the energy release stage is characterized by one or more foreshocks as well as by one or more mainshocks [4] [12] .

Citation: Giulio Riga, Paolo Balocchi (2016) Earthquake Forecast with the Seismic Sequence Hierarchization Method. Open Journal of Earthquake Research,05,79-96. doi:10.4236/ojer.2016.52006.

By using the graphic and quantitative analysis, we can learn the developmental stages of the branching structure, assess whether the phenomenon develops according to the expected standard scenario and know early the "hazard degree" (Short-Term Earthquake Probabilities) of the region examined, which heavily depends on space-time relationships between each event [13] . Therefore, the hazard degree is a function of the seismic sequence's developmental stage. If the area is located in the first developmental stages, the hazard degree of the area will be relatively low compared to the last developmental stages where it will feature a greater degree.

The analysis of the seismic sequence structure is performed using: 1) a datasets spatial coverage involving either parts of a region or several countries (from a few tens of km to thousands of km), 2) a magnitude values range of 2.0 - 10 m, 3) a depths range of 1 - 50 km.

BRANCHING STRUCTURE DEVELOPMENTAL PATTERN

The strongest earthquakes usually trigger aftershocks aimed at re-adjusting the new stress condition resulting from the strongest shock. The aftershocks show a peculiar pattern that develops over time following a hierarchization process identified as branching structure. Such structures are not born equal, but since the first developmental stage we can understand if they will generate a medium or strong earthquake. Indeed, the energy that a branching structure can release in the energy release stage, depends on the number and duration of the developmental stages and the average magnitude of the first developmental stage. The greater the duration, the number of the developmental stages and the average magnitude of the first developmental stage, the greater the released energy.

The entire branching structure preparation process can be divided into three phases: 1) energy accumulation; 2) triggering; 3) energy release (Figure 1) [4] .

1) Energy accumulation stage. It consists of five developmental stages related to temporal amplitude, magnitude and number of seismic events decreasing over time, which may even last for several years and in some cases constitutes 90% of the branching structure.

   The most energetic shocks take place during the first developmental stage.
2) Trigger stage. It consists of small magnitude shocks organized in an easily detectable micro-sequence called "Double Bottom-Three Small Earthquakes" (DB-3SE). This stage is easily recognizable when the last shock coincides with the point in which the elastic deformation probably reaches its peak local values (trigger point).
3) Energy release stage. During this stage, the energy is released immediately after the trigger point (Flash earthquakes type) or in one or more cycles of increasing amplitude and magnitude (Progressive earthquakes type), characterised by a seismic activity acceleration known as foreshocks [14] [15] . Foreshocks depend on different, still unclear physical mechanisms [16] - [18] and start to develop even hours/years before a strong earthquake

In summary, in the pattern of the branching structure the following apply:

1) The most energetic earthquakes occur, at irregular intervals, during the first part of the energy accumulation stage and in the final part of the energy release phase;
2) The amplitude of the branching structure's developmental stages tends to decrease over time in the energy accumulation stage and to increase in the energy release stage;
3) The energy can be abruptly (Flash earthquakes type) or gradually (Progressive earthquakes type) released, alternating standby with peak activity periods;
4) The energy release stage is activated by small magnitude earthquakes organized in simple DB-3SE micro- sequences wherein the last shock is the trigger point;
5) In the energy accumulation and release stages, multiple seismogenetic structures, at regional or interregional level, may be involved [19] - [21] ;
6) The aftershocks and foreshocks contribute to dissipate tectonic energy. Conversely, the cumulative effect of aftershocks increases it while foreshocks' lower it [22] .

Figure 1 shows the branching structure's pattern: we see that the first developmental stage (the widest) starts to develop after a strong earthquake, while the subsequent ones are related to the magnitude values decrease over time. The branching structure's trigger point determines the transition from the accumulation to the energy release stage that develops in a shorter time compared to the first.

METHODS

Linear Graph

The graph is one of the most important components in making both long- and short-term predictions process.

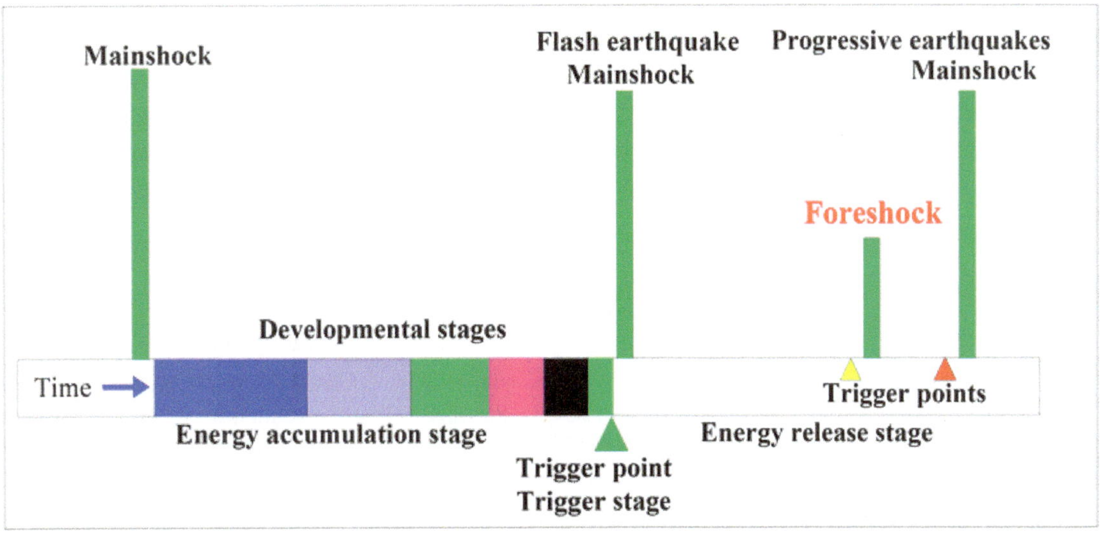

Figure 1. Branching structure's schematic representation. The variously colored rectangles indicate the developmental stages of the energy accumulation stage.

By representing the magnitude values as a linear graph, we can see how the magnitude values behave in the time window considered, identify the energy accumulation and release stage, detect the premonitory patterns [2] , set projections related to future magnitude values and strong earthquakes occurrence times.

The time frame used to process a short term graph can be daily/weekly, while for the medium/long-term is monthly/ yearly. Obviously, the wider the reference window, the greater the details shown in the graph.

The line segment that connects two points on the graph shows a local stage, which can be visually interpreted compared to other lines' by means of a coded color.

Figure 2 shows the linear graph of the magnitude values whose reading is quite simple. As we note, the seismic sequence develops by alternating energy release with adjustment stages. Black-colored circles indicate the magnitude points M in which the reverse of the cyclical adjustment (red-colored lines) and energy (green- colored lines) stages occur. Both the energy release and adjustment stages may consist of one or more seismic events.

Although this type of graph is perfect for a quick and immediate glimpse of the seismic sequence developmental structure, we need to examine it in depth to obtain more detailed information.

For an early identification of the energy accumulation and release stages, we need to join the relative maximum points with a straight line both in case of downward and upward magnitude values (Figure 3).

These straight lines are called dynamic trendlines (magnitude values vary with the time) and clearly indicate the direction in which the values develop, both in the forthcoming and in the long term period.

The upward breaking of the breakout trendline activates the energy release stage, which is characterized by one or more mainshoks (Flash earthquakes type) or by a first foreshock (Progressive earthquakes type) followed a first adjustment shock, which involves the magnitude values' return to somewhere close or below the energy accumulation trendline. Then the values begin to rise again until a second foreshock occurs, which allows us to understand the direction of the energy release trandline.

Typically, although with the Progressive earthquakes type we can record up to four foreshocks, most frequently the pattern consists of two foreshocks and one mainshock.

Besides indicating the ongoing trend, the trendlines temporarily prevent the magnitude values from rising by allowing, in some cases, the early detection of certain levels of magnitude that will be achieved in the future.

Actually, by analyzing the graph shown in Figure 3, we can see how the knowledge of the magnitude levels achieved in points 1 and 2 allows us to calculate the magnitude value in points 3 and 4 of the energy accumulation stage or the magnitude value in point 6 of the energy release stage after the the points 4 and 5 have formed. The point 4, which is common to the two phases, is the transition point from the energy accumulation and the energy release stage.

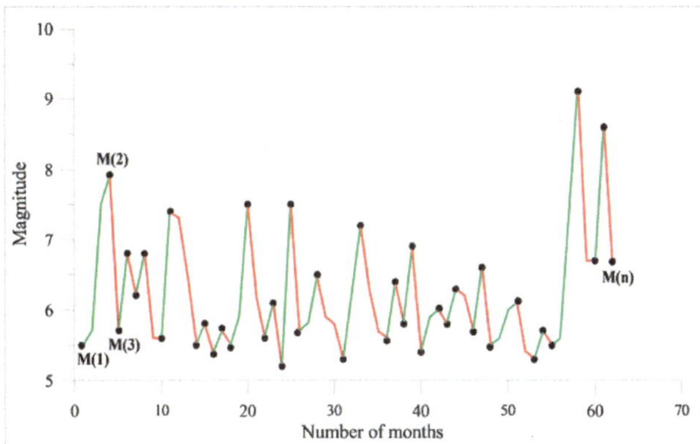

Figure 2. Earthquake linear graph, Japan, 2011.

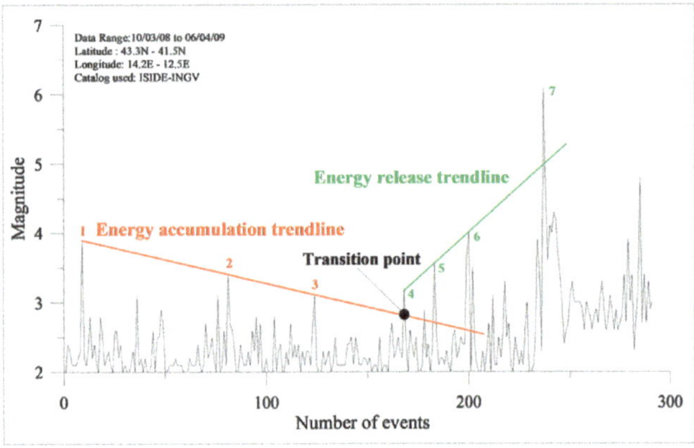

Figure 3. Earthquake linear graph, L'Aquila (Italy), 2009. The red-colored line indicates the energy accumulation trendline, while the green-colored one indicates the energy release trendline. The black-colored circles represent the transition point between the accumulation and the energy release stage.

DB-3SE Microsequence-Generated Trigger Point

Some studies performed on seismic sequences show how earthquakes of any magnitude can trigger other earthquakes [23] [24] that have different triggering properties [25] . Generally, low magnitude earthquakes are greater in number compared to those of higher magnitude: therefore, they collectively can trigger also big seismic events [24] [26] due to the stress in the surrounding area [27] . Through the study of seismic sequence, and with appropriate calculation algorithms it is possible to distinguish the seismic events triggered by the background activity [8] .

By analyzing the magnitude values trend occurring before the start of an energy release stage, we can identify a small shocks microsequence that may be used for the prediction of immediate and short term medium/high magnitude earthquakes. In fact, the energy release stage's trigger point coincides with the third shock in the DB-3SE microsequence, which consists of three, small magnitude shocks and is a very reliable stage-reverse signal.

Once completed the last phase of the branching structure's energy accumulation stage, in order to locate the first trigger point and the following ones, in the case where the energy release stage develops according to a Progressive earthquakes-type pattern, it is necessary that the magnitude values of the shocks that make up the temporary DB-3SE microsequence (Figure 4(a)) meet the following conditions:

$$M(n-1) \le M(n) > M(n+1) \quad (1)$$

$$M(n-1) < M(n) \ge M(n+1) \quad (2)$$

and

$$M(n) < M(n-2) \quad (3)$$

where M(n) is the second shock's magnitude value, while M(n − 1) and M(n + 1) are the magnitude values of the first and third shock, respectively.

For the completion of the DB-3SE microsequence, we need a subsequent shock whose magnitude must be:

$$M(n+2) \ge M(n)$$

(Figure 4(b)). (4)

Figure 4 shows both temporary and final DB-3SE microsequences, while the red-colored triangle indicates the the energy release stage's trigger point.

The trigger point coincides with the M(n + 1) magnitude shock in the DB-3SE microsequence, while the subsequent final shock often corresponds to a more energetic event (foreshock or mainshock).Figure 5 shows the four possible patterns of the final DB-3SE microsequence, depending on the magnitude values of the three shocks which it consists of (red-colored circles).

Seismic Sequence Hierarchization

A seismic sequence may be considered as a succession of cyclical shocks with variable amplitude over time where events occur following a particular hierarchization process during which each earthquake can generate several earthquakes [28] .

The herarchization process's basic concept is that a source event (strong earthquake), can generate seismic events of decreasing magnitude during the energy accumulation stage. In essence, these seismic events related to the source event and classified as secondary [29] , may also, in turn, generate other minor seismic events, thus triggering a process that continues up to the energy release phase's trigger point and may end after a strong event.

The number of seismic events included in a seismic cycle (adjustment and energy release stages) depends on the seismicity bottom value (i.e., the minimum value in the analyzed magnitude values' time series). The greater the bottom value, the smaller the number of events the cycle will consist of.

The use of a herarchization process in the seismic sequence analysis, allowed us to develop a graph pattern that effectively contributes to outline the past and future development of the seismic sequence. In fact, all the adjustment and energy release stages included in a seismic sequence constitute the "seismic grid", whose development is probably related to various mechanisms involving, among the others, the lithospheric rocks' deformation process and the interaction between the different seismogenetic structures on regional or interregional scale.

The seismic grid can be represented by some geometric entities, such as the "seismic nodes", i.e. the contact points between the two phases, the "reverse phase nodes" that are the transition points between the adjustment and the energy release stage, and, finally, the energy release (r) and accumulation (a) "seismic branches" that make up a seismic cycle (Figure 6).

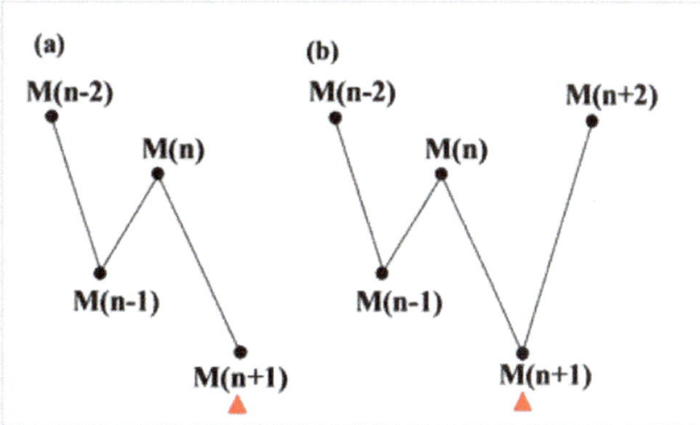

Figure 4. Schematic representation of DB-3SE microsequence: (a) temporary microsequence, (b) final microsequence.

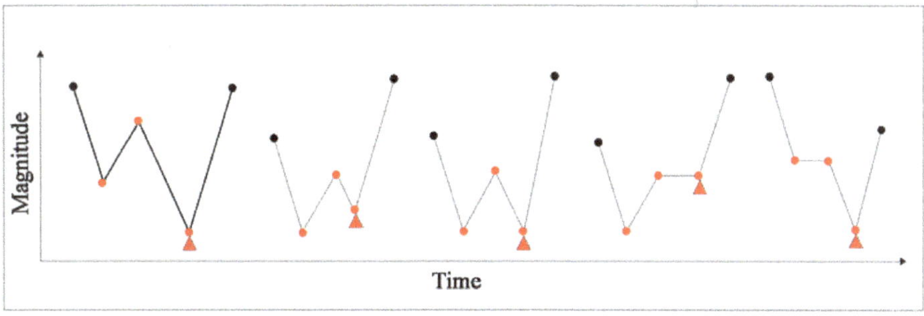

Figure 5. Schematic representation of microsequence DB-3SE. The red-colored circles indicate the "DB-3SE" microsequence while the red-colored triangle indicates the trigger point. The black-colored circles represent the microsequence's initial and final shocks.

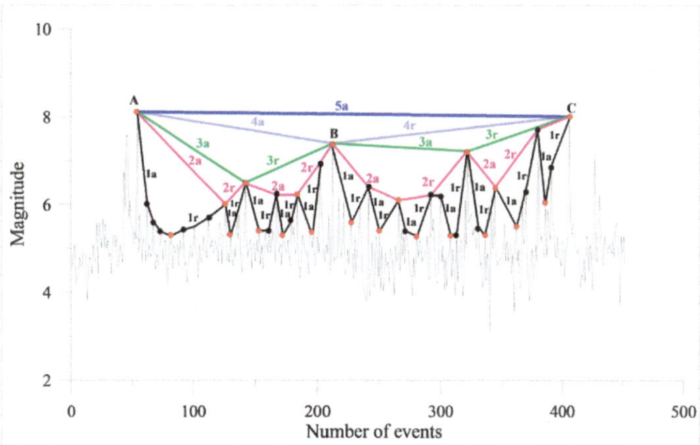

Figure 6. Schematic representation of the seismic sequence hierarchy. Each circle represents a seismic node, while the red-colored ones are the reverse phase nodes. The lines of different thickness and color are the earthquakes' branches. The higher order AC branch represents the higher order seismic cycle consisting of the energy accumulation (4a) and release (4r) stages. We see that the magnitude value is greater in the reverse nodes A, B and C related to higher order seismic branches.

As to the branches arrangement, the seismic grid can be arranged by applying a hierarchization method that consists in assigning an order number to the various branches that make up the seismic grid according to the following rules:

- One or more branches of n order energy accumulation followed by one or more n order energy release branches result in a n + 1 order branch (the hierarchical order increase occurs);
- The magnitude value of the "reverse phase nodes" increases as the cycle order increases;
- The higher order branch (5A) is the higher order seismic cycle.

In general, in the various cycles we can observe the proportionality principle, i.e. the higher is the order of the energy accumulation seismic branch, the greater the expected magnitude in the energy release stage.

Moreover, the smaller the inclination of the higher order energy accumulation branch, the greater the time or the number of shocks needed for the energy accumulation stage to be completed.

The end of a (n) order seismic cycle usually takes place after the amplitude of the (n − 1) order energy release branch has reached a value equal to the amplitude of the (n − 1) order energy accumulation branch plus 50% of it.

We can obtain more accurate data on the developmental level of a seismic grid by removing the seismic branches related to the energy release stage (Figure 7). The main branching structure obtained (S,A,B,C,D) retains the remote seismic events and the pre-earthquake seismic context's effects which we can obtain information from in order to determine its future development.

Note how the various orders seismic branches begin to form after a strong earthquake (source node S-source node). and are related to the magnitude values decrease over time. We also see how from the early stages of the seismic grid formation, it is possible to identify the branching structures that, depending on higher order seismic branch, will produce shocks of average magnitude or strong earthquakes in the energy release stage (the greater the order of the first seismic branch that originates from a source point, the greater the expected magnitude). Actually, the branching structures are not all the same but may take different appearances: their duration and the energy they release in the energy release stage, depend on the magnitude values of higher order branch and the time arrangement of low order subsequent seismic branches.

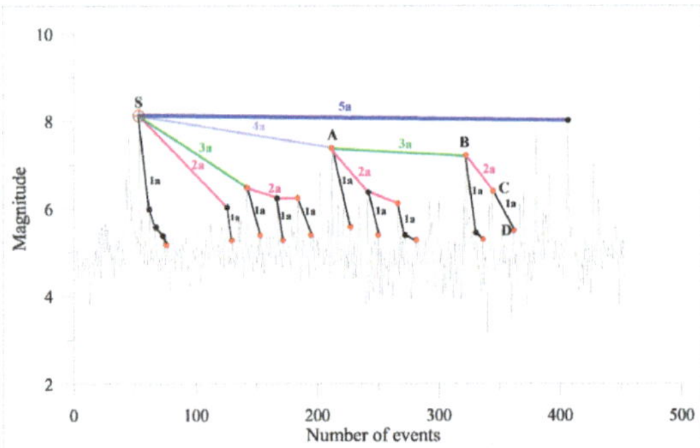

Figure 7. Schematic representation of the branching structure obtained by removing the energy release stage's seismic branches. The red-colored circles represent the reverse node of all the branching structure identified, while the letters S, A, B, C and D indicate those in the main branching structure (the point S is the source-point).

A branching structure is perfectly hierarchical when each (n) order branch is followed by another that features an immediately lower order (n – 1). If this is not the case, a seismic anomaly occurs. In addition, when the seismic branches' order decreases, the time or the number of shocks that we need to wait for the trigger stage to be formed, which precedes the energy release stage, decreases as well.

This means that only after the formation of the (1a) first order branch, the branching structure enters a critical stage where the DB-3SE microsequences generate the trigger point that can be profitably used for short term forecast.

Graphical Procedure for Branching Structures Identification

The long-term, monthly hierarchization of the seismic sequence (from 1973 to 2004) in Indonesia, which consists of any single maximum magnitude value monthly recorded by assuming a bottom value of 2.0 M, can be implemented by identifying the reverse nodes and following the multi-step procedure as follows:

1) Locate the first order reverse nodes P1(n) that meet the following condition:

$$M\left(n-1\right)<M1\left(n\right)\geq M\left(n+1\right) \qquad (5)$$

where, M1(n) is the value of the first order reverse node magnitude and M(n – 1) and M(n + 1) are the magnitude values of seismic events that precede and follow the first reverse seismic node (Figure 8), respectively;

2) Locate the second order reverse nodes P2(n) that meet the following condition:

$$M1\left(n-1\right)<M2\left(n\right)\geq M1\left(n+1\right) \qquad (6)$$

where, M2(n) is the magnitude value of the second order reverse node and M1(n – 1) and M1(n + 1) are the magnitude values of the first order reverse nodes that precede and follow the second order reverse node;

3) Draw the black-colored line joining the second order seismic nodes;

4) Repeat the same procedure to locate the second, third and fourth order reverse nodes (magenta-, green- and pastel blue-colored lines) (Figure 9);

5) Locate the line joining the second order reverse nodes (black-colored lines), P2m(n) second order minimum reverse nodes whose magnitude must comply with the following condition:

$$M2\left(n-1\right)\geq M2m\left(n\right)<M2\left(n+1\right) \qquad (7)$$

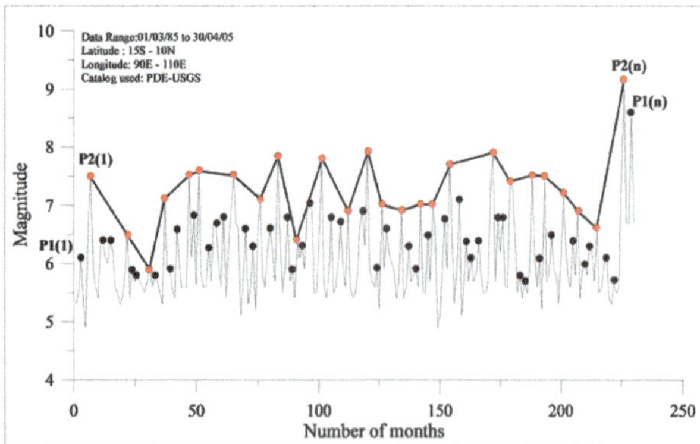

Figure 8. Earthquake's seismic sequence, Indonesia, 2004. Representation of the first order reverse nodes (black- and red- colored circles) and the second order reverse nodes (red-colored circles).

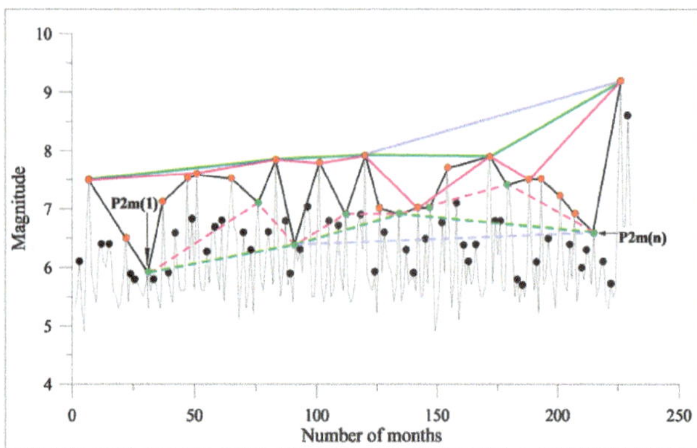

Figure 9. Fluctuation of various orders reverse nodes. The solid-colored line indicates various orders seismic branches that join the maximum reverse nodes, while the dashed colored lines indicate various orders seismic branches that join minimum reverse nodes.

where M2m(n) is the magnitude value of the second order, minimum reverse node and M2(n − 1) and M2(n + 1) are respectively the magnitude values of the second order reverse nodes that precede and follow the second order minimum reverse node (Figure 8);

6) Join the second order minimum reverse nodes with a line (magenta-colored dashed line);
7) Repeat the same procedure to draw higher order green, pastel blue and blue-colored dashed lines.

In order to identify various order branching structures we need to remove from the graph the energy release seismic branches as well as the descending seismic branches that join the minima (dashed lines).

Moreover, we need to identify the triggering stages (TS) of the branching structures, by drawing the line joining the lower extremity of the first order seismic branch with the trigger point generated by the DB-3SE microsequence (light green—colored line).

The colors assigned to the seismic branches and triggering stage let us determine the development status achieved by the various order branching structures.

Results

Figure 10 displays several various orders branching structures (BR1, BR2 and BR3) that have developed over time. The greatest (BR3) consists of an energy accumulation stage characterized by four developmental stages and a flash earthquake-type energy release stage featuring a mainshock with a magnitude of 9.1 Mw on 26-12-2004.

During the branching structures' early development and completion stages, several various orders trigger points were generated. In the graph, the yellow-colored triangles, placed at the extremities of the minimum ascending branches (magenta-colored dashed lines), represent the first order trigger point, while the red-colored triangles indicate the second order ones generated by the DB-3SE mirosequence.

In the lower branching structure (A,B and C), we note the abnormal formation of a fourth order branch that generated a trigger point in the node C (yellow triangle) just before the mainshock.

The same procedure can be used to locate the greater magnitude aftershocks in the energy accumulation phase.

Figure 11 shows the branching structures that have developed after the earthquake occurred Indonesia on 26 December 2004 (source node).

In particular, we observe a fourth order branching structure, which on 28 March 2005 the first significant aftershock developed from with a magnitude of 8.6 Mw, that originated the first, fourth order seismic branch.

Modified Aroon Oscillator

The Aroon oscillator [30] was modified so as to be able to obtain adequate information on the type of branching structure (primary or secondary) being formed and on the developmental stages' hierarchization degree (whether a branching structure is located in an accumulation or release stage).

The oscillator features a red line that indicates the energy accumulation stage and a green line indicating the energy release stage (Figure 12). The scale used by the oscillator ranges from 0% to 100%.

When the values are increasing, the oscillator indicates an energy release stage in progress, while when the values are decreasing, it indicates that an accumulation energy stage is being formed. The farther the oscillator from the value of 50%, the stronger the current phase. Usually, the oscillator is decreasing if seismic branching with the same order is subsequent to each other.

There are other basic situations for interpreting the indicator:

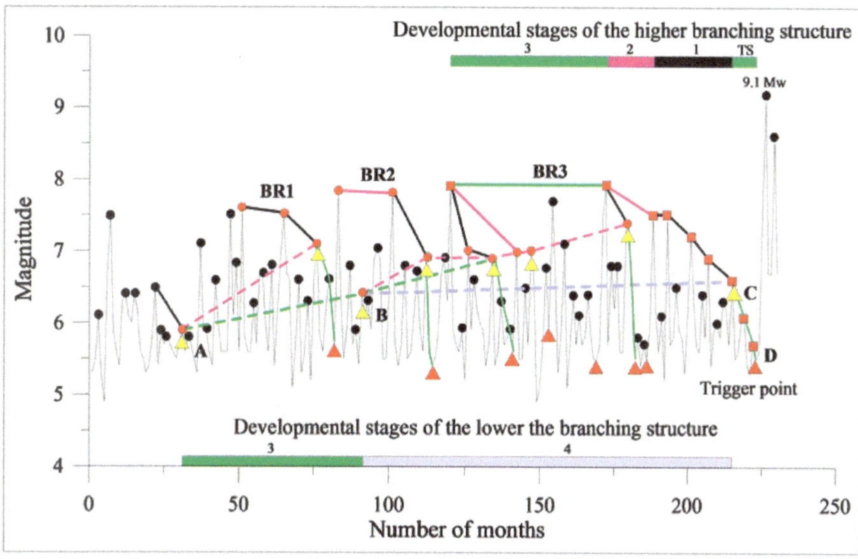

Figure 10. Earthquake's seismic sequence, Indonesia, 2004. The red colored squares show the higher order nodes of the branching structure, while the red-colored circles indicate the lower order ones, and the yellow- and red-colored triangles indicate the first and the second order trigger points, respectively. The upper part of the graph shows the developmental stages of the higher branching structure, while in the lower part the developmental stages of the lower branching structure less are reported.

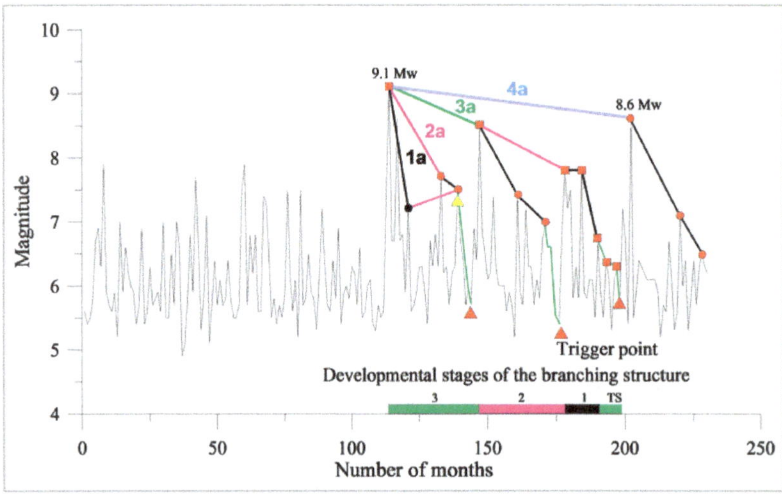

Figure 11. Earthquake's seismic sequence, Indonesia, 2004. Pattern of the branching structure related to the energy accumulation stage. The red-colored square indicates the nodes of the higher order branching structure, while the yellow and red triangles indicate the first and second order trigger points, respectively. The lower part of the graph shows the developmental stages of the branching structure associated with the first significant aftershock.

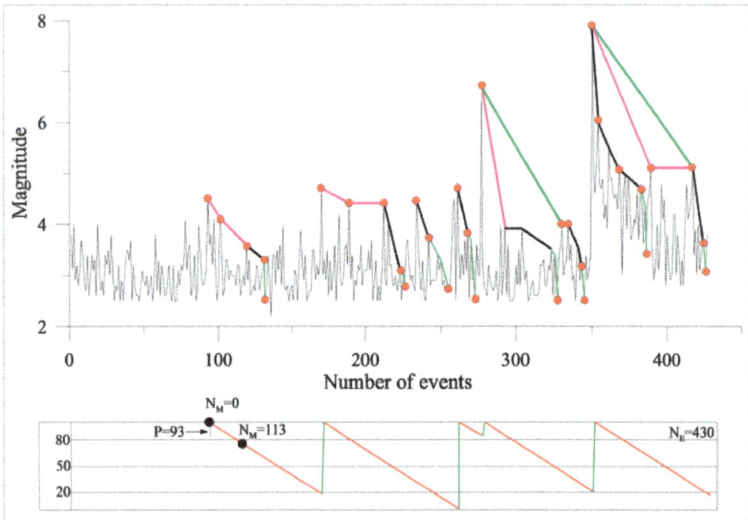

Figure 12. Schematic representation of modified Aroon oscillator. The red-colored lines indicate the energy accumulation stages, while the green lines indicate the energy release stages. The black-colored circles indicate the calculation points.

1) When the oscillator decreases from 100% to 80%, it indicates the beginning of an energy accumulation stage.
2) When the oscillator decreases 20% to 0%, it indicates that energy accumulation stage is being completed.
3) When the oscillator increases from 80% to 100%, it indicates that the energy release stage is being strengthened;

The formula to calculate the modified Aroon oscillator is the following:

$$AR = \left[\left(P - \left(N - N_M \right) \right) / P \right] \cdot 100 \quad (8)$$

where P (reference period) is the number of seismic events, days or months from the beginning of the seismic sequence to the first reference absolute or relative maximum value (mainshock or average magnitude shock), N is the oscillator's calculation point in relation to the beginning of the seismic sequence and N_M is the number of seismic events, days or months from the absolute or relative maximum value that has formed during the reference period with respect to the beginning of the sequence.

Usually the period P must be such that the already formed source points coincide with the points at which the oscillator reaches 100% (default p = 42).

The algorithm provides for a calculation cycle from N = P up to N_E (total number of events in the seismic sequence) in which we need to establish the position of the reference period maximum with respect to the origin of the seismic sequence through a second internal cycle of calculation from $N_M = N - (P - 1)$ up to N, and subsequently calculate the modified Aroon oscillator.

For example, if the period consists of 93 events and the maximum (N_M) and the calculation point (N) of the oscillator are equal to zero, we will have $AR = \left[(93 - (0 - 0))/93 \right] \cdot 100 = 100\%$. If the maximum ($N_M$) occurred in 93 past events and the oscillator's calculation point is N = 113, then the oscillator's value is $AR = \left[(93 - (113 - 93))/93 \right] \cdot 100 = 78\%$.

Maximum Magnitude Assessment Procedure

Figure 13 shows the branching structures' patterns, which have been identified in the Japan's monthly seismic sequence from 1973 to 31 August 2015.

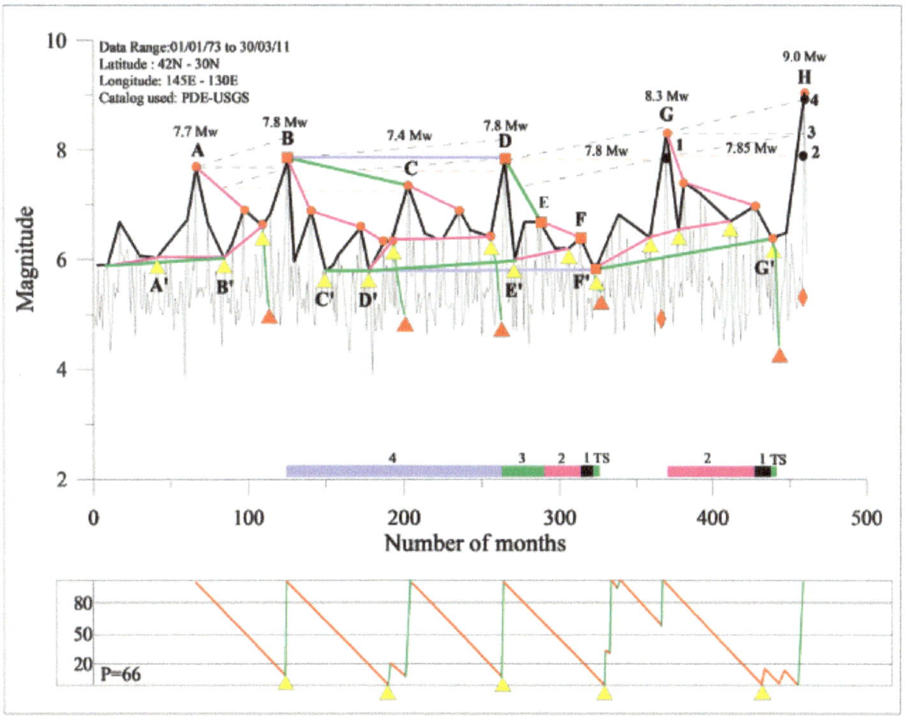

Figure 13. Japan's monthly sequence. The red-colored squares indicate the higher order nodes of the branching structure of higher order, the red circles the minor order ones, while the red and triangles and diamonds indicate the first, second and third order trigger points, respectively. In the lower part of the graph the modified Aroon oscillator is displayed.

The red squares indicate the main branching structure's nodes, while the red circles indicate the minor branching structures' ones.

The letters A, B, D, G and H denote the higher seismic nodes that correspond to the strongest earthquakes recorded in the time interval considered, while the letters A', B', C', D', E', F' and G' are the nodes relating to minor, second order ascending seismic branches included in the lower branching structure that correspond to first order trigger points (yellow-colored triangles).

The red-colored triangles and diamonds, generated by the DB-3SE microsequence upon the higher branching structure's completion and during the energy release stage, indicate the second and third order trigger points.

The development dynamics of the branching structure suggests to compute the magnitude value of the strongest earthquake in correspondence with the highest order branch.

From the mid-point of the higher order branch (point 1) included in the branching structure that preceded the earthquake occurred on 11 March 2011, we need to draw an horizontal, dashed line extending to the right up to the calculation point of the expected magnitude value (point 2). The intersection of this half line with the vertical line passing through the set calculation point, provides the minimum magnitude value (red-colored dashed line).

To calculate the maximum value we need to draw a horizontal dashed line from the point G that extends to the right up to the calculation point (point 3). Then we join the point 3 with the mid-point of the 2 order branch (point 1); from the point G we finally draw a parallel line to the segment 1 - 3 (black-colored dashed line).

The intersection of this parallel line with the vertical line passing through the calculation point, gives the maximum magnitude value (point 4).

We should however bear in mind that not always these minimum and maximum values are accurately reflected in the energy release stage. In the case where the range between the minimum and the maximum value is very wide, we can consider the mean value between the minimum and maximum values computed as the maximum value.

Based on Figure 14, which shows in details the energy release stage of the big earthquake occurred on 11 March 2011, with a magnitude of 9.0 Mw, we can formulate some interesting considerations about the characteristics of the higher and lower branching structures that preceded the earthquake.

First, we can observe that the first order trigger point coincided with the extremity of the third order lower seismic branch (green line) and that the second and third order trigger points (red box-included red triangle and diamond) that preceded the mainshock were generated by the DB-3SE microsequence.

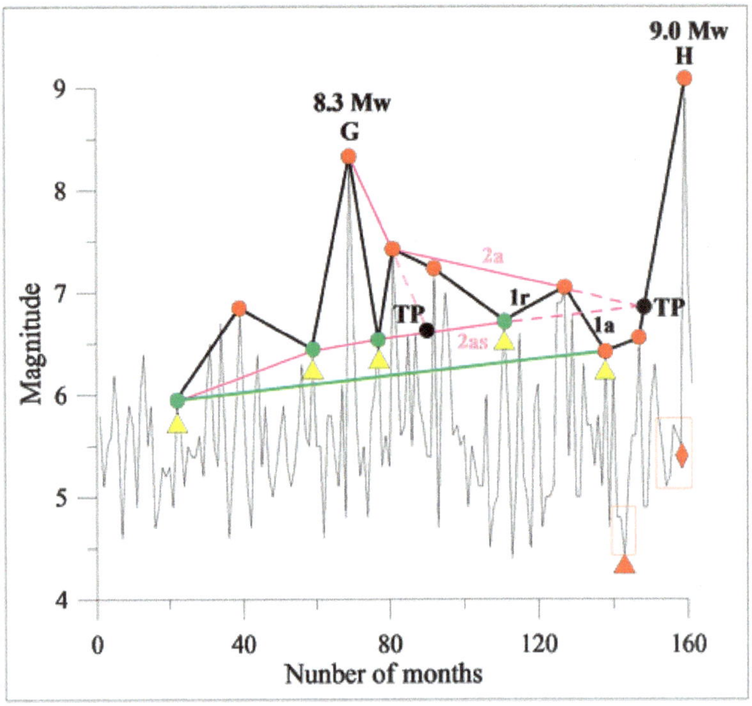

Figure 14. Energy release stage (Japan's monthly sequence). The red circles indicate the higher order nodes of the branching structure, the green circles represent the lower order ones, while the yellow and red triangles and the red diamonds indicate the first, second and third order trigger points. The black-colored circles indicate the trigger points generated by the intersection between the higher second order branches of energy accumulation (2a) lower and ascending magenta-colored ones (2as).

In addition, by means of a simple graphic procedure we identified a further trigger point (TP) generated by second order higher branches of the energy accumulation stage (2a) and lower ascending, magenta-colored branches (2as). In this case, the next DB-3SE microsequence (red diamond) represents a second order trigger point.

If the TP falls before the lower extremity of the first order energy accumulation branch (1a), we consider the latter as the trigger point.

Figure 15 shows the branching structures that have developed in the time interval between 1973 and 31-08-2015 in the Southern California's seismic sequence. In particular, we note the well hierarchized branching structure that has generated the earthquake occurred in Landers on 28-06-1992, with a magnitude of 7.3 Mw, which consisted of two developmental stages with the origin in the node B (source).

The TP was followed by first, second and third order trigger points. The maximum magnitude value calculated with the graphical method starting from the second order branch's mid-point was close to the one recorded.

The Landers earthquake has subsequently generated a third order branching structure (C,E,F,G,E') from which, after the second order trigger point, a hierarchical abnomaly has developed in the node H (seismic branches FH-HF') consistinting of a third order branch directly followed by a first order one. The energy release stage that followed was of Progressive earthquakes-type, where the strongest aftershocks, including the 7.2 Mw ones occurred on 04-04-2010 (Baja California, Mexico), were preceded by first and second order trigger points (yellow- and red-colored triangles). The TP was generated shortly before the strongest shock and after the second order trigger point. Subsequently, the earthquake on 04-04-2010 has activated a new branching structure characterized by a first, fourth order final seismic related to the energy accumulation (CD branch).

The minimum temporary magnitude value associated with this seismic bomb developmental phase is 7.15 M.

EXAMPLES OF BRANCHING STRUCTURES

The examples shown in Figure 16, processed from the data in Table 1, show the main branching structures

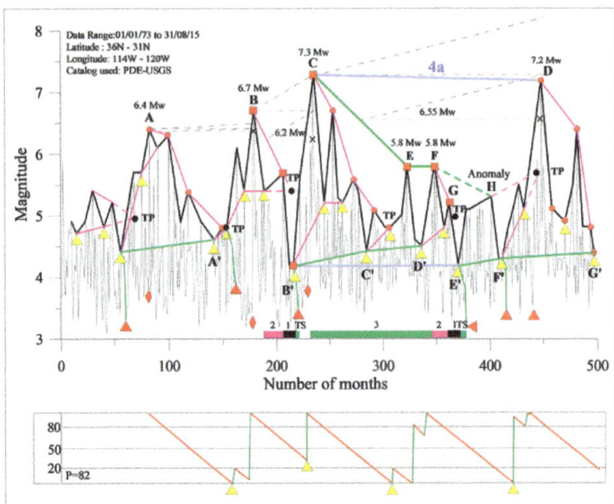

Figure 15. Southern California (US)'s monthly sequence. Red colored squares represent the higher order branching structure nodes, red circles the minor order ones, while the yellow and red triangles and the red diamonds indicate the first, second and third order trigger points, respectively. In the lower part of the graph developmental stages of the main branching structure are reported.

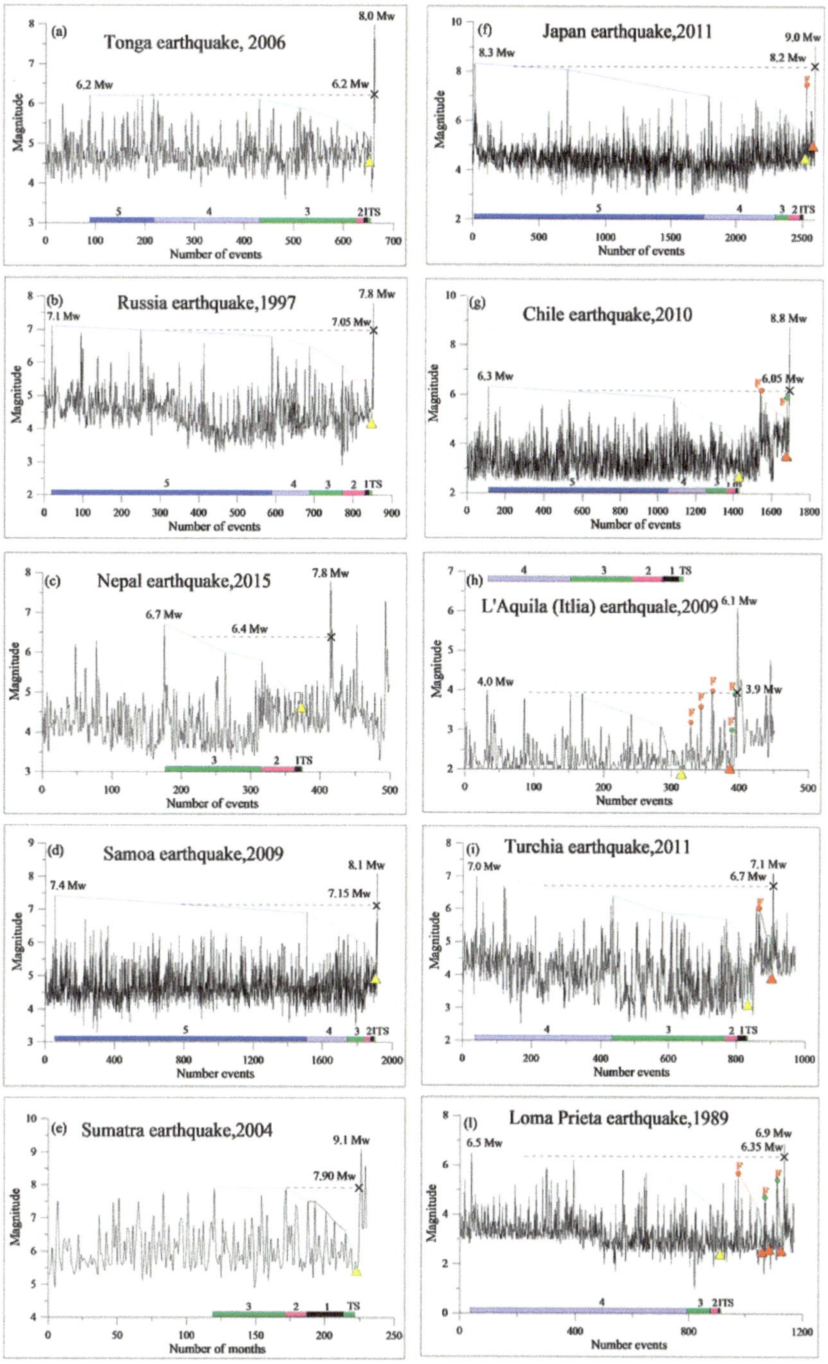

Figure 16. Examples of branching structure. The colored lines indicate the developmental stages, the letter "F" the foreshocks, the red circles the first order foreshocks and the green circles the second order foreshocks. The yellow and red triangles indicate the rigger points of main and secondary branching structures.

Table 1. Earthquakes, date range, coordinates of the region analyzed and catalogue used.

Earthquakes	Date Range	Latitude	Longitude	Catalog Used
L'Aquila—Italy	27/04/2007 - 06/04/2009	43.2 N - 41.5 N	14.2 E - 12.5 E	ISIDE-INGV
Loma Prieta—Ca, USA	24/09/1982 - 18/10/1989	38.0 N - 36.0 N	120.0 W - 123.0 W	PDE-USGS
Turkey	09/04/1988 - 24/10/2011	44.0 N - 36.0 N	44.0 E - 36.0 E	PDE-USGS
Japan	25/09/2003 - 11/03/2011	42.0 N - 30.0 N	130.0 E - 145.0 E	PDE-USGS

Chile	03/09/2006 - 27/02/2010	30.0 S - 40.0 S	63.0 W - 80.0 W	PDE-USGS
Samoa	09/01/1995 - 29/09/2009	12.0 S - 20.0 S	168.0 W - 176.0 W	PDE-USGS
Sumatra	01/01/1981 - 30/04/2005	90.0 S - 130.0 S	10.0 E - 15.0 E	PDE-USGS
Nepal	18/07/2002 - 12/05/2015	32.0 N - 26.0 N	88.0 E - 81.0 E	PDE-USGS
Tonga	22/03/2003 - 03/05/2006	16.0 S - 24.0 S	179.0 W - 178.0 W	PDE-USGS
Russia	28/06/1990 - 05/12/1997	59.0 N - 51.0 N	170.0 E - 154.0 E	PDE-USGS

that have developed before some strong earthquakes happened in the past in various parts of the world.

The branching structures, identified with the long term, seismic sequence hierarchization, show that big earthquakes can trigger additional earthquakes with decreasing magnitude during the energy accumulation stage. After the trigger point formation (yellow triangle), we notice some energy release stages of flash earthquakes (Figures 16 (a)-(e)) or progressive earthquakes type (Figures 16(f)-(l)). Comprehensive retrospective tests show that the proposed methodology is reasonably stable upon changes in the size of the seismic area and therefore it is able to locate all the branching structures that develop during the energy accumulation and release stages.

CONCLUSIONS

This study has the aim to propose a model for the graphical analysis of seismic sequences, starting from the assumption that the sequences do not develop in an unpredictable manner, but follow easily identifiable trends.

As a result, we can say that some developments recur cyclically, making it possible, if identified on time, the formulation of hypotheses on the future development of a seismic sequence.

The proposed model, based on the earthquakes time series' hierarchization, has shown that the seismic sequences develop according to similar developmental stages and that the mainshock magnitude value associated with a seismological branching sequence depends on the first developmental stage (higher order seismic branch).

This approach develops a dramatically innovative earthquake forecasting model starting from the seismicity analysis conducted on an area of different width, which provides a reasonable estimate of the expected magnitude.

REFERENCES

1. Ogata, Y. (1999) Seismicity Analysis through Point-Process Modeling: A Review. Pure and Applied Geophysics, 155, 471-507. http://dx.doi.org/10.1007/s000240050275
2. Helmstetter, A. and Sornette, D. (2002) Diffusion of Epicenters of Earthquake Aftershocks, Omori's Law, and Generalized Continuous-Time Random Walk Models. Physical Review E, 66, Article ID: 061104.http://dx.doi.org/10.1103/PhysRevE.66.061104
3. Gu, C., Schumann, A.Y., Baiesi, M. and Davidsen, J. (2013) Triggering Cascades and Statistical Properties of Aftershocks. Journal of Geophysical Research: Solid Earth, 118, 4278-4295. http://dx.doi.org/10.1002/jgrb.50306
4. Riga G. and Balocchi, P. (2016) Seismic Sequence Structure and Earthquakes Triggering Patterns. Open Journal of Earthquake Research, 5, No. 1.http://www.scirp.org/Journal/PaperInformation.aspx?PaperID=63538
5. Parsons, T. and Velasco, A.?A. (2009) On Near-Source Earthquake Triggering. Journal of Geophysical Research, 114, Article ID: B10307.http://dx.doi.org/10.1029/2008JB006277
6. Zhuang, J., Ogata, Y. and Vere-Jones, D. (2002) Stochastic Declustering of Space-Time Earthquake Occurrences. Journal of the American Statistical Association, 97, 369-380. http://dx.doi.org/10.1198/016214502760046925
7. Zhuang, J., Ogata, Y. and Vere-Jones, D. (2004) Analyzing Earthquake Clustering Features by Using Stochastic Reconstruction. Journal of Geophysical Research, 109, Article ID: B05301. http://dx.doi.org/10.1029/2003JB002879
8. Baiesi, M. and Paczuski, M. (2004) Scale-Free Networks of Earthquakes and Aftershocks. Physical Review E, 69, Article ID: 066106.http://dx.doi.org/10.1103/PhysRevE.69.066106
9. Baiesi, M. and Paczuski, M. (2005) Complex Networks of Earthquakes and Aftershocks. Nonlinear Processes in Geophysics, 12, 1-11. http://dx.doi.org/10.5194/npg-12-1-2005
10. Zaliapin, I., Gabrielov, A., Keilis-Borok, V. and Wong, H. (2008) Clustering Analysis of Seismicity and Aftershock Identification. Physical Review Letters, 101, Article ID: 018501.http://dx.doi.org/10.1103/PhysRevLett.101.018501
11. Marsan, D. and Lengliné, O. (2008) Extending Earthquakes' Reach through Cascading. Science, 319, 1076-1079. http://dx.doi.org/10.1126/science.1148783
12. Mogi, K. (1963) Some Discussions on Aftershocks, Foreshocks and Earthquake Swarms—The Fracture of a Semi- Infinite Body Caused by an Inner Stress Origin and Its Relation to the Earthquake Phenomena (Third

Paper). Bulletin of the Earthquake Research Institute, 41, 615-658.

13. Van Stiphout, T., Wiemer, S. and Marzocchi, W. (2010) Are Short-Term Evacuations Warranted? Case of the 2009 L'Aquila Earthquake. Geophysical Research Letters, 37, L06306. http://dx.doi.org/10.1029/2009gl042352
14. Jones, L.M. and Molnar, P. (1976) Frequency of Foreshocks. Nature, 262, 677-679.http://dx.doi.org/10.1038/262677a0
15. Abercrombie, R.E. and Mori, J. (1996) Occurrence Patterns of Foreshocks to Large Earthquakes in the Western United States. Nature, 381, 303-307.http://dx.doi.org/10.1038/381303a0
16. Keilis-Borok, V.I. and Malinovskaya, L.N. (1964) One Regularity in the Occurrence of Strong Earthquakes. Journal of Geophysical Research, 69, 3019-3024.
17. Jaumé, S.C. and Sykes, L.R. (1999) Evolving Towards a Critical Point: A Review of Accelerating Seismic Moment/ Energy Release Prior to Large and Great Earthquakes. In: Wyss, M., Shimazaki, K. and Ito, A., Eds., Seismicity Patterns, Their Statistical Significance and Physical Meaning, Birkhauser, Basel, 279-305. http://link.springer.com/chapter/10.1007%2F978-3-0348-8677-2_5http://dx.doi.org/10.1007/978-3-0348-8677-2_5
18. Sammis, S.G. and Sornette, D. (2002) Positive Feedback, Memory and the Predictability of Earthquakes. Proceedings of the National Academy of Sciences of the United States of America, 99, 2501-2508. http://arxiv.org/pdf/cond-mat/0107143v1.pdfhttp://dx.doi.org/10.1073/pnas.012580999
19. Ishibashi, K. (1988) Two Categories of Earthquake Precursors, Physical and Tectonic, and Their Roles in Intermediate- Term Earthquake Prediction. Pure and Applied Geophysics, 126, 687-700. http://dx.doi.org/10.1007/bf00879015
20. Shebalin, P. (2006) Increased Correlation Range of Seismicity before Large Events Manifested by Earthquake Chains. Tectonophysics, 424, 335-349. http://dx.doi.org/10.1016/j.tecto.2006.03.040
21. Mantovani, E., Babbucci, D., Tamburelli, C. and Viti, M. (2009) A Review on the Driving Mechanism of the Tyrrhenian-Apennines System: Implications for the Present Seismotectonic Setting in the Central-Northern Apennines. Tectonophysics, 476, 22-40.http://dx.doi.org/10.1016/j.tecto.2008.10.032
22. Ziv, A. and Schmittbuhl, J. (2003) The Seismic Cycle and the Difference between Foreshocks and Aftershocks in a Mechanical Fault Model. Geophysical Research Letters, 30, 2237.https://eost.unistra.fr/fileadmin/upload/EOST/Jean_Schmittbuhl/Publications/Articles/Zi-Sc03.pdf
23. Helmstetter, A. and Sornette, D. (2003) Foreshocks Explained by Cascades of Triggered Seismicity. Journal of Geo- bphysical Research: Solid Earth, 108, 2457.
24. Helmstetter, A., Kagan, Y.Y. and Jackson, D.D. (2005) Importance of Small Earthquakes for Stress Transfers and Earthquake Triggering. Journal of Geophysical Research: Solid Earth, 110, Article ID: B05S08.http://onlinelibrary.wiley.com/doi/10.1029/2004JB003286/full
25. Shearer, P.M. (2012) Space-Time Clustering of Seismicity in California and the Distance Dependence of Earthquake Triggering. Journal of Geophysical Research: Solid Earth, 117, B10306. http://dx.doi.org/10.1029/2012jb009471
26. Helmstetter, A. (2003) Is Earthquake Triggering Driven by Small Earthquakes? Physical Review Letters, 91, Article ID: 058501.http://dx.doi.org/10.1103/PhysRevLett.91.058501
27. Felzer, K.R. and Brodsky, E.E. (2006) Decay of Aftershock Density with Distance Indicates Triggering by Dynamic Stress. Nature, 441, 735-738.http://dx.doi.org/10.1038/nature04799
28. Marzocchi, W. and Lombardi, A.M. (2008) A Double Branching Model for Earthquake Occurrence. Journal of Geo- physical Research: Solid Earth, 113, Article ID: B08317.http://dx.doi.org/10.1029/2007JB005472http://onlinelibrary.wiley.com/doi/10.1029/2007JB005472/pdf
29. Werner, M.J. (2008) On the Fluctuations of Seismicity and Uncertainties in Earthquake Catalogs: Implications and Methods for Hypothesis Testing. University of California, Los Angeles. https://www.ethz.ch/content/dam/ethz/special-interest/mtec/chair-of-entrepreneurial-risks-dam/documents/dissertation/Max_Werner_thesis_final_Dec07.pdf
30. Chande, T. (1995) A Time Price Oscillator. Technical Analysis of Stocks & Commodities, 13, 369-374. ftp://80.240.216.180/Transmission/%D0%A4%D0%B0%D0%B9%D0%BB%D1%8B/S&C%20on%20DVD%2011.26/VOLUMES/V13/C09/THETIME.pdf

CHAPTER 3

MICRO-EARTHQUAKE MONITORING WITH SPARSELY SAMPLED DATA

Paul Sava

Keywords: Microearthquakes, Imaging, Interferometry

INTRODUCTION

Seismic imaging based on the single scattering assumption, also known as Born approximation, consists of two main steps: *wavefield reconstruction* which serves the purpose of propagating recorded data from the acquisition surface back into the subsurface, followed by an *imaging condition* which serves the purpose of highlighting locations where scattering occurs.

This framework holds both when the source of seismic waves is located in the subsurface and the imaging target consists of locating this source, as well as when the source of seismic waves is located on the acquisition surface and the imaging target consists of locating the places in the subsurface where scattering or reflection occurs. In this paper, I concentrate on the case of imaging seismic sources located in the subsurface, although the methodology discussed here applies equally well for the more conventional imaging with artificial sources.

An example of seismic source located in the subsurface is represented by micro-earthquakes triggered by natural causes or by fluid injection during reservoir production or fracturing. One application of micro-earthquake location is monitoring of fluid injection in brittle reservoirs when micro-earthquake evolution in time correlates with fluid movement in reservoir formations. Micro-earthquakes can be located using several methods including double-difference algorithms (Waldhauser and Ellsworth 2000), Gaussian-beam migration (Rentsch et al. 2004, 2007), diffraction stacking (Gajewski et al. 2007) or time-reverse imaging (Gajewski and Tessmer 2005; Artman et al. 2010).

Micro-earthquake location using time-reverse imaging, which is also the technique advocated in this paper, follows the same general pattern mentioned in the preceding paragraph: wavefield-reconstruction backward in time followed by an imaging condition extracting the image, i.e. the location of the source. The main difficulty with this procedure is that the onset of the micro-earthquake is unknown, i.e. time $t = 0$ is unknown, so the imaging condition cannot be simply applied as it is usually done in zero-offset migration. Instead, an automatic search needs to be performed in the back-propagated wavefield to identify the locations where wavefield energy focuses. This process is difficult and often ambiguous since false focusing locations might overlap with locations of wavefield focusing. This is particularly true when imaging using an approximate model which does not explain all random fluctuations observed in the recorded data. This problem is further complicated if the acquisition array is sparse, e.g. when receivers are located in a borehole. In this case, the sparsity of the array itself leads to artifacts in the reconstructed wavefield which makes the automatic picking of focused events even harder.

The process by which sampling artifacts are generated is explained in Fig. 1a–d. Each segment in Fig. 1a corresponds to a wavefront reconstructed from a receiver. For dense, uniform and wide-aperture receiver coverage and for reconstruction using accurate velocity, the wavefronts overlap at the source position (Fig. 1b). This idealized situation resembles the coverage typical for medical imaging, although the physical processes used are different. However, if the velocity used for wavefield reconstruction is inaccurate, then the wavefronts do not all overlap at the source position (Fig. 1c), thus leading to imaging artifacts. Likewise, if receiver sampling is sparse,

Citation: Paul Sava, Micro-earthquake monitoring with sparsely sampled data. doi:10.1007/s13202-011-0005-7.

reconstruction at the source position is incomplete (Fig. 1d), even if the velocity used for reconstruction is accurate. The cartoons depicted in Fig. 1a–d represent an ideal situation with receivers surrounding the seismic source, which is not typical for seismic experiments. In those cases, source illumination is limited to a range which correlates with the receiver coordinates.

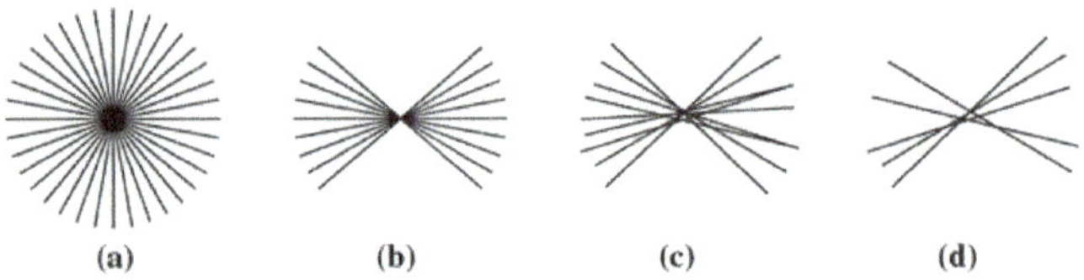

Figure 1: Schematic representation of focus constructions using time reversal. Each *line* in the plots represents a wavefront reconstructed at the source from a given receiver. The panels represent the following cases: **a** dense acquisition, complete angular coverage and correct velocity, **b** dense acquisition, partial angular coverage and correct velocity, **c** dense acquisition, partial angular coverage and incorrect velocity, and **d** sparse acquisition, partial angular coverage and incorrect velocity. Panel **d** represents the worst case scenario for micro-earthquake imaging

In general, artifacts caused by unknown velocity fluctuations and receiver sampling overlap and, although the two phenomena are not equivalent, their effect on the reconstructed wavefields are analogous. As illustrated in the following sections, the general character of those artifacts is that of random wavefield fluctuations. Ideally, the imaging procedure should attenuate those random wavefield fluctuations irrespective of their cause in order to support automatic source identification.

CONVENTIONAL IMAGING CONDITION

Assuming data D(x,t) acquired at coordinates xx function of time *t* (e.g. in a borehole) we can reconstruct the wave field V(x,y,t) at coordinates y in the imaging volume using an appropriate Green's function G(x,y,t) corresponding to the locations x and y (Fig. 2)

$$V(\mathbf{x},\mathbf{y},t) = D(\mathbf{x},t) *_t G(\mathbf{x},\mathbf{y},t), \tag{1}$$

where the symbol $*_t$ indicates time convolution. The total wavefield U(y,t) at coordinates y due to data recorded at all receivers located at coordinates xx is represented by the superposition of the reconstructed wavefields V(x,y,t):

$$U(\mathbf{y},t) = \int_{\mathbf{x}} \mathrm{d}\mathbf{x}\, V(\mathbf{x},\mathbf{y},t). \tag{2}$$

A conventional imaging condition (CIC) applied to this reconstructed wavefield extracts the image RCIC(y) as the wavefield at time $t = 0$

$$R_{\mathrm{CIC}}(\mathbf{y}) = U(\mathbf{y}, t=0). \tag{3}$$

This imaging procedure succeeds if several assumptions are fulfilled: first, the velocity model used for imaging has to be accurate; second, the numeric solution to the wave-equation used for wavefield reconstruction has to be accurate; third, the data need to be sampled densely and uniformly on the acquisition surface. In this paper, I assume that the first and third assumptions are not fulfilled. In these cases, the imaging is not accurate because contributions to the reconstructed wavefield from the receiver coordinates do not interfere constructively, thus leading to imaging artifacts. As indicated earlier, this situation is analogous to the case of imaging with an inaccurate velocity model, e.g. imaging with a smooth velocity of data corresponding to geology characterized by rapid velocity variations.

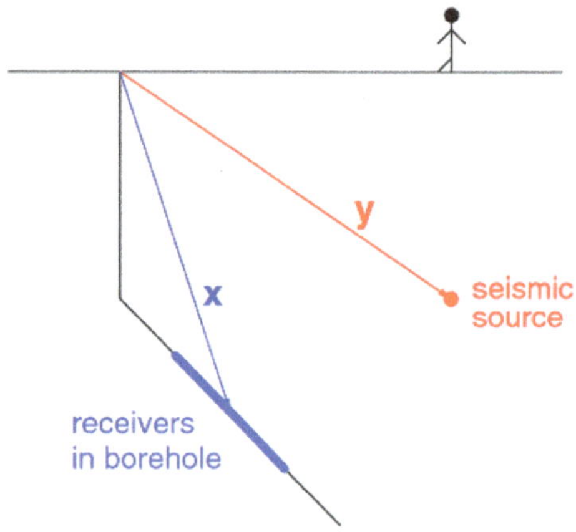

Figure. 2 Illustration of the variables x and y used for the description of the conventional and interferometric imaging procedures

Different image processing procedures can be employed to reduce the random wavefield fluctuations. The procedure advocated in this paper uses interferometry for noise cancellation. Interferometric procedures can be formulated in various frameworks, e.g. coherent interferometric imaging (Borcea et al. 2006) or wave-equation migration with an interferometric imaging condition (Sava and Poliannikov 2008).

Interferometric imaging condition

Migration with an interferometric imaging condition (IIC) uses the same generic framework as the one used for the conventional imaging condition, i.e. wavefield reconstruction followed by an imaging condition. However, the difference is that the imaging condition is not applied to the reconstructed wavefield directly, but it is applied to the wavefield which has been transformed using pseudo-Wigner distribution functions (WDF) (Wigner 1932). By definition, the zero frequency pseudo-WDF of the reconstructed wavefield U(y,t) is

$$W(\mathbf{y},t) = \int_{|t_h|\le T} \mathrm{d}t_h \int_{|\mathbf{y}_h|\le Y} \mathrm{d}\mathbf{y}_h U\left(\mathbf{y}-\frac{\mathbf{y}_h}{2}, t-\frac{t_h}{2}\right) U\left(\mathbf{y}+\frac{\mathbf{y}_h}{2}, t+\frac{t_h}{2}\right), \tag{4}$$

where *Y* and *T* denote averaging windows in space and time, respectively. In general, *Y* is three dimensional and *T* is one dimensional. Then, the image RIIC(y) is obtained by extracting the time $t = 0$ from the pseudo-WDF, W(y,t), of the wavefield U(y,t):

$$R_{\mathrm{IIC}}(\mathbf{y}) = W(\mathbf{y}, t=0) \tag{5}$$

The interferometric imaging condition represented by Eqs. 4 and 5 effectively reduces the artifacts caused by the random fluctuations in the wavefield by filtering out its rapidly varying components (Sava and Poliannikov 2008). In this paper, I use this imaging condition to attenuate noise caused by sparse data sampling or noise caused by random velocity variations. As suggested earlier, the interferometric imaging condition attenuates both types of noise at once, since it does not explicitly distinguish between the various causes of random fluctuations.

The parameters Y and T defining the local window of the pseudo-WDF are selected according to two criteria (Cohen 1995). First, the windows have to be large enough to enclose a representative portion of the wavefield which captures the random fluctuation of the wavefield. Second, the window has to be small enough to limit the possibility of cross-talk between various events present in the wavefield. Furthermore, cross-talk can be attenuated by selecting windows with different shapes, for example Gaussian or exponentially decaying. Therefore, we could in principle define the transformation in Eq. 4 more generally as

$$W(\mathbf{y},t) = \int_{|t_h|\le T} dt_h W_T(t,t_h) \int_{|\mathbf{y}_h|\le Y} d\mathbf{y}_h W_Y(\mathbf{y},\mathbf{y}_h) U\left(\mathbf{y}-\frac{\mathbf{y}_h}{2}, t-\frac{t_h}{2}\right) U\left(\mathbf{y}+\frac{\mathbf{y}_h}{2} \right. \quad (6)$$

where W_T and W_Y are weighting functions which could represent Gaussian, boxcar or any other local functions (Artman 2011, personal communication). For simplicity, in all examples presented in this paper, the space and time windows are rectangular with no tapering and the size is selected assuming that micro-earthquakes occur sufficiently sparse, i.e. the various sources are located at least twice as far in space and time relative to the wavenumber and frequency of the considered seismic event. Typical window sizes used here are 11 grid points in space and 5 grid points in time.

Example

I exemplify the interferometric imaging condition method with a synthetic example simulating the acquisition geometry of the passive seismic experiment performed at the San Andreas Fault Observatory at Depth (Chavarria et al. 2003; Vasconcelos et al. 2008). This numeric experiment simulates waves propagating from three micro-earthquake sources located in the fault zone (Fig. 3), which are recorded in a deviated well located at a distance from the fault. For the imaging procedure described in this paper, the micro-earthquakes represent the seismic sources. This experiment uses acoustic waves, corresponding to the situation in which we use the P-wave mode recorded by the three-component receivers located in the borehole (Figs. 4b,5b). The three sources are triggered 40 ms apart and the triggering time of the second source is conventionally taken to represent the origin of the time axis.

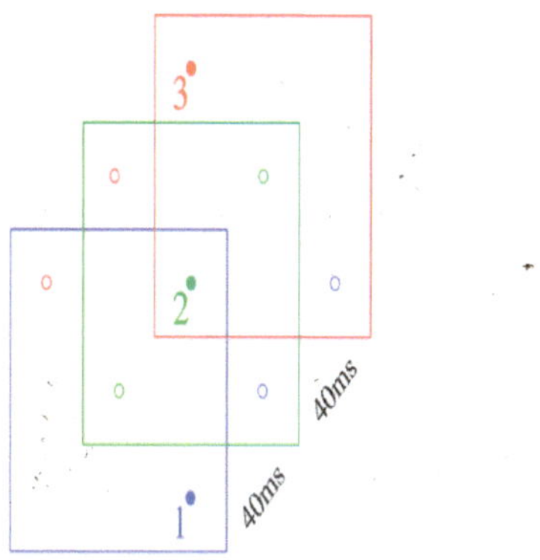

Figure. 3 Geometry of the sources used in the numeric experiment. The horizontal and vertical separation between sources is 250 m. The sources are triggered with 40 ms delays in the order indicated by their numbers. Time $t = 0$ is conventionally set to the triggering moment of source 2

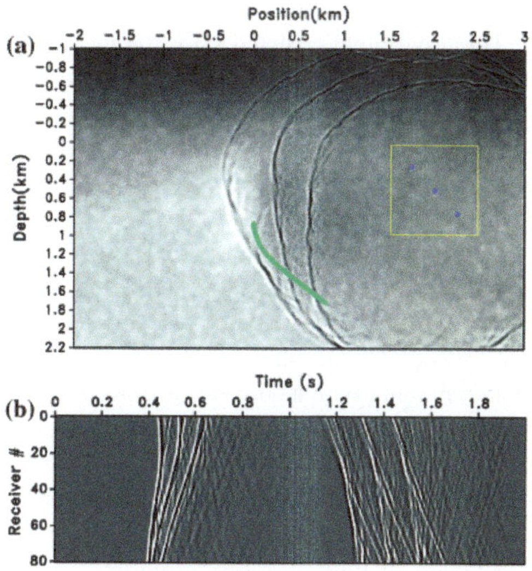

Figure. 4 a Wavefields simulated in random media and **b** data acquired with a dense receiver array. Overlain on the model and wavefield are the positions of the sources and borehole receivers. The *boxed area* corresponds to the images depicted in Fig. 6a and b

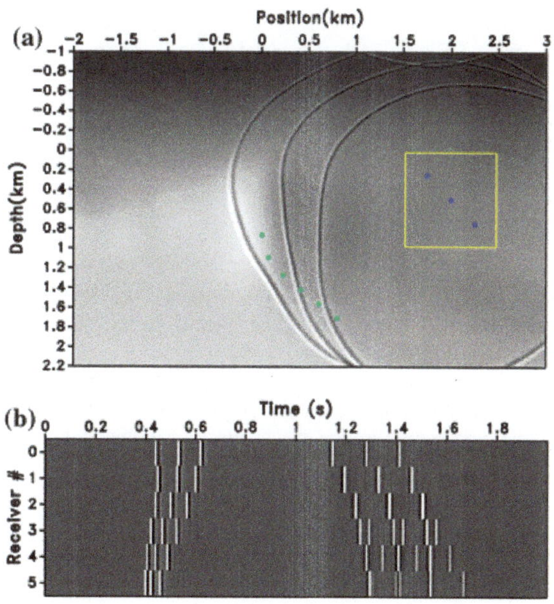

Figure. 5 a Wavefields simulated in smooth media and **b** data acquired with a sparse receiver array. Overlain on the model and wavefield are the positions of the sources and borehole receivers. The *boxed area* corresponds to the images depicted in Fig. 7a and b

The goal of this experiment is to locate the source positions by focusing data recorded using dense acquisition in media with random fluctuations or by focusing data recorded using sparse acquisition arrays in media without random fluctuations. In the first case, the imaging artifacts are caused by the fact that data are imaged with a velocity model that does not incorporate all random fluctuations of the model used for data simulation, while in the second case, the imaging artifacts are caused by the fact that the data are sampled sparsely in the borehole array. The third case is a combination of acquisition with two sparse arrays, and imaging with an inaccurate velocity model.

Figures 6a and b, 7a and b and 8a and b show the wavefields reconstructed in reverse time around the target location. From left to right, the panels represent the wavefield at different times. As indicated earlier, the time at which source 2 focuses is selected as time $t = 0$, although this convention is not relevant for the experiment and any other time could be selected as reference. The experiment depicted in Fig. 6a and b corresponds to modeling in a model with random fluctuations and migration in a smooth background model. In this experiment, the data used for imaging are densely sampled in the borehole, i.e. there are 81 receivers separated by approximately 12 m. In contrast, the experiment depicted in Fig. 7a and b corresponds to modeling and migration in the smooth background model. In this experiment, the data are sparsely sampled in the borehole, i.e. there are only six receivers obtained by selecting every 16th receiver from the original set. In all cases, panels (a) correspond to imaging with a conventional imaging condition, i.e. simply select the reconstructed wavefield at various times, and panels (b) correspond to imaging with the interferometric imaging condition, i.e. select various times from the wavefield transformed with a pseudo-WDF of 11 grid points in space and 5 grid points in time. For this example, WDF window corresponds to 44 m in space and 2 ms in time.

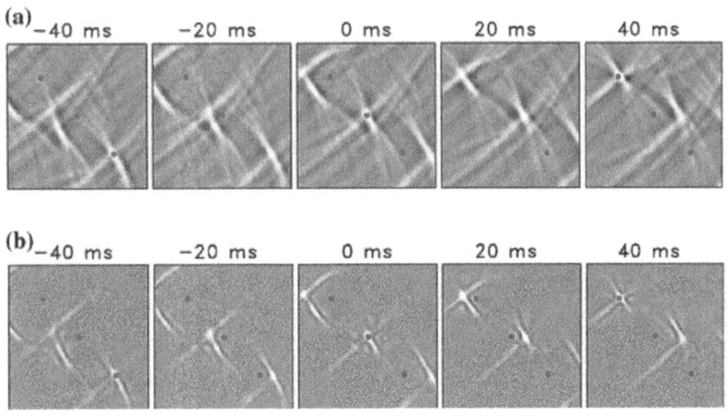

Figure. 6 Images corresponding to migration of the *densely sampled data* (Fig. 4b) modeled in the*random velocity* by **a** conventional IC and **b** interferometric IC using the background velocity. The *left-most panel* shows focusing at source 1, the *middle panel* shows focusing at source 2, and the *right-most panel* shows focusing at source 3. The *overlain dots*represent the exact source positions

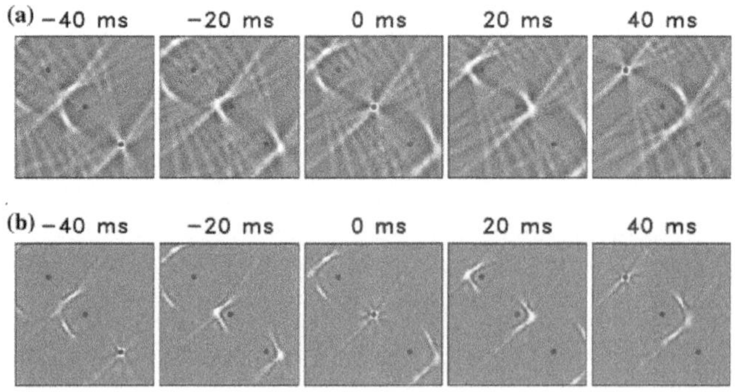

Figure. 7 Images corresponding to migration of the *sparsely sampled data* (Fig. 5b) modeled in the*background velocity* by **a** conventional IC and **b** interferometric IC using the background velocity. The *left-most panel* shows focusing at source 1, the *middle panel* shows focusing at source 2, and the *right-most panel* shows focusing at source 3. The *overlain dots*represent the exact source positions

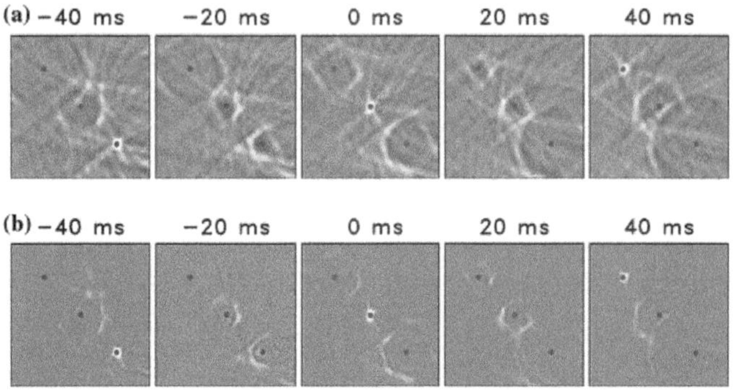

Figure. 8 Images corresponding to migration of the dual *sparsely sampled data* (Fig. 9b) modeled in the *random velocity* by **a** conventional IC and **b** interferometric IC using the background velocity. The *left-most panel* shows focusing at source 1, the *middle panel*shows focusing at source 2, and the *right-most panel* shows focusing at source

THE*OVERLAIN DOTS* REPRESENT THE EXACT SOURCE POSITIONS

Figure 6a shows significant random fluctuations caused by wavefield reconstruction using an inaccurate velocity model. The fluctuations caused by the random velocity and encoded in the recorded data are not corrected during wavefield reconstruction and they remain present in the model. Likewise, Fig. 7a shows significant random fluctuations caused by reconstruction using the sparse borehole data. However, the pseudo-WDF applied to the reconstructed wavefields attenuates the rapid wavefield fluctuations and leads to sparser, better focused images that are easier to use for source location. This conclusion applies equally well for the experiments depicted in Fig. 6a and b or 7a and b.

The final example corresponds to the case of acquisition with two separate sparse arrays (Fig. 9a, b). As expected, the wavefields are far less noisy after the application of the WDF, and the focusing is increased due to the larger array aperture. This facilitates an automatic procedure for focusing identification, since most of the spurious noisy is eliminated from the image.

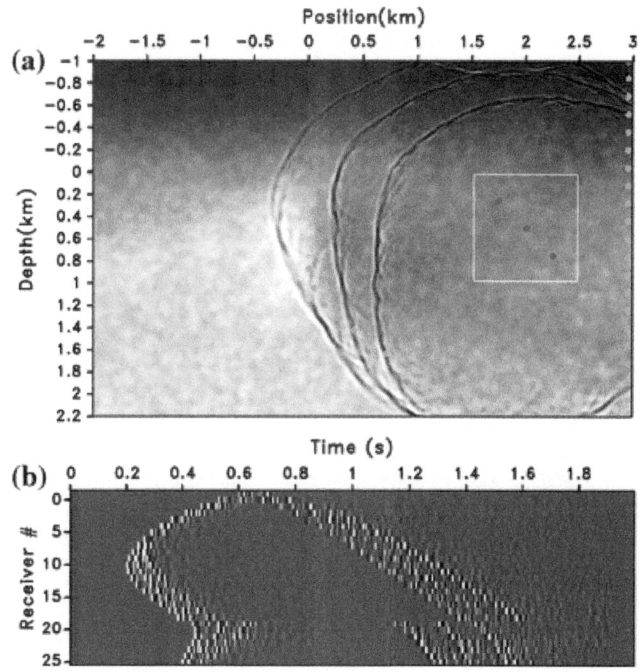

Figure. 9 **a** Wavefields simulated in random media and **b** data acquired with two sparse receiver arrays. Overlain on the model and wavefield are the positions of the sources and borehole receivers. The *boxed area* corresponds to the images depicted in Fig. 8a and b. The *top traces* in **b** correspond to the vertical array, and the other traces correspond to the sparse deviated array

Finally, I note that the 2D imaging results from this example show better focusing than what would be expected in 3D. This is simply because the 1D acquisition in the borehole cannot constrain the 3D location of the micro-earthquakes, i.e. the azimuthal resolution is poor, especially if scatterers are not present in the model used for imaging. This situation can be improved using data acquired in several boreholes or using additional information extracted from the wavefields, e.g. polarization of multicomponent data.

CONCLUSIONS

The interferometric imaging condition used in conjunction with time-reverse imaging reduces the artifacts caused by random velocity fluctuations that are unaccounted-for in imaging and by the sparse wavefield sampling on the acquisition array. The images produced by this procedure are crisper and support automatic picking of micro-earthquake locations. Imaging with sparse arrays allows increased aperture for identical acquisition cost with that of a narrower but denser array. At the same time, a larger aperture improves focusing of the events, thus facilitating automatic event identification. The interferometric imaging procedure has a similar structure to conventional imaging and the moderate cost increase is proportional to the size of the windows used by the pseudo-Wigner distribution functions. The source positions obtained using this procedure can be used to monitor fluid injection or for studies of naturally occurring earthquakes in fault zones.

ACKNOWLEDGMENTS

This work is supported by the sponsors of the Center for Wave Phenomena at Colorado School of Mines and by a research grant from ExxonMobil. The reproducible numeric examples in this paper use the Madagascar open-source software package freely available from

REFERENCES

1. Artman B, Podladtchikov I, Witten B (2010) Source location using time-reverse imaging. Geophys Prospect 58:861–873
2. Borcea L, Papanicolaou G, Tsogka C (2006) Coherent interferometric imaging in clutter. Geopysics 71:SI165–SI175
3. Chavarria J, Malin P, Shalev E, Catchings R (2003) A look inside the San Andreas Fault at Parkfield through vertical seismic profiling. Sci Agric 302:1746–1748
4. Cohen L (1995) Time frequency analysis: Signal processing series. Prentice Hall, Englewood Cliffs
5. Gajewski D, Anikiev D, Kashtan B, Tessmer E, Vanelle C (2007) Localization of seismic events by diffraction stacking, In: 76th annual international meeting, SEG, Expanded Abstracts, pp 1287–1291
6. Gajewski D, Tessmer E (2005) Reverse modelling for seismic event characterization. Geophys J Int 163:276–284
7. Rentsch S, Buske S, Luth S, Shapiro SA (2004) Location of seismicity using Gaussian beam type migration. In: 74th annual international Meeting. Society of Exploration Geophysicists, pp 354–357
8. Rentsch S, Buske S, Luth S, Shapiro SA (2007) Fast location of seismicity: a migration-type approach with application to hydraulic-fracturing data. Geophysics 72:S33–S40
9. Sava P, Poliannikov O (2008) Interferometric imaging condition for wave-equation migration. Geophysics 73:S47–S61
10. Vasconcelos I, Snieder R, Sava P, Taylor T, Malin P, Chavarria A (2008) Drill bit noise illuminates the San Andreas fault. EOS. Trans Am Geophys Union 89:349
11. Waldhauser F, Ellsworth W (2000) A double-difference earthquake location algorithm: method and application to the northern Hayward fault. Bull Seismol Soc Am 90:1353–1368
12. Wigner E (1932) On the quantum correction for thermodynamic equilibrium. Phys Rev 40:749–759

CHAPTER 4

G: FRACTURE ENERGY, FRICTION AND DISSIPATION IN EARTHQUAKES

S. Nielsen, E. Spagnuolo, M. Violay, S. Smith, G. Di Toro, A. Bistacchi

Keywords: Earthquake scaling, Fracture energy, Laboratory experiments, High velocity friction

INTRODUCTION

Seismic rupture is controlled by an energy balance involving elastic work, dissipation by anelastic processes (friction, damage, and plastic strain) and wave radiation. Creation or reactivation of rupture within a solid requires an energy input G (J m^{-2}) per unit surface, called material toughness or fracture energy (G may be envisioned, for example, as the energy spent in severing atomic bonds). Griffith (1921) originally proposed a rupture propagation criterion based on the balance, during rupture advancement, between the energy made available by release of elastic stress and the surface energy necessary to create fresh crack face. Irwin (1957) later proposed to add plastic strain energy around the crack tip in the balance. Since then, the energy balance concept was expanded and developed in different forms (Rice 1968) and applied to a wide variety of problems including engineering applications (Broberg 1999), rock mechanics (Atkinson and Meredith 1987), and earthquake physics (Ida 1972; Andrews 1976; Aki 1979; Wong 1982; Ohnaka et al. 1986; Andrews 2005; Lancieri et al. 2012; Malagnini et al.1994).

During earthquakes, stress is rapidly released because of a drop in friction, causing high (≈1 m/s) slip velocity, high (several km/s) rupture velocity, and wave radiation. Friction drops with slip and slip velocity, a process known as weakening. The simplest and most widely adopted model for frictional weakening, introduced by Ida (1972), consists in a linear decrease of the friction coefficient with slip from peak μ_p to a dynamic, steady-state μ_{ss}, over a slip distance D_c. (The corresponding shear stress drops from $\tau_p = \mu_p\sigma_n$ to $\tau_{ss} = \mu_{ss}\sigma_n$, where σ_n is the fault-normal stress). The weakening distance D_c has been widely used in fault studies and often imposed as a constant in fault models (Andrews 1976, and references therein; Mai et al. 2006). However, several authors have argued that D_c should be treated as a variable (Cocco and Bizzarri 2002; Nielsen et al. 2010), since the effective weakening distance depends on slip history, loading conditions, and is likely to become dynamically determined during rupture itself. It is recognized (Palmer and Rice 1973) that frictional work (product of shear stress and slip) during the weakening process equates to a particular form of fracture energy, such that many earthquake studies have adopted the simple form

$$G = \frac{1}{2}(\tau_p - \tau_{ss})\, D_c \ , \tag{1}$$

assuming Ida's slip weakening law, whereas the slightly more general form

$$G(u) = \int_0^u (\tau(u') - \tau(u))\, du' \ , \tag{2}$$

can take into account forms of non-linear friction decay and considers the possibility that final slip $u < D_c$ and the shear stress remains higher than the lowest possible dynamic value (Abercrombie and Rice 2005).

Laboratory measures of D_c obtained under different conditions of sliding velocity and normal stress show very large variations. On the one hand, rate and state friction (RAS) parameters (Dieterich 1979; Ruina 1983) obtained by velocity-stepping at sub-seismic slip rates indicate an evolution distance $10^{-6} < L < 10^{-4}$ m, and a modest (a few percent or less) friction

---

Citation: S. Nielsen, E. Spagnuolo, M. Violay, S. Smith, G. Di Toro, A. Bistacchi, G: Fracture energy, friction and dissipation in earthquakes, doi:10.1007/s10950-016-9560-1.

drop. Rupture modeling shows (Guatteri et al. 2001; Tinti et al. 2004) that, due to the complexity in the slip history, a fixed parameter L in the RAS law can result in different effective slip weakening distances D_c. However, the difference between D_c and L is at most 50 % and both well below 1 mm. On the other hand, rotary friction experiments performed at seismic (≈1 m/s) slip rates yield weakening distances in the order of meters, substantial friction drops (between 50 and 95 %) and fracture energies arguably in the same order as the seismological estimates (see discussion in further sections).

Seismological estimates of D_c suffer from poor resolution and from a fundamental indetermination: rupture is mainly sensitive to G (Peyrat et al. 2004; Spudich and Guatteri2000) which is in essence the product of stress drop ($\tau_p - \tau_{ss}$) and D_c, as shown in Eq.1, so that both cannot be determined independently. Exceptional circumstances may allow an unbiased constraint of D_c in earthquakes, like the recording of radiation from a patch of fault where rupture is propagating at supershear velocity (i.e., exceeding shear wave velocity), as argued by Cruz-Atienza and Olsen (2010).

Ultimately, fault properties which control rupture propagation are better represented by fracture energy G rather than by individual aspects of the weakening process. In addition, Gappears in the energy balance (see later discussion around Eq. 15) against radiated energy (which can be estimated from detected waves) and elastic energy release (which can be estimated from final fault slip). As a consequence, under a series of assumptions, it is possible to obtain estimates of G for earthquakes and its variation with seismic moment and slip, as attempted by several authors (Aki 1979; Abercrombie and Rice 2005; Tinti E. et al. 2005; Malagnini et al. 1994; Causse et al. 2014; Viesca and Garagash 2015).

The strain energy release rate GG (J m^{-2}) is defined as the energy (originating from elastic strain release) made available per unit of newly created fracture surface area. In dynamic rupture propagation, GG depends on rupture velocity V_r, and $G(V_r)$ can be defined analytically for simple cases of cracks or slip pulses propagating at constant velocity (Kostrov 1974; Nielsen and Madariaga 2003; Rice et al. 2005). The simplest example is for the anti-plane (Mode II) crack where $\mathcal{G} = \mathcal{G}_0\sqrt{1-(V_r/V_S)^2}$, g_0 is the elastic strain energy release for a quasi-static crack (in the limit of vanishing small rupture velocity V_r) and V_S is the shear wave velocity. The square root term is the well-known Lorentz contraction encountered also in special relativity. Though no closed mathematical expression for GG exists for general cases, $G(V_r)$ is generally a monotonously decreasing function of V_r for sub-Rayleigh rupture velocities ($V_r < V_{Ral}$), but can be more complex at intersonic rupture velocities ($V_S < V_r < V_P$) (Broberg 1999); G is maximized and equates to G0G0 at low rupture velocity. For the purpose of earthquake studies, the criterion for rupture advancement is $G(V_r)$=G, where G is the material toughness (or fracture energy) of the material. When a rupture stops (e.g., at the end of earthquake rupture propagation), the maximum available strain energy release does not suffice to compensate the dissipation through fracture energy G; as a consequence, the condition G≥G0 is verified. For a shear crack of radius a with static stress drop $\Delta\sigma = \tau_0 - \tau_1$ (where τ_0 is initial and τ_1 final shear stress) and a shear stiffness μ', the static fracture energy release rate is

$$\mathcal{G}_0 = \frac{\pi\,\Delta\sigma^2 a}{\mu'} \tag{3}$$

Relation 3 was used by Aki (1979) to derive for the first time a lower bound estimate of material toughness G for an earthquake fault barrier (the radius a was inferred from the perimeter of aftershock activity around an asperity). On the other hand, the well-known scaling of slip with fault size and stress drop is described by

$$u = C_1\,\frac{\Delta\sigma}{\mu'}\,a \tag{4}$$

(C_1 is a dimensionless constant of order unity, depending on the shape of the ruptured area). Replacing a from Eq. 4 into 3, and using $C = \pi/C_1$, we obtain that G_0=C Δ_σ u. Given that the stress drop Δ_σ, though highly variable in earthquakes, shows no general dependence on u, and that G at rupture arrest scales with G0G0, it follows that fracture energy should scale linearly with slip, i.e.,

$$G \propto u \tag{5}$$

to allow the arrest of earthquakes of all sizes. This preliminary result indicates that fracture energy in earthquakes is not necessarily related to an intrinsic material property such as material toughness, unless we assume that separate magnitude earthquakes occur on separate faults with arresting barriers of distinct properties at fault ends.

We discuss fracture energy in terms of frictional weakening, since both can be related as illustrated in Eq. 2. In high-velocity friction experiments, the observed abrupt dynamic weakening is clearly related to thermal decomposition, phase changes, breakdown reactions, etc. under intense frictional heating (Han et al. 2007; Di Toro et al. 2011).

Thermal pressurization of fluids trapped in the fault zone has also been invoked as a possible mechanism, and modeled under a series of assumptions, in particular about fault zone width and permeability (Rice 2006; Viesca and Garagash 2015). Under favorable circumstances, thermal pressurization may effectively preclude thermal decomposition in seismogenic crust faults by limiting coseismic temperature rise.

We remark that in high slip velocity experiments conducted in the presence of undrained fluid (water), a minor degree of pressurization is observed only in the later phases of slip (Violay et al. 2015). In both dry and wet experiments, decomposition weakening mechanisms appear to be triggered extremely early and efficiently (see description of weakening phases in Section 2), buffering the temperature rise and reducing the range of conditions where TP may be important on natural faults.

The experimental setup of Violay et al. (2015) allowed only to explore pressurization for a limited set of conditions: bare rock (no pressurized gouges were explored) and fixed fluid volumes. More efficient TP was predicted only through extrapolation of the experimental results to a different set of conditions. These suggested that an efficient TP would require that small volumes of fluids be trapped in a low permeability fault zone—such that a reduced heat capacity would maximize fluid temperature increase, and that diffusion may not buffer the pressure rise (for a complete discussion, see Violay et al. 2015). Otherwise, TP may become important mostly after large slip amounts (hence on mature faults generating large earthquakes, a point also suggested by modeling Viesca and Garagash 2015).

We will not develop here the physical interpretation of thermal decomposition weakening, nor endeavour in the design of a general constitutive relation for high-velocity friction. These will be the the topics of a separate dedicated study.

Instead, we analyze in detail the features of the experimental weakening curves and provide a general fit which is purely empirical, with the synthesis of a large number of experiments and their result in terms of frictional breakdown energy G_f. We attempt to reconcile seismological studies and laboratory measurements of dissipation and friction.

It is striking that previous laboratory experiments (Wong 1982, and reference therein) repeatedly obtained markedly lower values of fracture energy (by at least three orders of magnitude) than those estimated for earthquakes from simple scaling relationships (Aki 1979), at least for intermediate to large magnitudes.

Here, the laboratory measurements show a remarkable compatibility with the seismological estimates of fracture energy. This is arguably due to the experimental conditions, which cover an extended range here to approach those of natural earthquakes. In particular, high slip velocity (>1 m/s), large amounts of cumulative slip (0.001–10 m), and intermediate normal stress conditions (10–30 MPa) are combined.

G_f can be obtained directly by integrating the experimentally measured shear stress curve $\tau(u)-\tau(U)$ with respect to slip u according to Eq. 2. Since the earthquake estimates of G and the laboratory measures of G_f are conceptually and dimensionally equivalent quantities, we proceed to the comparison of our experimental data with both newly derived and previously published seismological estimates.

We discuss the similarities and the differences between G' and G_f, and argue that dissipation on faults does include friction but also more general damage forms occurring off-fault (Andrews2005). In recent years, it has been argued that an anelastic (off-fault plastic) damage zone forms adjacent to the fault, owing to the large stress concentration in the vicinity of the rupture tip (Reches 1999; Dor et al. 2006; Mitchell et al. 2011; Fondriest et al. 2015). In addition to the on-fault friction, which is measured in the experiments herein, during rupture propagation on natural faults the off-fault damage process may exert important controls on rupture, including reduction in rupture velocity and peak slip velocity (Andrews 2005), rupture arrest (Hok et al.2010), influence on rupture directivity, and on the formation of short slip pulses (Xu et al.2015). The width of the damage band generally increases with cumulative fault slip (Shipton et al. 2006b), increasing the dissipated energy in off-fault plastic strain, with the consequence that apparent fracture energy increases with rupture length. Damage can be induced by stress concentration in the vicinity of the propagating rupture tip (Andrews 2005). In addition, we discuss possible forms of dissipation arising from distributed deformation occurring in the accommodation of slip onto fault surfaces which are non-planar and present roughness at all scales.

DYNAMIC WEAKENING IN HIGH VELOCITY ROCK FRICTION EXPERIMENTS

Following Nielsen et al. (2015), we report the general trend of friction weakening, and the corresponding frictional fracture energy, for a representative sample of 28 experiments selected from a larger catalogue (experiments with low signal-to-noise ratio were chosen for a variety of conditions and lithologies, from more than 1000 experiments in total). All experiments were performed on a high-velocity, rotary shear apparatus (SHIVA) installed at INGV-Roma (Di Toro et al. 2010). The experiments were performed on cohesive, pre-cut rock samples representative of crustal seismic environments within the basement: silicate-rich rocks (microgabbro, Niemeijer et al. 2011; basalt, Violay et al. 2014) and calcite-rich rocks (Carrara marble with 99 % calcite, Violay et al. 2013); methods are described in detail in the references above. Experimental conditions varied, with imposed normal stress in the range 5–40 MPa, under velocity control with maximum sliding velocity 1–6.5 m/s, and slip accelerations (at the start of the experiments) of 3–6.5 m/s^2. Samples were exposed to conditions of room humidity and temperature throughout the experiments (although during the high velocity slip intervals, temperatures in the vicinity of the slip surface increased in excess of 1000°C under the effect of frictional heating). As observed in previous studies of friction at seismic slip velocity (≈1 m/s), frictional melting develops during the experiment for silicate-rich rocks (microgabbro and basalt, Tsutsumi and Shimamoto 1997; Hirose and Shimamoto 2005; Violay et al. 2014), but not on calcite-rich rocks (Han et al. 2007).

Dynamic weakening measured during high-velocity friction experiments under high normal stress shows substantial variations. However, when the experimental procedure is accurate enough to allow good signal-to-noise ratio and a high degree of repeatability, several systematic features are observed—for an example of high-velocity rotary friction (HVRF) experimental repeatability, see Violay et al. (2015). We will now analyze in detail the anatomy of the weakening curve in a couple of representative examples.

As illustrated in Fig. 1b, upon rapid loading, shear stress rises linearly with strain as the sample is elastically loaded previous to the start of slip (phase 0). A short slip (phase I) then occurs under high, almost constant or slightly strain-hardening friction (compatible with Byerlee's law with a friction coefficient >0.5). Phase I gives way very early (<1 cm of slip, which corresponds roughly to slip rates of the order of 10–20 cm/s under high imposed accelerations) to (phase II) when the high frictional power triggers efficient lubrication processes (Di Toro et al. 2011), and thus abrupt weakening is initiated. During all of phase II, friction drop versus slip follows a linear trend in a log-log diagram (Fig. 1b). Steady-state (phase III) in silicate-rich samples is achieved almost immediately after the target slip velocity has been reached (end of acceleration phase), where a low sliding friction value (<0.1) is maintained, with minor fluctuations, as long as slip rate is not modified. Note that in experiments under lower normal stress on gabbro, the steady-state is achieved much later during the steady-velocity sliding, and a second peak of strengthening is observed (Hirose and Shimamoto 2005). These differences are due to the larger normal stress, slip velocity, and the resulting larger frictional power dissipated in the experiments discussed here, whereby weakening is accelerated (Nielsen et al. 2010; Di Toro et al. 2011).

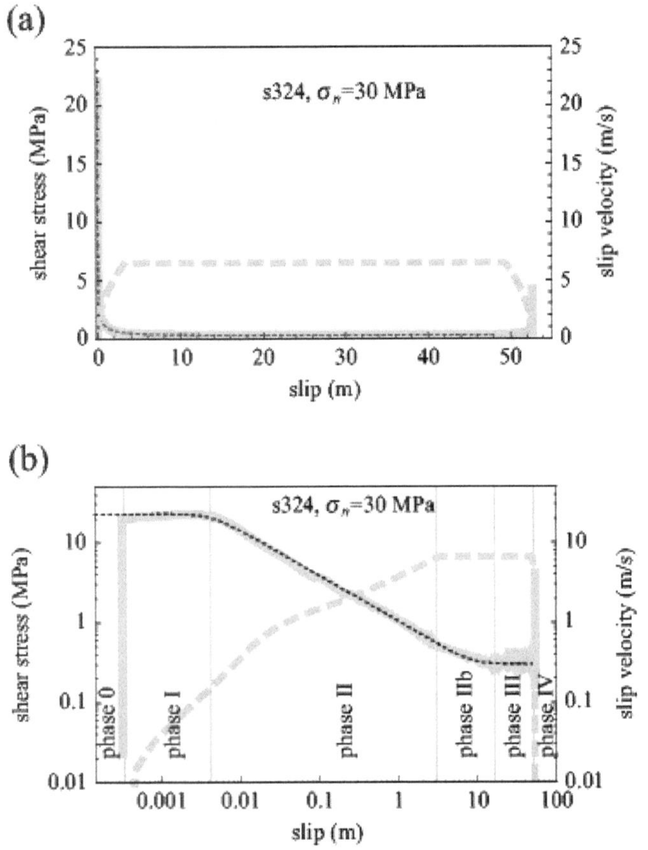

Figure. 1 Shear stress (experimental weakening curve in solid grey), power law fit (dashed black), and slip velocity (dashed grey) for experiment S234, performed on Carrara marble at normal stress 30 MPa. **a** Linear plot. **b** Log-log scale plot (different phases are indicated as defined in text). The power law fit is according to Eq. 9, with parameters $u_w = 0.0043$ m, $\tau_p = 23$ MPa, $\tau_d = 0.3$ MPa, $\alpha = 0.57$. In phase II–IIb, the weakening can be approximated by $\tau \propto u^{-0.57}$

In the case of calcite-rich samples, an intermediary (phase IIb) is observed during which a slight weakening continues even after the end of slip acceleration; in this case, phase III is reached only after several meters of slip. During phase IIb, friction versus slip still appears as log-log linear but with a reduced slope (Fig. 2).

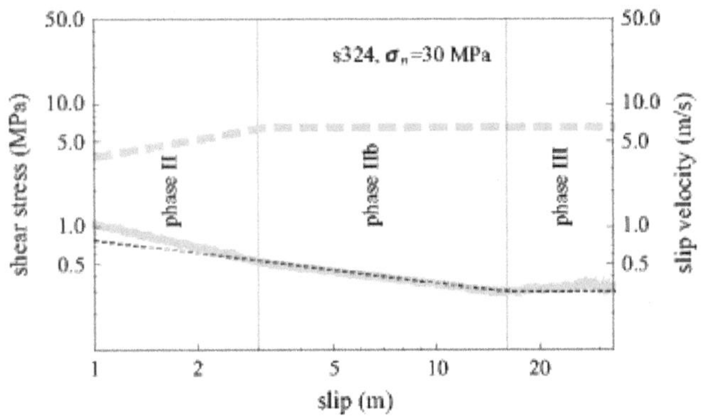

Figure. 2 Detail of phase IIb of experiment S324 (same as Fig. 1). Phase IIb shows the continuation of weakening under constant slip rate; shear stress decay is best fit by $\tau \propto u^{-0.35}$ (black dashed curve) in phase IIb (as opposed to $\tau \propto u^{-0.57}$ in phase II)

Deceleration of slip (phase IV) is characterized by a rapid recovery of friction; in many experiments, the recovery reaches about 20 % of the peak stress. This value obviously depends on the imposed deceleration rate: friction is strongly velocity-weakening but not instantaneously related to slip velocity, so that a faster deceleration results in reduced recovery. A discussion on friction recovery is found in Del Gaudio et al. (2009); however, in that case, the deceleration was not imposed by the control system as in the experiments discussed here, but resulted from rapid dissipation of the moment of inertia in the rotary shear apparatus.

During the abrupt weakening phase, shear stress is best described by a power law of the form $\tau \propto u^{-\alpha}$ where $0.5 < \alpha < 0.6$ for phase II ($\alpha \approx 0.35$ for phase IIb). Given that the weakening phase is tapered at its beginning by an approximately constant peak value τ_p and at its end by the steady-state, dynamic sliding value τ_d, the three different stages can be described by:

$$u \ll u_w, \quad \tau \approx \tau_p \tag{6}$$

$$u_w < u < u_w(\tau_p/\tau_{ss} - 1)^{1/\alpha}, \quad \tau \approx \tau_p \left(\frac{u}{u_w}\right)^{-\alpha} \tag{7}$$

$$u \gg u_w(\tau_p/\tau_{ss} - 1)^{1/\alpha}, \quad \tau \approx \tau_{ss} \tag{8}$$

where u_w is the slip value at which weakening is initiated; τ will have dropped to steady state value τ_{ss} when slip has reached $u = u_w(\tau_p/\tau_{ss}-1)^{1/\alpha}$. An empirical fit for the entire frictional phases I–III (excluding phase IIb and the recovery phase IV) can be obtained in the form

$$\tau = \left(\frac{\tau_p^n}{1 + \left(\frac{u}{u_w}\right)^{\alpha n}} + \tau_{ss}^n \right)^{1/n} \tag{9}$$

where n is an ad hoc integer introduced for convenience, to match the curvature of the shear stress function at start and end of the weakening phase (here we used n = 8). The fit corresponding to Eq. 9 is represented as a dashed curve along with the experimental data in Figs. 1, 2, and 3. The exponent 0.57 was obtained by least-squares minimization during phase II for all the experiments together; the tapering (τ_p, τ_{ss}) and the slip for onset of weakening (u_w) are introduced for each individual curve by trial-and-error.

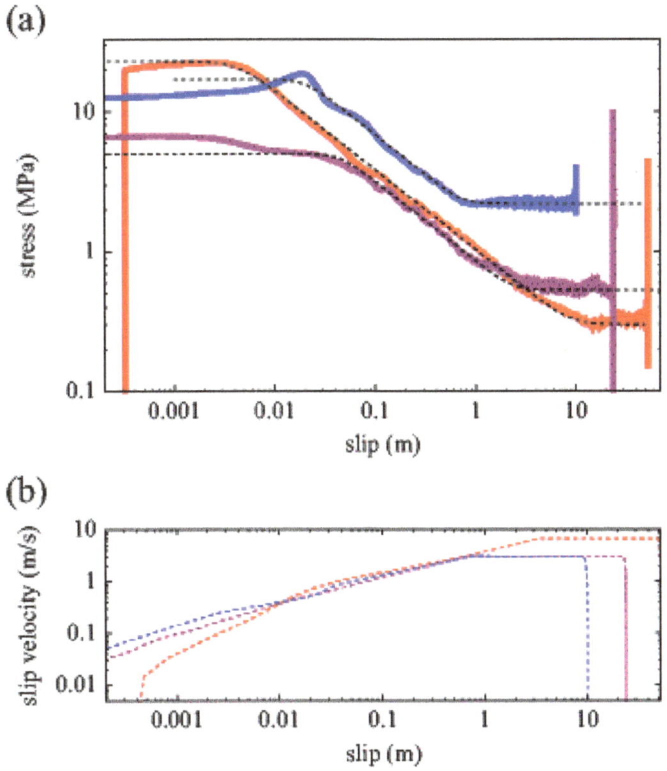

Figure. 3 Comparison of experiments S324 (red) and S330 (purple), performed on calcite (Carrara marble), and S543 performed on gabbro (blue). Acceleration/deceleration is 6.5 m/s^2 in all cases. **a** Shear stress as a function of slip; the thin dashed lines indicate the analytical fit according to Eq. 9, with parameters u_w = 0.0043 m, τ_p = 23 MPa, τ_d = 0.3 MPa, α = 0.57 (S324); u_w = 0.045 m, τ_p = 5 MPa, τ_d = 0.53 MPa, α = 0.57 (S330); u_w = 0.023 m, τ_p= 17 MPa, τ_d = 2.2 MPa, α = 0.57 (S543). Shear stress decay in phase II is always close to$\tau \propto u^{-0.57}$. **b** Slip velocity curves for the three experiments (same color code)

Under approximately constant slip acceleration V˙, we shall have $V = \sqrt{2\,u\,\dot V}$. Assuming that weakening is triggered once a critical slip velocity V_w has been reached, then the corresponding slip is

$$u_w \approx \frac{1}{2}\frac{V_w^2}{\dot V}. \tag{10}$$

For experiment S324 (Fig. 1), we measure $V_w \approx 0.16$ m/s and $\dot V \approx 3\,\{\mathrm{m\backslash\ s}\}^{-2}$, resulting in $u_w \approx 0.0043$ m. The exponent α is close to 0.5 in phase II, with a best fit value is α = 0.57. Using the rough approximation that $\tau \propto 1/V$ during the weakening phase, and noting that $V = \sqrt{2\,u\,\dot V}$ under constant acceleration, one would obtain $\tau \propto \frac{1}{\sqrt{2\dot V}}u^{-0.5}$, where the exponent value is fairly close to the experimental fit of α = 0.57 (this value holds for many experiments during phase II).

A comparison of two experiments performed on calcite under different maximum slip velocity and normal stress shows that phase II is very similar (Fig. 3); the change in the loading conditions mainly affects the values of τ_p, τ_d, and u_w.

In case of silicate-rich rocks, no phase IIb is observed, and steady-state is reached much earlier, as soon as the slip acceleration phase is over. Frictional melt and extrusion of melt out of the slipping zone takes place in silicate-rich rocks with relatively rapid ($v \approx 1$ mm/s or more under our experimental conditions) shortening of the sample. As discussed by (Nielsen et al. 2008,2010), the melting front advances into the sample catching-up with the thermal diffusion, creating a thermal boundary layer of finite thickness in such a way that a steady-state is rapidly reached. The steady-state thickness of the boundary layer is $b = \kappa/v$ where κ is the thermal diffusivity and v is the melting front velocity (or sample shortening half-velocity). Dimensional arguments indicate that steady-state will be reached within a time $t \approx 2\kappa/v^2$ (less than 1 s and less than 1 m slip at our experimental conditions). With increased frictional power dissipation (hence with increased slip velocity, acceleration, and normal stress), sample shortening velocity v is faster and steady-state is achieved in reduced time and slip amounts. One example for gabbro (experiment S543) is shown in Figs. 3 and 4; the main differences with calcite experiment of Fig. 1 are the slight slip-hardening phase before weakening, the absence of phase IIb and the faster achievement of steady-state. Besides these differences, the same general fit (Eq. 9) and exponent as for the calcite example applies to cases of frictional melting.

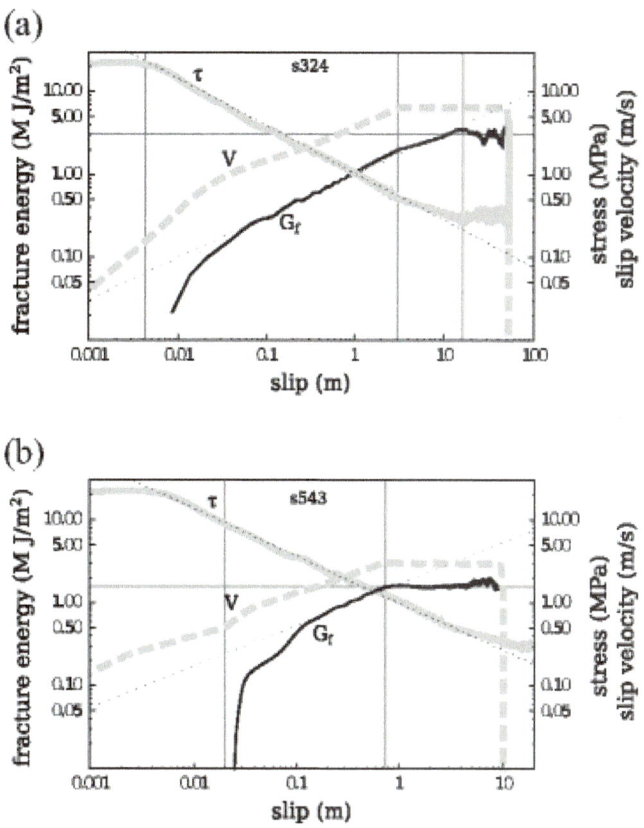

Figure. 4 Frictional equivalent fracture energy G_f (solid black curve), obtained by integration of the data according to Eq. 2. Stress (solid grey) and velocity (dashed grey) are also represented. (Dotted and dashed black lines are indicative log slopes of −0.57 and +0.5, respectively). **a** Experiment S324 performed on calcite. **b** Experiment S543 performed on gabbro. Note that the gabbro experiment (**b**), reaches steady-state much earlier because of frictional melting, thus G_f saturates much earlier, than the calcite experiments (slip about 1m as opposed to 10 m). See text for further details

In conclusion, the sliding shear stress during the weakening interval (phase II–IIb) is reasonably well described by a power law in slip, tapered by peak stress and steady-state dynamic stress on either end. The steady-state is achieved at shorter slip for melt experiments, where it is further anticipated by higher normal stress and slip rate. However, steady state is not anticipated by normal stress increase in calcite experiments, where no melting occurs, although this impression is conveyed when inspecting the shear stress curves on linear axes plots due to the higher peak stress value.

Measures of a slip-weakening distance D_c have been attempted previously in high velocity friction experiments; D_c was defined as the distance over which $\tau-\tau_{ss}$ drops down to $0.05\ (\tau_p-\tau_{ss})$ (Han et al. 2010). It has been argued in the case of frictional melt (Nielsen et al. 2008,2010) that the apparent slip-weakening distance scales as Dc∝1/(Vσ2n)Dc∝1/(Vσn2) (inverse of velocity and inverse normal stress square). However, in the light of the fact that, under different experimental conditions, the weakening part of the curves collapse into a single powerlaw, where changes mostly affect initial (τ_p) and steady-state (τ_p) values, the significance of D_c defined as a characteristic distance becomes less clear.

FRACTURE ENERGY AND FRICTIONAL WEAKENING

Using Eqs. 6 and 7, we can obtain a theoretical prediction for fracture energy $G_f(u) = u_w\tau_p\left(\frac{\alpha}{\alpha-1} + (\frac{u}{u_w})^{-\alpha} + \frac{1}{1-\alpha}(\frac{u}{u_w})^{1-\alpha}\right)$ by integrating the frictional weakening according to Eq. 2. Given that $\alpha < 1$, in the case that $u \gg u_w$, we may neglect first two terms to obtain:

$$G_f(u) \approx \frac{1}{1-\alpha}\, u_w\tau_p\left(\frac{u}{u_w}\right)^{1-\alpha}, \text{ for } u < u_w(\tau_p/\tau_{ss}-1)^{1/\alpha} \qquad (11)$$

$$G_f(u) = \text{const.} \approx \tfrac{1}{1-\alpha}\, u_w\tau_p(\tau_p/\tau_{ss}-1)^{\frac{1-\alpha}{\alpha}}, \quad \text{for } u \geq u_w(\tau_p/\tau_{ss}-1)^{1/\alpha} \qquad (12)$$

the second equation indicates that G_f saturates after reaching the steady-state with τ_d at $u = u_w(\tau_p/\tau_{ss}-1)^{1/\alpha}$.

In a similar fashion, we may compute the discrete summation equivalent to Eq. 2 using τ and umeasured in the experiments, to obtain an experimental curve $G_f(u)$.

The latter is represented as a solid black curve in Fig. 4 for two examples (calcite S324 and gabbro S543), and an average G_f for the 28 different experiments mentioned in the previous section is shown in (Fig. 5). Note that the friction recovery phase during the deceleration takes place within a very small slip amount, so that the frictional energy related to recovery is negligible and is not taken into account here in the definition of G_f. A consequence is that $G_f(u)$ obtained at any partial slip value $u < u_{final}$ can be considered as a reliable prediction of the finalG_f that would have resulted for an experiment arrested earlier at $u_{final} = u$ (see Appendix for details on the computation of experimental G_f). On the other hand, we anticipate that the presence of a strong recovery will introduce a non negligible difference between G_f and the seismological estimate G' which is obtained under the assumption of no undershoot (no recovery), as discussed further below.

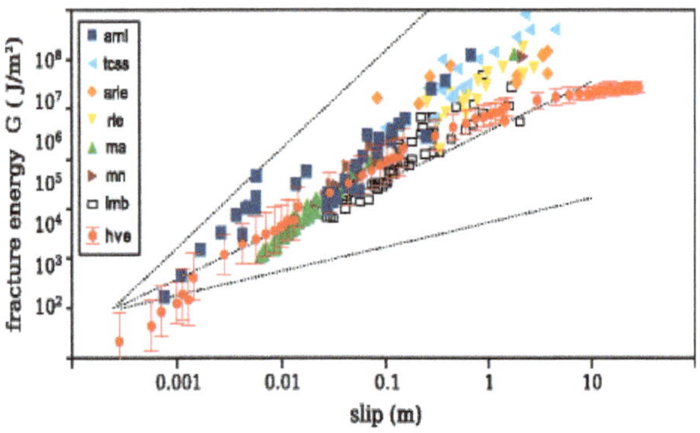

Figure. 5 Experimental (G_f, in red) and seismological (G', other colors) estimates of fracture energy under coseismic slip conditions. Dashed lines indicate exponent 0.5, 1, and 2 for reference. Red disks correspond to values of G_f averaged for 28 different experiments, at various slip amounts. arnl: Northridge aftershocks (Abercrombie and Rice 2005); arle: large earthquakes (Abercrombie and Rice 2005); tcss: numerical simulations (Tinti E. et al. 2005); rle: large earthquakes (Rice 2006); ma: L'Aquila (Malagnini et al. 1994); mn: Northridge sequence (Malagnini et al. 1994); lmb: Tocopilla, Chile (Lancieri et al. 2012); hve: high velocity friction experiments

COMPARISON OF G_F WITH VALUES ESTIMATED FROM SEISMOLOGICAL DATA

The scaling shown by Eq. 7 is, in some aspects, in agreement with independent estimates of earthquake fracture energy. Equating the fracture energy to the difference between the available strain energy (in excess of the minimum sliding friction) and the radiated energy (as quantified from kinetic energy in the far-field), Abercrombie and Rice (2005) (hereafter AR) proposed an estimate of G', the apparent fracture energy, noting that G' and the actual G will coincide if the final stress is equal to the minimum sliding shear stress (no undershoot, i.e., final stress higher than the sliding stress, neither overshoot, i.e., final stress lower than the sliding stress). Based on the best-fit of a number of earthquake data AR proposed the empirical relation:

$$G' = 5.25\ 10^6\ u^{1.28} \qquad (13)$$

(we use G' for fracture energy values resulting from seismological estimates, following AR, as opposed to G_f resulting from experimental measurements). In addition, Tinti E. et al. (2005) proposed $G_k \propto u^{1.81}$. Their estimate was obtained by imposing the slip retrieved from kinematic inversion as a boundary condition in dynamic simulations, and the resulting stress evolution was computed.

In Fig. 5, G' from a literature compilation (Nielsen et al. 2015) and the average G_f value for all 28 high-velocity experiments are shown together. G_f and G' are comparable for slips of about u= 1 cm ($G \approx 10^4$ J/m²), as reported by Nielsen et al. (2015), with the addition here of the Tocopilla earthquake sequence (Lancieri et al. 2012). G' and G_f both increase with slip up to about 10^6 J/m², however, while G_f saturates at u ≈ 10 m, it appears that G' continues to increase, with the result that for some of the reported earthquake sequences G' is substantially larger that G_f at large magnitudes (u≥10m). The exponent in G_f ($1-\alpha \approx 0.5$) is lower than in G'($1-\alpha \approx 1.28$ as reported by AR or slightly more for the earthquake sequences reported here). We discuss further below the possible causes of such discrepancy.

Assuming that G'(u) and shear stress τ(u) are related through Eq. 2, and that G' has the form of Eq. 13, AR took the derivative of both with respect to u and after some algebra obtained a stress evolution of the form:

$$\tau(u) = C_0 - 4.8\ 10^6\ u^{0.28}. \tag{14}$$

C_0 can then be assimilated to the peak shear stress τ_p at u = 0. If this form were used with very low values of τ_p (e.g., 5 MPa) and relatively high values of u (e.g., 10 m), although it is not a very likely combination of parameters for earthquake source, it may result in the unphysical feature of a negative friction. On the other hand, if the exponent of 1.28 in (Eq. 13) is replaced by any positive value smaller than 1 (for example 0.5), then retrieving τ from expression (2) results in a quite different expression for shear stress evolution, of the form $\tau = C + Bu^{-0.5}$ (where C and Bare constants to be determined by limit conditions), which can produce a weakening law without incurring into negative friction. (By setting $C = 0$ and $B = \tau_p\sqrt{u_w}$, we do obtain a solution of the type of Eq. 7). Thus, a value greater than 1 in the exponent in Eq. 13 resulting from seismological estimates produces an unphysical friction law if we assume that G' is related to frictional dissipation alone.

On the other hand, assuming that a single powerlaw scaling such as Eq. 13, and a single associated dissipation mechanisms, would represent earthquake mechanics over a seven magnitude-range or more is an oversimplification. It has been pointed out before (Rice 2006; Viesca and Garagash 2015) that Eq. 13 fails to capture the trend for earthquakes in the larger magnitude range. As magnitude increases, transitions in the weakening mechanisms and in the rupture mode (from crack-like to pulse-like, Viesca and Garagash (2015)) may be possible causes of the observed changes in G'(u). In addition, uncertainty in G' for large earthquakes is due to the difficulty of including large near-field radiation and/or the use of finite-source kinematic inversions with non-unique solutions. Indeed, a large scatter in fracture energy G' is observed at large magnitudes (Fig. 5), with the lower end members (lmb, Tocopilla sequence) being broadly compatible with the experimental measurements of G_f.

The above comments suggest either (1) that G' estimates may be affected by uncertainty or bias in the large magnitude; or (2) that G' represents a measure of other dissipation processes that friction alone. In the latter case, friction needs to be redefined in an equivalent, broader sense suitable for the scale of earthquake faults. A transition from friction-dominated dissipation to diffusive-dominated processes may occur in larger earthquakes. For example, thermal pressurization mechanisms may become more effective at larger slip; diffuse anelastic co-seismic strain may affect a wider fault zone on mature faults and become important in the energy budget of rupture.

In the following sections (Sections 5–9), we consider different sources of mismatch between G'and G_f at large magnitudes, due to either a bias in the seismological estimates G', or to the limitations of using experimental measures of friction to account for large earthquake dissipation.

First, we will consider whether a more complex loading history than that used in the experiments (constant acceleration ramp) may radically alter the G_f trend and whether it may approach that of G' under realistic conditions (Section 5). Subsequently, we discuss the effect of normal stress increase in the experiments, showing that it has no significant influence on the fracture energy, except for slip <10 cm (Section 6). Then, we discuss whether a bias due to undershoot in the G' estimates may explain the difference with G_f (Section 7). Finally, we consider possibilities of G' being the signature of other dissipation processes than friction: off-fault plasticity (Section 8) and non-planar fault topography, discussing possible scaling of G'with u due to fault non-planarity (Section 9).

POSSIBLE OUTCOMES OF A DIFFERENT LOADING HISTORY

As noted in Section 2, using the rough approximation that during the weakening phase friction is inversely proportional to slip velocity such that τ∝1/V, and noting that $V = \sqrt{2\,u\,\dot{V}}$ under constant acceleration, one would obtain $\tau \propto \frac{1}{\sqrt{2\dot{V}}}u^{-0.5}$, a power law of slip with an exponent fairly close to the experimental fit α = 0.57. We remark that it is a limitation of the experiments that a constant acceleration $\dot{V}$ is used, followed by a constant "target" velocity. Given the close link between τ(t) and V(t), it is expected that the scaling τ versus u may differ to some extent if the loading time history was altered. Nonetheless, most dynamic rupture models produce a very abrupt acceleration phase in the vicinity of the rupture tip (corresponding more or less to the frictional weakening phase), which we assume here can be reasonably linearized by assuming constant V. More marked differences in the slip velocity are expected after the acceleration phase. Indeed, models (Kostrov 1974; Andrews 1976; Broberg 1999) predict a peak velocity near the end of the weakening phase followed by a relatively rapid decay in V, whereas the current experiments impose a constant slip rate; however, this later evolution would not alter the fracture energy which mostly results from the stress evolution within the weakening phase only.

If we now assume that the approximation τ∝1/V holds but drop the assumption of constant acceleration and set $\tau = \tau_p - C_1u^{\beta-1}$, where—compatibly with G' estimates and Eq. 2—we have $C_1 \approx 4.8\ 10^6$ and $\beta-1 \approx 0.28$. Using an indicative value τ_p = 100 MPa, this would require that slip velocity during the weakening phase follows . It is clear that in this case, velocity is increasing unrealistically slowly with increasing slip, as this contradicts (1) the abrupt acceleration resulting from the elastodynamic solution systematically observed in rupture propagation models and (2) the observation that wave radiation is mostly generated in the vicinity of the rupture front, which again supposes an abrupt acceleration of slip.

As a consequence, we think it is unlikely that the difference between G′ and G may be reconciled by assuming a different slip velocity profile, as the latter would hardly be compatible with fundamental results of elastodynamics and with seismological observations.

INFLUENCE OF NORMAL STRESS

We now discuss whether the normal stress does have an influence on fracture energy, and whether it may justify the difference between the experimental measures and the seismological estimates of fracture energy. While the normal stress in the experiments discussed here is in the range 10–40 MPa, we expect it to be higher on natural faults at seismogenic depths. Normal stress resolved on the fault will depend on the deviatoric stress and the type of faulting (Malagnini et al. 2010), but it will on average be of the same order as the lithostatic load. At the base of the seismogenic crust (≈10–15 km), the lithostatic load can be up to ≈300 MPa assuming no pore pressure, ≈200 MPa under hydrostatic pressure and considerably less in case of overpressure. The arithmetic average in the ten topmost kilometers may therefore be up to about 150 MPa.

Normal stress influences the peak and the steady-state shear stress, and the apparent weakening distance (Nielsen et al. 2010). However, previous theoretical modeling of frictional melt and related experimental data obtained in rotary shear experiment showed no systematic variation of fracture energy with normal stress. Here, we confirm these preliminary results by extending them to a wider range of normal stress and to either cases where frictional melt is taking place (gabbro, basalt) or not (calcite experiments).

Figure 6 shows a collection of fracture energy measured under different normal stress and at three different values of slip. For the case of gabbro, at slip of 1 and 10 cm, there appears to be an order magnitude increase in G_f as normal stress varies from 10 to 40 MPa. However, it is to note that at 1 and 10 cm experiments indicate a degree of variability of the same order of magnitude (experiments at 20 MPa were repeated). As a consequence, such trend is not significantly above the experimental error. When the weakening is about to be complete (1 m slip), the data variability is reduced and no systematic trend is observed in G_f with normal stress. The data for calcite, on the other hand, do not show any significant trend above the variability.

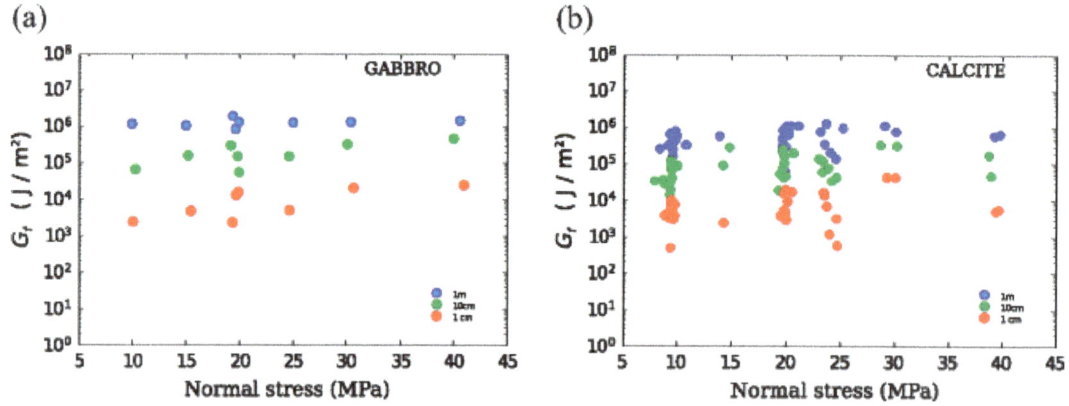

Figure. 6 Measure of fracture energy for experiments performed under different normal stress (in the range ~10 to ~40 MPa) and three different slip amounts (red = 1 cm, green = 10 cm,blue = 1 m). No significant trend is observed with normal stress. The increase with slip is replicates the observation of Figs. 4 and 5. **a** Gabbro experiments; **b** calcite experiments

SYSTEMATIC BIAS DUE TO UNDER- OR OVERSHOOT

Seismological estimates of G′ are based on the assumption that there is no under- or overshoot in earthquakes (i.e., the assumption that final stress τ_1 is equal to the minimum dynamic sliding stress τ_d, see Fig. 7). Some consequences of such assumption are discussed in AR and (Viesca and Garagash 2015). Here, we explore the possibility the same assumption could cause the hiatus between the experimental and seismological fracture energies.

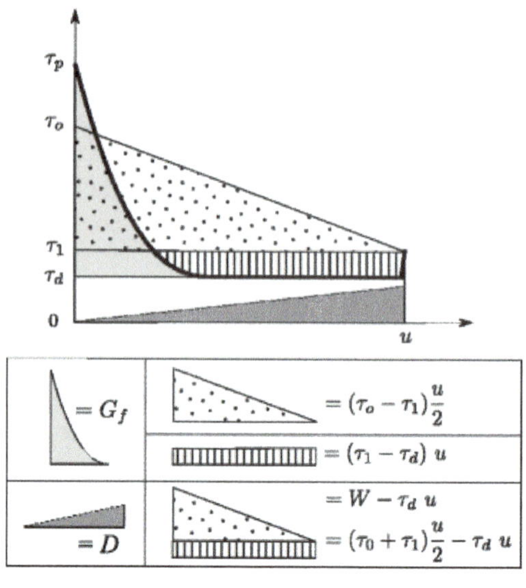

Figure. 7 Schematic energy partition as a function of slip and initial, dynamic, and final stress. The total area below the diagonal line

connecting $(0, \tau_0)$ and (u, τ_1) is the available elastic strain energy W. Frictional dissipation is the total area below the shear stress curve (thick black curve). Light grey area indicates the fracture energy due to friction, G_f. Thehatched rectangular area indicates the energy error incurred by neglecting $\tau_1 - \tau_d$. Thedark grey triangle below the dashed line (named D) indicates a general dissipation process (not friction) increasing with slip (in this case linearly). See text for futher details

Because only the static (final) stress drop $\Delta\sigma = \tau_0 - \tau_1$ (τ_0 is initial stress, see Fig. 7) can be estimated through seismology, but not the absolute stress or the initial, final and dynamic stress values, the estimate from earthquake data is $G' = (\tau_0 - \tau_1)\frac{u}{2} - \frac{\mu' E_s}{M_o} u$ (under the assumption$\tau_1 = \tau_d$) based on average slip, stress drop, radiated energy E_s, and seismic moment M_0. Recalling the energy balance in the earthquake process already discussed elsewhere (Scholz1990; Rivera and Kanamori 2005; Abercrombie and Rice 2005) and illustrating schematically stress and frictional work flow as of Fig. 7, we can write:

$$
\begin{aligned}
G &= W - \frac{E_s}{A}\,, \\
&= (\tau_0 + \tau_1)\,\frac{u}{2} - \tau_d\, u - \frac{E_s}{A}\,, \\
&= (\tau_0 + \tau_1)\,\frac{u}{2} - \tau_d\, u - \frac{\mu' E_s}{M_o}\, u\,.
\end{aligned}
$$

(15)

(where W is the work of the elastic strain release, μ' is the shear stiffness, and A is the area of the fault). As a consequence $G = G' + (\tau_1 - \tau_d)$ u. If the actual G was smaller than the G'approximation, and possibly in agreement with the experimental measure G_f, this would imply that $\tau_1 < \tau_d$ (overshoot). However, only a weak overshoot is generated by crack-like ruptures under favorable circumstances (Madariaga 1976). On the other hand, pulse-like ruptures which provide a better fit to large eathquakes kinematic inversions (Heaton 1990), systematically generate undershoot (Nielsen and Madariaga 2003).

As a consequence, the assumption that $\tau_d \approx \tau_1$ for G' estimates is most likely to underestimate the actual fracture energy than otherwise. Therefore, this cannot reconcile the difference $G_f < G'$at large magnitudes.

OFF-FAULT ENERGY SINKS

As noted by Shipton et al. (2006a), the effective G' observed at the seismological scale should implicitly incorporate energy sinks other than frictional dissipation alone. On a planar fault, the stress level off-fault is sufficiently high to induce anelastic damage on a thickness increasing with length propagation, as initially proposed by Poliakov (2002). In a model by Andrews (2005), the dissipation due to plastic strain results in G scaling with slip so that $G \propto u^{1.0}$, an exponent which is also compatible with the barrier toughness model for the arrest of crack growth (discussed in Eqs. 3, 4, and 5). The exponent 1.0 is intermediate between that estimated from frictional dissipation and that estimated from seismological data.

Indicatively, a trend representing a general dissipation growing with slip is shown in Fig. 7 as the grey area the bottom of the diagram (we represent a case with $D \propto u^{1.0}$ compatibly with models of off-fault plastic dissipation, but more general trends may be speculated, see section below on non-planar fault surfaces). This ulterior energy sink D has to be subtracted from the radiated energy. The balance of energy Eq. 15 then becomes:

$$
\begin{aligned}
G + D &= W - \tau_d\, u - \frac{E_s}{A}\,, \\
&= (\tau_0 + \tau_1)\,\frac{u}{2} - \tau_d\, u - \frac{\mu\, E_s}{M_o}\, u\,.
\end{aligned}
$$

(16)

The above shorthand form, where values are averaged on the fault rupture area A, can be equated to the more general integral form (Kostrov 1974; Rivera and Kanamori 2005):

$$
\begin{aligned}
E_s &= \tfrac{1}{2}\textstyle\int_A (\tau_1 + \tau_0)\, u\, dA - \int_A 2\,\gamma\, dS \\
&\quad - \textstyle\int_{t0}^{t_1} dt \int_A \tau\, \dot{u}\, dA \\
&= \tfrac{1}{2}\textstyle\int_A (\tau_1 + \tau_0)\, u\, dA \\
&\quad - \textstyle\int_{t0}^{t_1} dt \int_A \tau_d\, \dot{u}\, dA \backslash - \int_A 2\,\gamma\, dA \\
&\quad - \textstyle\int_{t0}^{t_1} dt \int_A (\tau - \tau_d)\, \dot{u}\, dA
\end{aligned}
$$

(17)

(where, for simplicity, we assumed that slip u is in a fixed direction (no rake rotation) to avoid vectorial notation). Here, we may equate $W = \frac{1}{2A}\int_A(\tau_1 + \tau_0)\, u\, dA$, , $G = G_f = \frac{1}{A}\int_{t0}^{t_1} dt \int_A (\tau - \tau_d)\, \dot{u}\, dA$ and $\tau_d\, u = \frac{1}{A}\int_{t0}^{t_1} dt \int_A \tau_d\, \dot{u}\, dA$. . The remaining term $\int_A 2\,\gamma\, dA$, was introduced to account for the energy sink in the singular stress terms ahead of the crack tip, considering the case where the process zone at the crack tip is vanishingly small. If we assume that for earthquakes there is no singularity strictly speaking and the process zone is finite, we may instead equate the γ term to diffuse off-fault dissipation in the vicinity of the crack tip, and ideally equate $D = \frac{1}{A}\int_A 2\,\gamma\, dA$.

In addition, natural fault surfaces are rough at many different scales and the slipping zone is often made of non-cohesive rock (gouge), whereas experiments are performed on smooth, machined surfaces (wavelength $<10^{-3}$ m) and, often, on cohesive rocks. We discuss possible effects of roughness and thickness of damage zone in the next section.

NON-PLANAR FAULT SURFACES

The non-planarity of natural faults creates additional stress and strain during slip (Chester and Chester 2000; Sagy and Brodsky 2009; Griffith et al. 2010). If sufficiently strong, such stress may result in asperity abrasion (especially at small wavelengths) and off-fault damage which will introduce energy sinks in addition to that of friction on the sliding surface itself (Dunham et al. 2011). Moreover, in the presence of fault gouge (non-cohesive rock), typical at depths <6–7 km, a relevant portion of energy may be dissipated during the slip-localizing process. An example of 600 m wide damage structure associated to ancient seismic faulting at depth, has been reported in detail by Smith et al. (2013). Since G' is obtained by estimating the amount of dissipation with respect to strain energy and radiated energy, it will implicitly incorporate the sum of all dissipative processes due to rupture propagation and fault slip, necessarily larger or equal to the laboratory, friction-only related measurements.

With increasing fault slip, the size of the topographical asperities which are removed, the wavelength of fault bends which are involved, and the thickness of the damage zone increases. As a consequence, the apparent G' from seismological estimates, implicitly accounting for the effects of off-fault damage, may be larger than G_f and have a different slip dependence than the experimentally measured G_f, which accounts for small-scale dissipation by friction alone.

Natural fault surface roughness can be described by a distribution with power spectrum $P(k) = k^{-1-2H}$ where k is wavenumber and H is the Hurst (or Hausdorff) exponent (Schmittbuhl et al.1993; Bistacchi et al. 2011; Candela et al. 2012). Examples of synthetic topographic profiles are shown in Fig. 8 with H = 0.8 (and indicatively, though unrealistic, with H = 1.14). If a fraction of the elastic strain work is dissipated while the asperities are cyclically strained, such process may yield a dissipation increasing as a power of slip due to the increasingly large scale roughness involved.

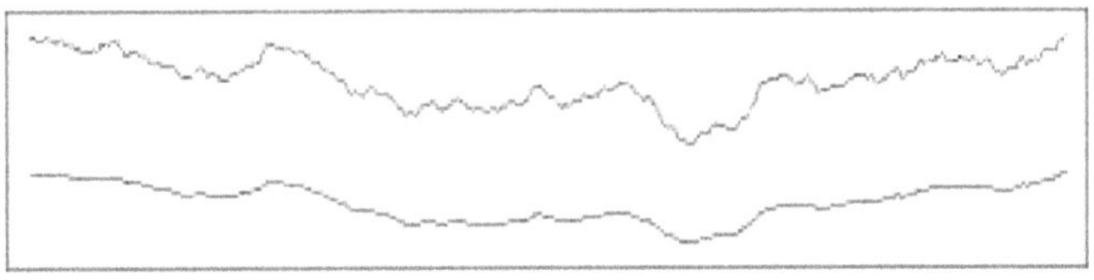

Figure. 8 Comparison of self-affine randomly generated topographies with Hurst exponent H = 0.8 (top curve) within the range of values observed on fossil seismic faults exhumed from depths of ≈9 km (Bistacchi et al. 2011) and H = 1.14 (bottom curve). The larger H = 1.14 value, implying relatively larger fault topography at larger scales, but relatively smooth fault at smaller scales, would generate a strain energy per unit fault area which scales as$G' \propto u^{1.28}$ as estimated by AR. The self-similar case corresponding to H = 1 (not shown) produces an intermediate roughness

In the following paragraphs, we provide estimated scaling of strain and dissipation induced by slip on a non-planar fault. Although a number of simplifying assumptions are made, the dimensional arguments and the resulting scaling should not be altered by the assumptions. Since the estimates rely on a model of infinitesimal strain and elastic stress, we note that our approximate model should produce an upper bound estimate. In future studies, a verification may be attempted by using numerical tools which possibly allow for finite strain analysis, a Lagrangian approach, and the implementation of anelastic strain.

We start our analysis using a simple sinusoidal topography $y = h \sin(2\pi x/\lambda)$ for the fault surface. Assuming that the fault surface itself is lubricated, elastic stress and strain induced by topography in the fault vicinity can be computed using the well-known analytical solution based on Flamant problem, under the assumption that amplitude h is small with respect to the wavelength λ (see for example Saucier et al. (1992)).

As slip increases on the fault, we may assume that a finite fraction of the elastic work spent in the cyclic compression and decompression of the sinusoidal bends is dissipated in anelastic strain and permanent damage. Damage in the vicinity of natural faults is consistently observed, and in some instances, it has been shown to correlate with the fault bends (Griffith et al. 2010).

Fault with a Single Wavelength in Topography

We consider two ideal models of dissipation due to strain on a non-planar fault. The first model assumes that the stress distribution around then fault can be approximated by an elastic model, illustrated in Fig. 9, but that a portion of elastic strain energy is dissipated through each deformation cycle. The second model assumes that asperities are abraded by irreversible anelastic strain occurring in a thickness which is approximately that of fault topography elevation h(λ) at the wavelength λ ≈ u. Neither model is producing an exponent in dissipation compatible with the seismological estimates. To avoid the assumptions and approximations used in this analysis, a full numerical simulation could be performed which is beyond the scope of this study, for example, using methods illustrated by Dunham et al. (2011).

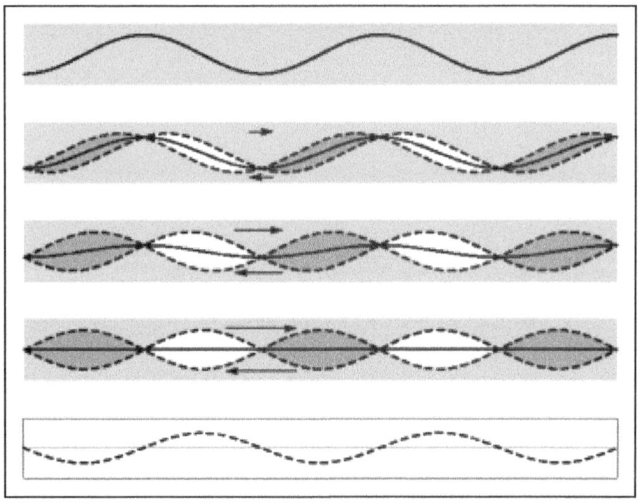

Figure. 9 Wavy fault (black thick curve) after offsets of 15,2515,25, and 1212; λ. Dashed curves indicate the virtual boundaries of top and bottom sides if they were offset without strain (virtual overlap and underlap). At u=12λu=12λ, the single wavelength topography has flattened-out. For reference, the normal stress corresponding to u=12λu=12λ is shown (bottom dotted curve)

Elastic stress and strain energy in cyclic deformation

We first discuss the elastic stress perturbation resulting from slip on a wavy fault of simple topography represented by $y = h\sin(2\pi x/\lambda)$. As discussed in Appendix, when the fault slips an amount u it will undergo $n = u/\lambda$ strain cycles, resulting in an absolute value of elastic strain work which scales as $w \propto u\,(h^2/\lambda^2 + Ah/\lambda)$ where A is a dimensionless ratio of prestress versus elastic stiffness (see Appendix). Assuming that the topographical elevation h can be related to wavelength such that

$$h = \phi\,\lambda^H, \tag{18}$$

where ϕ is a normalizing factor, we may write $w \propto u\phi^2\lambda^{2H-2} + u\phi A\lambda^{H-1}$. For fault slip u ≈ λ, we obtain:

$$w \propto \phi^2 \lambda^{2H-1} + A \phi \lambda^H \quad (19)$$

which, given the dimensions of ϕ as of Eq. 18, yields correct dimensions of J m^{-2}.

For slip u on the fault, strain due to non-planarity will involve predominantly wavelengths $\lambda \approx$ uand below.

Although Eq. 19 is derived from the description of a purely elastic stress/strain relation, we may assume that if during each deformation cycle a fraction of work is dissipated through visco-elastic or plastic processes, the scaling may remain close to that of Eq. 19. Hence, we may consider w as an upper-bound estimate of the roughness-induced dissipation, and assuming that the strain is dominated by the upper wavelength $\lambda \approx$ u, we find

$$G_r \propto u^{2H-1} + A\, u^H, \quad (20)$$

which may contribute to the apparent fracture energy G′ in the earthquake balance. Using an indicative value H = 0.8, we obtain a scaling exponent of about 0.6 (if A≪1 in Eq. 20) or 0.8 (ifA≫1). In both cases, the exponent is quite lower than the seismological estimates. Considering the effect of smaller wavelengths in the above computation, though not developed here, leads to similar results.

Finite brittle or ductile strain

We consider another simplified model where we assume that slip u results in the irreversible anelastic abrasion of asperities in the scale $\lambda \approx$ u, through brittle, ductile, or plastic distributed deformation processes in the vicinity of the fault. An ideal sketch can be found in Fig. 10. In our ideal model, shear deformation during abrasion of asperities at scale λ takes place within a band of thickens approximately $h(\lambda)$, the elevation of fault topography at a given wavelength. As slip progresses, wavelengths involved in the abrasion process become larger and larger, and the thickness of the deformation band increases. In Fig. 10, two different wavelengths are shown, together with the increasing area of shear. The deformation within the shear band is of the order of $\varepsilon_p \approx h/u$; If shear occurs at an average plastic yield stress σ_y, over a thickness h, we will have w(u) = σ_y (h^2/u) per unit fault area. Using $h = \lambda^H = u^H$, we obtain $w \approx u^{2H-1}$. In this case, too, the exponent of the dissipated work is quite smaller (≈0.6) than the exponent (≈1.28 obtained fog G′ from seismological estimates. Furthermore, we note that since the strain $\varepsilon_p \approx h/\lambda \approx \lambda^{H-1}$ decreases with λ for H < 1, a limit where infinitesimal strain can be accommodated elastically will be reached as the slip increases, thus inducing a saturation of dissipated energy. Assuming a limit of elastic strain at 1 %, and $h = 0.05\, \lambda^H$, dissipation would saturate at slip u ≈ 3.6 m. Such an amount of slip lies within the large end-members of the seismological catalog of Fig. 5; therefore, the lack of larger events would hinder the observation of any change in behavior.

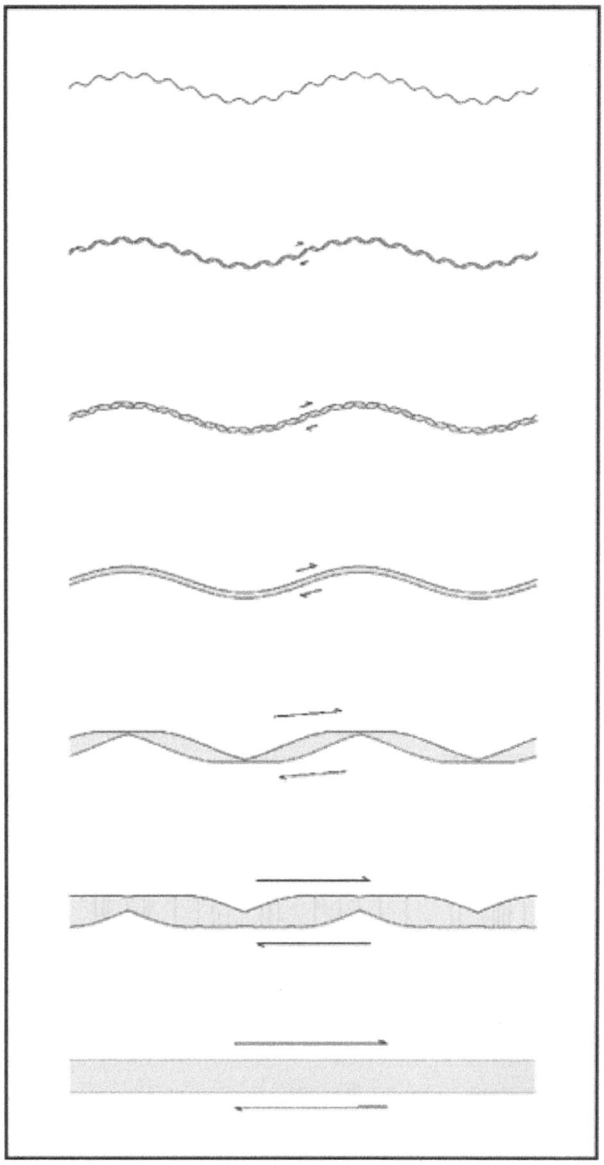

Figure. 10 Schematic evolution of fault surface waviness and associated damage with increasing slip. A simple model with only two different wavelengths is shown

Fault segmentation

According to the above discussion, fault roughness described by H ≈ 0.8 would yield off-fault dissipation scaling with a much lower exponent (≈0.6 versus ≈1.3) than that captured by the seismological estimates of G′.

However, power-law roughness of faults is measured on individual, continuous segments (Bistacchi et al. 2011; Candela et al. 2012). Further complexity is induced by fault discontinuities, i.e., jogs or fault segmentation. In particular for large earthquakes, seismic ruptures have been known to span more than one fault segments, for example, in the well-documented Landers earthquake (Aochi et al. 2003, and references therein).

A fault jog is associated with the termination of an individual fault segment, and the transition of slip onto another fault segment within an echelon-type structure. As a consequence, stress concentration in the vicinity of jogs is similar to that at the tips of an isolated rupture (albeit generally larger owing to constructive interference of both segments), inducing substantial damage and dissipation.

Indeed, documented km-scale contractional jogs show an increase of up to five times in the thickness of the damage zone (Bistacchi et al. 2010, reported on the Sprechenstein fault system, Eastern Italian Alps). This additional source of dissipation and its scaling are not quantitatively explored here, but may constitute an interesting case for further modelling and for a systematic census of jog-related anelastic strain in faults.

CONCLUSION

In this study, we report detailed evolution of shear stress in high slip velocity experiments on calcite-rich and silica-rich rocks which are representative end-members of the Earth crust and some, but not all, earthquake fault environments.

We show that, under the experimental conditions explored here (slip rate V = 3–6.5 m/s, slip acceleration V˙=6.5 m~s−2V˙=6.5 m~s−2 m/s², normal stress 10–30 MPa): (1) frictional weakening with slip is compatible with a power law of the type $\tau \propto u^{-\alpha}$ with $\alpha \approx 0.57$. (2) Weakening is typically triggered at slip velocities of $V_w \approx 0.4$ m/s and $V_w \approx 0.1$ m/s at normal stresses of 10 and 30 MPa, respectively; extrapolating this trend V_w will decrease even further at higher normal stresses. (3) Constant acceleration results in the relation between slip u and slip rate $V = \sqrt{2u\dot{V}}$, with the consequence that the weakening typically starts after slip of $0.004\ \text{m} < u_w < 0.04$ m given the above V_w and V. (4) Weakening is complete after slip of the order of $u \approx 1$ m for silica-rich rocks, as soon as the acceleration phase ends, but up to $u \approx 10$–15 m for calcite, well into the interval of constant slip velocity. The difference is ascribed to the frictional melting and fast sample shortening in the case of silicate-rich rocks, resulting in a rapid achievement of steady state.

We compute the equivalent fracture energy G_f resulting from the frictional weakening and compare it with estimates G' obtained for earthquakes from seismology estimates. Both are in the form of a power law of slip; however, the exponents are substantially different ($\tau \propto u^{0.5}$ for G_f versus $\tau \propto u^{1.28}$ for G'). Though G' and G_f values are compatible for earthquakes with slip in the range of centimeter to meter of slip (approximate magnitudes $3 < M_w < 7$ and $10^4 < G < 10^6 \text{J.m}^{-2}$), they do diverge considerably for larger earthquakes, owing to the different exponent.

To explain the large exponent from seismological estimates, we attempt a preliminary analysis of off-fault dissipation due to fault non-planarity. The latter yields a scaling of dissipation with slip with an exponent between 0.6 and 0.8, which is much smaller that the seismological estimates (1.28). On the other hand, we may assume that the main energy sink is due to plastic strain associated to the stress concentration at the tip of a propagating rupture. In that case, it has been shown (Andrews 2005) that dissipation (per unit fault area) increases proportionally to fault length. According to the elementary seismological scaling relations, slip is proportional to fault length, in this case the dissipation yields a scaling exponent of 1, which is close to and possibly compatible with the seismological estimate of 1.28, given its reduced precision.

The dissipation related to rupture tips can be multiplied and amplified by the presence of fault jogs along the seismic rupture, which act as multiple stress concentrators within the main rupture area.

The striking conclusion is that, because faults are rapidly and efficiently lubricated upon fast slip initiation, the dominant dissipation mechanism in large earthquakes may not be friction but be the plastic, off-fault strain due to fault segmentation and stress concentrations in a growing region around the fracture tip.

ACKNOWLEDGMENTS

We acknowledge funding provided by the European Research Council projects 205175 (USEMS) and 614705 (NOFEAR), and NERC capital grant CC019 (HiFAST). S.A.F.S. acknowledges funding from the Marsden Fund Council (project UOO1417 to Smith), administered by the Royal Society of New Zealand.

APPENDIX: COMPUTATION OF G_F

The experimental value of G_f is obtained by discretizing Eq. 2. Using the recorded data points, which are sampled at a time interval dt = 40 10^{-6} s, and using n, m to represent the time indexes, then at time t = ndt we have:

$$G_f(n) = \sum_{m=1}^{n} (\tau(m) - \tau(n))\,(u(m) - u(m-1))\,.$$

Note that in case of slip-hardening (transient increase of τ), instead of subtracting $\tau(n)$ in the above equation, we subtract the minimum observed τ value within the interval [1, n].

Elastic stress and strain around wavy fault

The elastic stress and strain distribution for slip on an interface described by $y = h \sin(2\pi x/\lambda)$(assuming that $u \ll \lambda$ and $h \ll \lambda$) is of the form:

$$\begin{aligned}
\sigma_{ij}(x,y) &= E\left(a_{ij} + b_{ij}\tfrac{2\pi y}{\lambda}\right)\tfrac{h\,u}{\lambda^2}\; g(2\pi x/\lambda) \\
&\quad \times \exp(-2\pi|y|/\lambda) + \sigma^0_{ij} \\
\varepsilon_{ij}(x,y) &= \left(c_{ij} + d_{ij}\tfrac{2\pi y}{\lambda}\right)\tfrac{h\,u}{\lambda^2}\; g(2\pi x/\lambda) \\
&\quad \times \exp(-2\pi|y|/\lambda) + \varepsilon^0_{ij}
\end{aligned} \tag{21}$$

where g is a function of periodicity 2π (e.g., sine or cosine), E is Young modulus, and u is slip. (The zero superscripts indicate background initial values of stress and strain). After Saucier, $g = \cos(2\pi x/\lambda)$ for diagonal elements of the stress and strain tensors, and $\sin(2\pi x/\lambda)$ for off-diagonal terms. (a_{ij}, b_{ij}) take the values $(1,1), (1,-1), (0,1)$ for $ij = xx, yy,$and $xy,$ respectively, and (c_{ij}, d_{ij}), take the values take the values $(1-\nu,1+\nu), (1-\nu,-1-\nu), (0,1+\nu)$ for $ij = xx, yy,$ and xy, respectively. These general properties apply either to the study of lubrified interfaces in Hertzian contact with partial opening (Dundurs, Block), to the model proposed by Saucier for offset on lubrified sinusoidal interface (Saucier) or to the model including friction at the interface proposed by Chester and Chester (2000).

If we release the assumption that $u \ll \lambda$, when slip u is exactly equal to one wavelength, the two half-spaces on each side of the fault are exactly in phase again, so that strain and stress perturbation cancel. As a consequence, a periodic topography will sustain a cyclic stress fluctuation which can be obtained by replacing $u/\lambda \to \frac{1}{2\pi}\sin(2\pi u/\lambda),$ and we obtain:

$$\begin{aligned}
\sigma_{ij}(x,y) &= E\left(a_{ij} + b_{ij}\tfrac{2\pi y}{\lambda}\right)\tfrac{h}{2\pi\lambda}\; \sin(2\pi u/\lambda) \\
&\quad \times g(2\pi x/\lambda)\exp(-2\pi|y|/\lambda) + \sigma^0_{ij} \\
\varepsilon_{ij}(x,y) &= \left(c_{ij} + d_{ij}\tfrac{2\pi y}{\lambda}\right)\tfrac{h}{2\pi\lambda}\; \sin(2\pi u/\lambda) \\
&\quad \times g(2\pi x/\lambda)\exp(-2\pi|y|/\lambda) + \varepsilon^0_{ij}
\end{aligned} \tag{22}$$

(in the limit $u \ll \lambda$ we have $u/\lambda \approx \frac{1}{2\pi}\sin(2\pi u/\lambda)$ so that we retrieve Eq. 21. Work density dw induced by elementary strain dε is $dw = \sigma_{ij} d\varepsilon_{ij},$ with implicit summation over repeated indexes. In terms of slip, we can write $dw = \sigma_{ij}(u)\, \varepsilon'_{ij}(u)\, du,$ where the prime denotes the first derivative with respect to u. Here, σ_{ij} refers to the stress variation with respect to initial condition.

Integrating the absolute value of dw over u and averaging over x, then integrating over all y, we obtain the same scaling for for both expressions (21) and (22), namely, $w = f_{ij}\, E\, h^2/\lambda + g_{ij}\, \sigma^0_{ij} h$ (where $f_{ij} = \frac{1}{128\,\pi^3}(\frac{1}{2}a_{ij}c_{ij} + bc_{ij} + a_{ij}d_{ij} + b_{ij}d_{ij})$ and $g_{ij} = \frac{1}{2\,\pi^3}(c_{ij} + d_{ij})$) for each cycle of strain, where w is the absolute value of elastic strain work performed for each cycle of strain ($u = \lambda$). After $n = u/\lambda$ cycles, the total work will be:

$$w = \Sigma_{ij}\, f_{ij}\, E\, u\, h^2/\lambda^2 + \Sigma_{ij} g_{ij} \sigma^0_{ij}\, u\, h/\lambda.$$

$$w \propto u\left(\frac{h^2}{\lambda^2} + A\,\frac{h}{\lambda}\right) \tag{23}$$

Since $h \ll \lambda$ but likely $E \gg g_{ij}\, \sigma^0_{ij},$ we do neglect neither term in Eq. 23, and define the dimensionless ratio $A = (E\, \Sigma_{ij} g_{ij} \sigma^0_{ij})/(\sigma^0_{ij} \Sigma_{ij}\, f_{ij}).$

REFERENCES

1. Abercrombie R, Rice J (2005) Can observations of earthquake scaling constrain slip weakening?. Geophys J Int 162:406–424
2. Aki K (1979) Characterization of barriers on a earthquake fault. J Geophys Res 84:6140–6148
3. Andrews DJ (1976) Rupture propagation with finite stress in antiplane strain. J Geophys Res 81:3575–3582
4. Andrews DJ (2005) Rupture dynamics with energy loss outside the slip zone. J Geophys Res Solid Earth 110(B1). JB003,191, doi:10.1029/2004JB003191, b01307
5. Aochi H, Madariaga R, Fukuyama E (2003) Constraint of fault parameters inferred from nonplanar fault modeling. Geochem Geophys Geosyst 4(1020). doi:10.1029/2001GC000207
6. Atkinson BK, Meredith PG (1987) Experimental fracture mechanics for rocks and minerals. Fracture Mechanics of Rock:477–525. doi:10.1016/b978-0-12-066266-1.50016-8
7. Bistacchi A, Massironi M, Menegon L (2010) Three-dimensional characterization of a crustal-scale fault zone: the pusteria and sprechenstein fault system (eastern alps). J Struct Geol 32(12):20222041. doi:10.1016/j.jsg.2010.06.003
8. Bistacchi A, Griffith W, Smith S, Di Toro G, Jones R, Nielsen S (2011) Fault roughness at seismogenic depths from LIDAR and photogrammetric analysis. Pure Appl Geophys 168:2345–2363. doi:10.1007/s00024-011-0301-7
9. Broberg KB (1999) Cracks and Fracture. Academic Press
10. Candela T, Renard F, Klinger Y, Mair K, Schmittbuhl J, Brodsky EE (2012) Roughness of fault surfaces over nine decades of length scales. J Geophys Res 117(B8):jb009,041, doi:10.1029/2011jb009041
11. Causse M, Dalguer LA, Mai PM (2014) Variability of dynamic source parameters inferred from kinematic models of past earthquakes. Geophys J Int 196(3):1754–1769. doi:10.1093/gji/ggt478
12. Chester M, Chester JA (2000) Stress and deformation along wavy frictional faults. J Geophys Res 105(B10):23,421–23,430
13. Cocco M, Bizzarri A (2002) On the slip-weakening behavior of rate- and state dependent constitutive laws. Geophys Res Lett 29(11):11-1–11-4. doi:10.1029/2001GL013999
14. Cruz-Atienza V, Olsen K (2010) Supershear mach-waves expose the fault breakdown slip. Tectonophysics 493:285–296. doi:10.1016/j.tecto.2010.05.012
15. Del Gaudio P, Di Toro G, Han R, Hirose T, Nielsen S, Shimamoto T, Cavallo A (2009) Frictional melting of peridotite and seismic slip. J Geophys Res 114(B06):306. doi:10.1029/2008JB005990
16. Di Toro G, Niemeijer A, Tripoli A, Nielsen S, Di Felice F, Scarlato P, Spada G, Alessandroni R, Romeo G, Di Stefano G, Smith S, Spagnuolo E, Mariano S (2010) From field geology to earthquake simulation: a new state-of-the-art tool to investigate rock friction during the seismic cycle (shiva). Rendiconti Lincei 21:95–114. doi:10.1007/s12210-010-0097-x
17. Di Toro G, Han R, Hirose T, De Paola N, Nielsen S, Mizoguchi K, Ferri F, Cocco M, Shimamoto T (2011) Fault lubrication during earthquakes. Nature 471:494–499. doi:10.1038/nature09838
18. Dieterich JH (1979) Modeling of rock friction. I: experimental results and constitutive equations. J Geophys Res 84:2161–2168
19. Dor O, Ben-Zion Y, Rockwell TK, Brune J (2006) Pulverized rocks in the Mojave section of the San Andreas fault zone. Earth Planet Sci Lett 245 (3-4):642–654. doi:10.1016/j.epsl.2006.03.034
20. Dunham EM, Belanger D, Cong L, Kozdon JE (2011) Earthquake ruptures with strongly rate-weakening friction and off-fault plasticity, part 2: Nonplanar faults. Bull Seismol Soc Am 101 (5):2308–2322. doi:10.1785/0120100076
21. Fondriest M, Aretusini S, Di Toro G, Smith SA (2015) Fracturing and rock pulverization along an exhumed seismogenic fault zone in dolostones: The foiana fault zone (southern alps, Italy). Tectonophysics 654:56–74. doi:10.1016/j.tecto.04.015
22. Griffith AA (1921) The phenomena of rupture and flow in solids. Philosophical Transactions of the Royal Society (London) A221:163–198
23. Griffith WA, Nielsen S, Di Toro G, Smith FAS (2010) Rough faults, distributed weakening, and off-fault deformation. J Geophys Res 115(B08):409. doi:10.1029/2009JB006925
24. Guatteri M, Spudich P, Beroza GC (2001) Inferring rate and state friction parameters from a rupture model of the 1995 Hyogo-ken Nanbu (Kobe) Japan earthquake. J Geophys Res Solid Earth 106 (B11). 26,511–26,521 doi:10.1029/2001JB000294
25. Han R, Shimamoto T, Hirose T, Ree JH, Ando J (2007) Ultralow friction of carbonate faults caused by thermal decomposition. Science 316:878–881. doi:10.1126/science.1139763
26. Han R, Hirose T, Shimamoto T (2010) Strong velocity weakening and powder lubrication of simulated carbonate faults at seismic slip rates. J Geophys Res Solid Earth 115(B3). B03,412. doi:10.1029/2008JB006136
27. Heaton TH (1990) Evidence for and implications of self-healing pulses of slip in earthquake rupture. Phys Earth Planet Inter 64:1–20. corso
28. Hirose T, Shimamoto T (2005) Growth of molten zone as a mechanism of slip weakening of simulated faults in gabbro during frictional melting. J Geophys Res 110:B05202 doi:10.1029/2004JB003207
29. Hok S, Campillo M, Cotton F, Favreau P, Ionescu I (2010) Off-fault plasticity favors the arrest of dynamic ruptures on strength heterogeneity: two-dimensional cases. Geophys Res Lett 37(2). L02,306. doi:10.1029/2009GL041888, l02306
30. Ida Y (1972) Cohesive force across the tip of a longitudinal-shear crack and Griffith's specific surface energy. J Geophys Res 77:3796–3805

31. Irwin GR (1957) Analysis of stresses and strains near the end of a crack traversing a plate. J Appl Mech 24:361–364

32. Kostrov BV (1974) Crack propagation at variable velocity. J Appl Math Mech 38:511–519

33. Lancieri M, Madariaga R, Bonilla F (2012) Spectral scaling of the aftershocks of the Tocopilla 2007 earthquake in northern Chile. Geophys J Int:1–12. doi:10.1111/j.1365-246X.2011.05327.x

34. Madariaga R (1976) Dynamics of an expanding circular fault. Bull Seismol Soc Am 66:639–666

35. Mai PM, Somerville P, Pitarka A, Dalguer L, Song S, Beroza G, Miyake H, Irikura K (2006) On scaling of fracture energy and stress drop in dynamic rupture models: consequences for near-source ground-motions. Geophysical Monograph Series:283–293. doi:10.1029/170gm28

36. Malagnini L, Nielsen S, Mayeda K, Boschi E (2010) Energy radiation from intermediate- to large-magnitude earthquakes: implications for dynamic fault weakening. J Geophys Res 115. B06,319. doi:10.1029/2009JB006786

37. Malagnini L, Munafò I, Cocco M, Nielsen S, Mayeda K, Boschi E (1994) Gradual fault weakening with seismic slip: inferences from the seismic sequences of L'Aquila, 2009, and Northridge. Pure Appl Geophys 171(10):2709–2730. doi:10.1007/s00024-013-0752-0

38. Mitchell T, Ben-Zion Y, Shimamoto T (2011) Pulverized fault rocks and damage asymmetry along the Arima-Takatsuki tectonic line, Japan. Earth Planet Sci Lett 308(3-4):284–297. doi:10.1016/j.epsl.2011.04.023

39. Nielsen S, Madariaga R (2003) On the self-healing fracture mode. Bull Seismol Soc Am 93 (6):2375–2388. doi:10.1785/0120020090

40. Nielsen S, Di Toro G, Hirose T, Shimamoto T (2008) Frictional melt and seismic slip. J Geophys Res 113. B01,308. doi:10.1029/2007JB005122

41. Nielsen S, Mosca P, Giberti G, Di Toro G, Hirose T, Shimamoto T (2010) On the transient behavior of frictional melt during seismic slip. J Geophys Res 115. B10,301. doi:10.1029/2009JB007020

42. Nielsen S, Spagnuolo E, Smith S, Violay M, Toro GD, Bistacchi A (2015) Scaling in natural and laboratory earthquakes. in press,. Geophys Res Lett. doi:10.1002/2015GL067490

43. Niemeijer A, Di Toro G, Nielsen S, Di Felice F (2011) Frictional melting of gabbro under extreme experimental conditions of normal stress, acceleration and sliding velocity. J Geophys Res 116. B07,404. doi:10.1029/2010JB008181

44. Ohnaka M, Kuwahaua Y, Yamamoto K, Hirasawa T (1986) Dynamic breakdown process and the generating mechanism for high frequency elastic radiation during stick-slip instabilities. In: Das S, Boatright J, Scholz C (eds), vol 37. AGU Geophys. Mono, Earthquake Source Mechanics, pp 13– 24

45. Palmer A, Rice J (1973) The growth of slip surfaces in the progressive failure of over-consolidated clay. Proc Roy Soc Lond 332:527–548

46. Peyrat S, Olsen KB, Madariaga R (2004) Which dynamic rupture parameters can be estimated from strong ground motion and geodetic data ?. Pure Appl Geophys 161:2155–2169

47. Poliakov ANB (2002) Dynamic shear rupture interactions with fault bends and off-axis secondary faulting. J Geo phys Res 107:B11. doi:10.1029/2001jb000572

48. Reches Z (1999) Mechanisms of slip nucleation during earthquakes. Earth Planet Sci Lett 170 (4):475–486. doi:10.1016/S0012-821X(99)00122-3

49. Rice JR (2006) Heating and weakening of faults during earthquake slip. J Geophys Res 111:B05311 doi:10.1029/2005JB004006

50. Rice JR, Sammis CG, Parsons R (2005) Off-fault secondary failure induced by a dynamic slip pulse. Bull Seismol Soc Am 95(1):109–134. doi:10.1785/0120030166

51. Rice JR (1968) A path independent integral and the approximate analysis of strain concentration by notches and cracks. J Appl Mech 35:379–386

52. Rivera L, Kanamori H (2005) Representations of the radiated energy in earthquakes. Geophys J Int 162(1):148–155. doi:10.1111/j.1365-246x.2005.02648.x

53. Ruina A (1983) Slip instability and state variable friction laws. J Geophys Res 88(10):359–10, 370

54. Sagy A, Brodsky EE (2009) Geometric and rheological asperities in an exposed fault zone. J Geophys Res 114(B2). doi:10.1029/2008jb005701

55. Saucier F, Humphreys E, II RW (1992) Stress near geometrically complex strike-slip faults: application to the San-Andreas fault at cajon pass, Southern California. J Geophys Res 97(B4):5081–5094

56. Schmittbuhl J, Gentier S, Roux S (1993) Field measurements of the roughness of fault surfaces. Geophys Res Lett 20:639–641

57. Scholz CH (1990) The mechanics of earthquakes and faulting. Cambridge University Press

58. Shipton ZK, Evans JP, Abercrombie RE, Brodsky EE (2006a) The missing sinks: Slip localization in faults, damage zones, and the seismic energy budget Earthquakes: Radiated energy and the physics of faulting: American Geophysical Union, Geophysical Monograph Series, vol 170, pp 217–222

59. Shipton ZK, Soden AM, Kirkpatrick JD, Bright AM, Lunn RJ (2006b) How thick is a fault? Fault displacement-thickness scaling revisited. Geophysical Monograph Series:193–198. doi:10.1029/170gm19

60. Spudich P, Guatteri M (2000) What can strong-motion data tell us about slip-weakening fault-friction laws?. Bull Seismol Soc Am 90:98–116

61. Smith SA, Bistacchi A, Mitchell TM, Mittempergher S, Di Toro G (2013) The structure of an exhumed intraplate seismogenic fault in crystalline basement. Tectonophysics 599:29–44. doi:10.1016/j.tecto.2013.03.031

62. Tinti E, Bizzarri A, Piatanesi A, Cocco M (2004) Estimates of slip weakening distance for different dynamic rupture models. Geophys Res Lett 31(2). L02,611. doi:10.1029/2003GL018811.l02611

63. Tinti E, Spudich P, Cocco M (2005) Earthquake fracture energy inferred from kinematic rupture models on extended faults. J Geophys Res 110. B12,303. doi:10.1029/2005JB003644

64. Tsutsumi A, Shimamoto T (1997) High-velocity frictional properties of gabbro. Geophys Res Lett 24:699–702

65. Viesca RC, Garagash DI (2015) Ubiquitous weakening of faults due to thermal pressurization. Nature Geosci. doi:10.1038/ngeo2554, advance online publication

66. Violay M, Nielsen S, Spagnuolo E, Cinti D, Di Toro G, Di Stefano GD (2013) Pore fluid in experimental calcite-bearing faults: abrupt weakening and geochemical signature of co-seismic processes. Earth Planet Sci Lett 361(0):74–84. doi:10.1016/j.epsl.2012.11.021

67. Violay M, Di Toro G, Gibert B, Nielsen S, Spagnuolo E, Del Gaudio P, Azais P, Scarlato PG (2014) Effect of glass on the frictional behavior of basalts at seismic slip rates. Geophys Res Lett 41(2):348–355. doi:10.1002/2013gl058601

68. Violay M, Di Toro G, Nielsen S, Spagnuolo E, Burg J (2015) Thermo-mechanical pressurization of experimental faults in cohesive rocks during seismic slip. Earth Planet Sci Lett 429:1–10. doi:10.1016/j.epsl.2015.07.054

69. Wong TF (1982) Shear fracture energy of westerly granite from post-failure behavior. J Geophys Res 87(B2):990– 1000

70. Xu S, Ben-Zion Y, Ampuero JP, Lyakhovsky V (2015) Dynamic ruptures on a frictional interface with off-fault brittle damage: feedback mechanisms and effects on slip and near-fault motion. Pure Appl Geophys 172(5):1243–1267. doi:10.1007/s00024-014-0923-7

CHAPTER 5

LOCATING THE FOCUS OF A STARTING EARTHQUAKE

Alexander Ivanchin

Keywords: Earthquake Forecasting, Locating the Focus of Earthquake, Stresses of Earthquake

INTRODUCTION

An earthquake is always an unexpected phenomenon. Modern science is not able to predict the time or the place or the earthquake strength. The problem of locating the focus of a starting earthquake has not even been set due to the poor level of understanding the processes preceding its start. At present the main earthquake hypothesis is the "explosive" relaxation of the high elastic stresses accumulated in the lithosphere. A serious objection to the above hypothesis is the stress relaxation caused by the plastic flow whose rate grows exponentially with temperature and stress [1] . In the lithosphere the temperature and the pressure increase with the depth and, therefore, models of the earthquake focus with no elastic stress concentrations, such as the Inertial Earthquake Focus (IEF) model, are more probable [2] - [4] .

Nowadays, geophysical measurements can be performed only on the surface or in the thin subsurface layer of the Earth, so all the earthquake hypotheses are purely theoretical. Although some data can be obtained from seismic waves, their amount is far from sufficient, since the waves are long and provide little information. Therefore, there remains only one way of locating the focus of a starting earthquake—to develop a method using the elastic stress field of a starting earthquake focus. This method is suggested in the present work.

Citation: Alexander Ivanchin (2014) Locating the Focus of a Starting Earthquake. International Journal of Geosciences,05,1137-1148. doi: 10.4236/ijg.2014.510096.

INERTIAL EARTHQUAKE FOCUS

In the lithosphere tectonic plates move and turn relative to each other. Due to the Earth's rotation each plate possesses the moment of momentum $\mathbf{M}$ parallel to the axis of rotation of the Earth's body [5]

$$\mathbf{M} = I\mathbf{\Omega}.$$

Here I is the moment of inertia, $\mathbf{\Omega}$ is the vector of the angular velocity of the Earth's rotation. There can be two types of change of the angular velocity of rotation illustrated in Figure 1 by the example of a rotating cylinder. One is precession, when by the action of the external moment, that is perpendicular to the axis of rotation , the orientation of the axis of rotation in space changes, whereas the angular velocity value does not change in magnitude. The direction of the axis of rotation changes from the vertical orientation in Figure 1(a) into the horizontal one, as is shown inFigure 1(b), its position relative to the material points of the cylinder does not change, and the external moment does not perform work. The other type of change of the angular velocity of rotation is shown in Figure 1(c), where the position of the axis of rotation relative to the material points of the cylinder changes.

For a cylinder to change the orientation of the rotation axis relative to the points of the body, it is necessary, first, in the horizontal rotation position of a cylinder (b) to apply the external moment of force directed opposite to the angular velocity vector $\mathbf{\Omega}_1$ and stop the rotation. Then, it is necessary to apply the external moment of force directed along the axis and make the cylinder rotate around and along the axis (Figure 1(c)).

For simplicity, let us consider the IEF in the form of a sphere. As a result of tectonic movement, the direction of axis will deviate by an angle ξ from the Earth's axis of rotation.

If the turn occurred in a vacuum, the direction of the angular velocity would deviate by the angle ξ and its value would be Ω_1. At the same time, the equality $\Omega_1 = \Omega$ would be satisfied. However, the IEF is located in the solid lithosphere and rotates together with the Earth, so the angular velocity cannot differ in value or direction from the Earth's angular velocity Ω. Here there occurs a change in the position of the axis of rotation relative to the IEF points. Therefore, first we stop the IEF rotation applying elastic stresses with the moment of force $\bar{\bar{\mathbf{K}}}$ inclined at the angle of $\pi+\xi$ to the Earth's rotation axis, with the IEF kinetic energy wholly turning into the potential energy of the elastic stresses occurring around it in the lithosphere. Then we make the IEF rotate reaching the angular velocity Ω applying to it the elastic stress with the moment of force $\bar{\mathbf{K}}$ directed along the Earth's rotation axis. The kinetic energy of rotation completely turns into the potential energy of the elastic stresses and vice versa, which means that the following equality will be satisfied

$$\bar{K} = \bar{\bar{K}} \tag{2.1}$$

The resulting vector is

$$\mathbf{K} = \bar{\mathbf{K}} - \bar{\bar{\mathbf{K}}} \tag{2.2}$$

hence

$$K = 2\bar{K}\sin\frac{\xi}{2} \tag{2.3}$$

Thus, the IEF gets its own moment of force $\mathbf{K}$. As is known, the condition of equilibrium in the elasticity theory requires that all the concentrated moments be zero [6] . However, the above condition applies only to in-

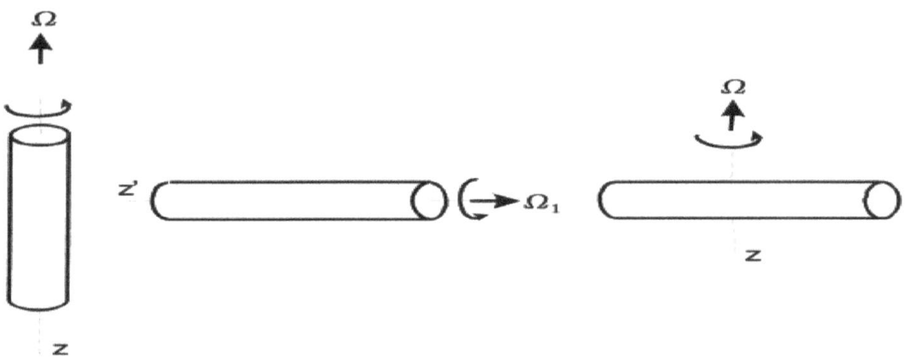

(a) (b) (c)

Figure 1. The orientation change of the rotating axis is displayed.

ertial systems, whereas the rotating Earth is not such a system. If the Earth did not rotate, there would be no moment of force .

The stress field of the moment of force $\mathbf{K}$ cannot be reduced by plastic relaxation without turning the IEF as a whole returning to the initial position. As the angle ξ increases, the stresses grow and can achieve a critical value, the IEF turns abruptly decreasing the angle ξ and the stress field, and an earthquake takes place.

THE ELASTIC FIELD OF THE MOMENT OF FORCE

The equation of elastic equilibrium for the displacement vector $\bar{\mathbf{U}}$ is [6]

$$\nabla(\nabla\cdot\bar{\mathbf{U}}) - \frac{1-2\nu}{2(1-\nu)}\nabla\times(\nabla\times\bar{\mathbf{U}}) = 0 \tag{3.1}$$

The deformation tensor is

$$\bar{\varepsilon}_{ij} = \frac{1}{2}\left(\frac{\partial\bar{U}_i}{\partial x_j} + \frac{\partial\bar{U}_j}{\partial x_i}\right)$$

and the stress tensor is

$$\bar{\sigma}_{ij} = 2\mu\left[\bar{\varepsilon}_{ij} + \frac{\nu}{1-2\nu}\left(\bar{\varepsilon}_{xx} + \bar{\varepsilon}_{yy} + \bar{\varepsilon}_{zz}\right)\delta_{ij}\right] \tag{3.2}$$

Here μ is the shear modulus, ν is the Poisson coefficient, δ_{ij} is the Kronekker symbol. For the sphere S of the radius R with the center in the origin of coordinates the force is

$$\bar{\mathbf{F}} = \int_S \bar{\sigma}\mathbf{dS} \tag{3.3}$$

and the moment of force is

$$\bar{\mathbf{K}} = \int_S \mathbf{r}\times(\bar{\sigma}\mathbf{n})\mathbf{d}S \tag{3.4}$$

Here $\mathbf{r}$ is the radius-vector, $\mathbf{n}$ is the unitary normal to S. If the moment (3.4) is not zero, the region will be referred to as the Torque Region (TR). Substituting (3.2) into (3.4) we derive

$$\bar{\mathbf{K}} = 2\mu\int_S \mathbf{r}\times(\bar{\varepsilon}\mathbf{n})\mathrm{d}S + 2\mu\frac{\nu}{1-2\nu}\int_S(\nabla\cdot\bar{\mathbf{U}})(\mathbf{r}\times\mathbf{n})\mathrm{d}S = 2\mu\int_S \mathbf{r}\times(\bar{\varepsilon}\mathbf{n})\mathrm{d}S \tag{3.5}$$

since for a sphere with the center in the origin of coordinates the vector $\mathbf{r}$ and $\mathbf{n}$ are parallel. Suppose that $\bar{\mathbf{U}}$ is a solenoidal vector, that is, there exists the vector $\mathbf{P}$ such that

$$\bar{\mathbf{U}} = \nabla\times\mathbf{P} \tag{3.6}$$

Then (3.1) is reduced to the equation

$$\nabla\times\nabla\times\nabla\times\mathbf{P} = \mathbf{0} \tag{3.7}$$

since the first term in (3.1) becomes zero. For the moment of force $\bar{K}$ in (3.5) to be independent of R, it is necessary that $\bar{\mathbf{U}} \sim 1/r^2$. Therefore, let us take the vector $\mathbf{P}$ in the form

$$\mathbf{P} = A\left\{0,0,\frac{1}{r}\right\} \tag{3.8}$$

Here A is the arbitrary constant. The vector (3.8) is the solution (3.7), then

$$\bar{\mathbf{U}} = \frac{A}{r^3}\{-y,x,0\} = \frac{A}{r^2}\{0,0,\sin\vartheta\} \tag{3.9}$$

Here the displacement vector is written first in the Cartesian system of coordinates and then in the spherical one. The sequence of writing the spherical vector components is as follows: radial $\bar{U}_r$ zenith $\bar{U}_\vartheta$ and asymuthal $\bar{U}_\varphi$. The Cartesian coordinates are related to the spherical ones in the following way

$$x = r\sin\vartheta\cos\varphi, \quad y = r\sin\vartheta\sin\varphi, \quad z = r\cos\vartheta$$

Differentiating (3.9) we obtain that

$$\nabla\cdot\bar{\mathbf{U}} = 0 \tag{3.10}$$

The potential vector $\bar{\mathbf{U}}$ produces a deformation tensor having two components not equal to zero in the sphe- rical coordinates

$$\bar{\varepsilon}_{r\varphi} = \bar{\varepsilon}_{\varphi r} = -\frac{3A}{2r^3}\sin\vartheta \tag{3.11}$$

The stress tensor has the same

$$\bar{\sigma}_{r\varphi} = \bar{\sigma}_{\varphi r} = 2\mu\bar{\varepsilon}_{r\varphi} = -3\frac{\mu A}{r^3}\sin\vartheta \tag{3.12}$$

The values (3.9), (3.11), (3.12) are not zero only outside the IEF. Let us choose a system of coordinates so that the moment of force had only one component along the axis z not equal to zero, then

$$\bar{K} = \int_0^{2\pi} d\varphi \int_0^{\pi} r\left(\sigma_{r\varphi} r^2 \sin\vartheta\right) d\vartheta = -3\pi^2\mu A \tag{3.13}$$

hence

$$A = -\frac{K}{3\pi^2\mu} \tag{3.14}$$

If a source is produced by inertial effects, that is the IEF, then multiplying (3.14) by $2\sin(\xi/2)$ we derive

$$K = -6\pi^2\mu A\sin\frac{\xi}{2} \tag{3.15}$$

Similarly,

$$\mathbf{U} = \frac{2A}{r^3}\sin\frac{\xi}{2}\{-y,x,0\} = \frac{2A}{r^2}\sin\frac{\xi}{2}\{0,0,\sin\vartheta\} \tag{3.16}$$

$$\varepsilon_{r\varphi} = \varepsilon_{\varphi r} = -\frac{3A}{r^3}\sin\vartheta\sin\frac{\xi}{2} \tag{3.17}$$

$$\sigma_{r\varphi} = \sigma_{\varphi r} = 2\mu\varepsilon_{r\varphi} = -6\frac{\mu A}{r^3}\sin\vartheta\sin\frac{\xi}{2} \tag{3.18}$$

The elastic energy density will be written as

$$w = 2\mu\varepsilon_{r\varphi}^2 = 18\mu A^2\frac{\sin^2\vartheta}{r^6}\sin^2\frac{\xi}{2}$$

The elastic energy of the stresses of the moment of force (2.3) is

$$W = \iiint_V w dV = \int_R^{\infty} dr \int_0^{2\pi} d\varphi \int_0^{\pi} wr^2 \sin\vartheta d\vartheta = 16\pi\mu\frac{A^2}{R^3}\sin^2\frac{\xi}{2} \tag{3.19}$$

The unknown constant A can be determined from two conditions. The first condition is when the moment of force is specified, and then the value of A is found according to (3.14). The second condition is when A is determined according to a specified kinetic energy of rotation. The energy density of the elastic field (3.9) is as follows

$$\bar{w} = \frac{9}{2}\mu A^2\frac{\sin^2\vartheta}{r^6}$$

and the energy will be

$$\bar{W} = \iiint_V \bar{w} dV = 4\pi\frac{A^2}{R^3} \tag{3.20}$$

The constant A is determined from the condition of the complete change of the kinetic energy

$$W_k = \frac{I\Omega^2}{2} = \frac{4}{15}\pi\rho R^5\Omega^2$$

into the potential energy of the elastic stresses (3.12). The inertial moment of a sphere is

$$I = \frac{8\pi\rho R^5}{15} \tag{3.21}$$

As a result, we obtain

$$A = \Omega R^4\sqrt{\frac{\rho}{15\mu}} \tag{3.22}$$

Substituting (3.22) into (3.15) we derive the value of the moment of force

$$K = -6\pi^2\Omega R^4\sqrt{\frac{\rho\mu}{15}}\sin\frac{\xi}{2} \tag{3.23}$$

The stress field according to (3.18) looks like

$$\sigma_{r\varphi} = \sigma_{\varphi r} = 2\mu\varepsilon_{r\varphi} = -\frac{\Omega R^4}{r^3}\sqrt{\frac{12\rho\mu}{5}}\sin\vartheta\sin\frac{\xi}{2}.$$

Substituting (3.22)into (3.19) we obtain an expression for the IEF

$$W = \frac{16}{15}\pi R^5\Omega^2\rho\sin^2\frac{\xi}{2} \quad (3.24)$$

On the IEF surface the stresses are

$$\sigma_{r\varphi} = -\Omega R\sqrt{\frac{12\rho\mu}{5}}\sin\vartheta\sin\frac{\xi}{2} \quad (3.25)$$

As is shown in [7] , in order to derive a general solution for the elasticity problem, a nonpotential solution should be added the potential solution (3.16). However, in the Appendix it is shown that the interaction energy of the potential and nonpotential solutions is zero. Addition of the nonpotential solution to the potential one (3.16) will increase the system energy, and for this reason it is excluded from consideration.

INTERACTION OF FOCUSES. LOCATING AN EARTHQUAKE FOCUS

If there exist two torque regions, there is energy of elastic interaction between them that can be used to locate the focus of a starting earthquake. For this purpose, it is necessary to create an artificial TR in the lithosphere applying shear stresses to the surface of the artificial cavity. Let us designate the values relating to the artificial TR by one point at the top (further referred to as the first TR), and the IEF by two points (further referred to as the second TR). The centres of these focuses are located on the abscissa at point $-a$ for the first TR and $+a$ for the second TR. The plane is vertical and passes through the Earth's center, y -coordinate is normal to it and forms the angle $\pi/2-\psi$ with the Earth's axis. The direction of y -coordinate is chosen so that the system of coordinates was right. The vector potentials of the moment of force will be

$$\dot{\mathbf{P}} = \frac{\dot{A}}{\dot{r}}\left\{\sin\dot{\vartheta}\cos\dot{\varphi}, \sin\dot{\vartheta}\sin\dot{\varphi}, \cos\dot{\vartheta}\right\}, \quad \ddot{\mathbf{P}} = \frac{\ddot{A}}{\ddot{r}}\left\{\sin\ddot{\vartheta}\cos\ddot{\varphi}, \sin\ddot{\vartheta}\sin\ddot{\varphi}, \cos\ddot{\vartheta}\right\} \quad (4.1)$$

Here

$$\dot{r} = \sqrt{(x-a)^2 + x^2 + y^2}, \quad \ddot{r} = \sqrt{(x+a)^2 + x^2 + y^2}.$$

The zenith angles $\dot{\vartheta}$ and $\ddot{\vartheta}$ are measured from z axis, and the angles $\dot{\varphi}$ and $\ddot{\varphi}$ from that of abscissa. The moments $\dot{\mathbf{K}}$ and $\ddot{\mathbf{K}}$ coincide in direction with $\dot{\mathbf{P}}$ and $\ddot{\mathbf{P}}$. The laboratory Cartesian system with the origin in the center of the first focus is oriented in the following way (the variables related to it are designated by the sign ^): the axis $\hat{z}$ coincides with the vertical at the given point and is directed upward, the axis $\hat{x}$ is directed along the meridian to the north and the axis $\hat{y}$ is directed along the latitude to the west. The plane $\hat{x}\hat{z}$ is meridianal, the angle between the normals of the planes $\hat{x}\hat{z}$ and xz , that is between the axes y and $\hat{y}$, is ψ . The change to the laboratory Cartesian system of coordinates is implemented by turning around axis $\hat{z}$ by the angle ψ and is described by the matrix

$$\mathbf{\Lambda} = \begin{pmatrix} \cos\psi & -\sin\psi & 0 \\ \sin\psi & \cos\psi & 0 \\ 0 & 0 & 1 \end{pmatrix}$$

The displacement vectors in the laboratory system of coordinates can be represented like this

$$\dot{\mathbf{U}} = \nabla\times\left(\mathbf{\Lambda}\dot{\mathbf{P}}\right) = \frac{A}{r^3}\begin{Bmatrix} -y\cos\dot{\vartheta} + z\sin\dot{\vartheta}\sin\left(\dot{\varphi}+\psi\right) \\ (x+a)\cos\dot{\vartheta} - z\sin\dot{\vartheta}\cos\left(\dot{\varphi}+\psi\right) \\ \sin\dot{\vartheta}\left[y\cos\left(\dot{\varphi}+\psi\right) - x\sin\left(\dot{\varphi}+\psi\right)\right] \end{Bmatrix},$$

$$\ddot{\mathbf{U}} = \nabla\times\left(\mathbf{\Lambda}\ddot{\mathbf{P}}\right) = \frac{\ddot{A}}{\ddot{r}^3}\sin\frac{\xi}{2}\begin{Bmatrix} -y\cos\ddot{\vartheta} + z\sin\ddot{\vartheta}\sin\left(\ddot{\varphi}+\psi\right) \\ (x-a)\cos\ddot{\vartheta} - z\sin\ddot{\vartheta}\cos\left(\ddot{\varphi}+\psi\right) \\ \sin\ddot{\vartheta}\left[y\cos\left(\ddot{\varphi}+\psi\right) - x\sin\left(\ddot{\varphi}+\psi\right)\right] \end{Bmatrix}. \quad (4.2)$$

The displacements (4.2) determine the deformation tensors

$$\dot{\varepsilon}_{xx} = 3\frac{\dot{A}}{\dot{r}^5}(x+a)\left[y\cos\dot{\vartheta} - z\sin\dot{\vartheta}\sin(\dot{\varphi}+\psi)\right],$$
$$\dot{\varepsilon}_{yy} = 3\frac{\dot{A}}{\dot{r}^5}y\left[-(x+a)\cos\dot{\vartheta} + z\sin\dot{\vartheta}\cos(\dot{\varphi}+\psi)\right],$$
$$\dot{\varepsilon}_{zz} = 3\frac{\dot{A}}{\dot{r}^5}z\sin\dot{\vartheta}\left[-y\cos(\dot{\varphi}+\psi)+(x+a)\sin(\dot{\varphi}+\psi)\right],$$
$$\dot{\varepsilon}_{xy} = \frac{3\dot{A}}{2\dot{r}^5}\left\{\left[-(x+a)^2+y^2\right]\cos\dot{\vartheta} + z\sin\dot{\vartheta}\left[(x+a)\cos(\dot{\varphi}+\psi) - y\sin(\dot{\varphi}+\psi)\right]\right\},$$
$$\dot{\varepsilon}_{xz} = \frac{3\dot{A}}{2\dot{r}^5}\left\{yz\cos\dot{\vartheta} + \sin\dot{\vartheta}\left[y(x+a)\cos(\dot{\varphi}+\psi)+\left((x+a)^2-z^2\right)\sin(\dot{\varphi}+\psi)\right]\right\},$$
$$\dot{\varepsilon}_{yz} = \frac{3\dot{A}}{2\dot{r}^5}\left\{-(x+a)z\cos\dot{\vartheta} + \sin\dot{\vartheta}\left[(z^2-y^2)\cos(\dot{\varphi}+\psi)+y(x+a)\sin(\dot{\varphi}+\psi)\right]\right\}.$$

(4.3)

$$\ddot{\varepsilon}_{xx} = 3\frac{\ddot{A}}{\ddot{r}^5}\sin\frac{\xi}{2}(x-a)\left[y\cos\ddot{\vartheta} - z\sin\ddot{\vartheta}\sin(\ddot{\varphi}+\psi)\right],$$
$$\ddot{\varepsilon}_{yy} = 3\frac{\ddot{A}}{\ddot{r}^5}\sin\frac{\xi}{2}y\left[-(x-a)\cos\ddot{\vartheta} + z\sin\ddot{\vartheta}\cos(\ddot{\varphi}+\psi)\right],$$
$$\ddot{\varepsilon}_{zz} = 3\frac{\ddot{A}}{\ddot{r}^5}\sin\frac{\xi}{2}z\left[-y\cos(\ddot{\varphi}+\psi)+(x-a)\sin(\ddot{\varphi}+\psi)\right],$$
$$\ddot{\varepsilon}_{xy} = \frac{3\ddot{A}}{2\ddot{r}^5}\sin\frac{\xi}{2}\left\{\left[-(x-a)^2+y^2\right]\cos\ddot{\vartheta} + z\sin\ddot{\vartheta}\left[(x-a)\cos(\ddot{\varphi}+\psi) - y\sin(\ddot{\varphi}+\psi)\right]\right\},$$
$$\ddot{\varepsilon}_{xz} = \frac{3\ddot{A}}{2\ddot{r}^5}\sin\frac{\xi}{2}\left\{yz\cos\ddot{\vartheta} + \sin\ddot{\vartheta}\left[(x-a)y\cos(\ddot{\varphi}+\psi)-\left((x-a)^2-z^2\right)\sin(\ddot{\varphi}+\psi)\right]\right\},$$
$$\ddot{\varepsilon}_{yz} = \frac{3\ddot{A}}{2\ddot{r}^5}\sin\frac{\xi}{2}\left\{-(x-a)z\cos\ddot{\vartheta} + \sin\ddot{\vartheta}\left[(z^2-y^2)\cos(\ddot{\varphi}+\psi)+(x-a)y\sin(\ddot{\varphi}+\psi)\right]\right\}.$$

(4.4)

The total deformation tensor is

$$\varepsilon = \dot{\varepsilon} + \ddot{\varepsilon}$$

(4.5)

The total elastic energy density is written as

$$w = \frac{2\mu\nu}{1-2\nu}\left(\varepsilon_{xx}+\varepsilon_{yy}+\varepsilon_{zz}\right)^2 + \mu\left(\varepsilon_{xx}^2+\varepsilon_{yy}^2+\varepsilon_{zz}^2+2\varepsilon_{xy}^2+2\varepsilon_{xz}^2+2\varepsilon_{yz}^2\right)$$

(4.6)

Substituting here (4.5) we find that the total density energy is the sum of the following components: the density of the energy of the first focus

$$\dot{w} = \frac{2\mu\nu}{1-2\nu}\left(\dot{\varepsilon}_{xx}+\dot{\varepsilon}_{yy}+\dot{\varepsilon}_{zz}\right)^2 + \mu\left(\dot{\varepsilon}_{xx}^2+\dot{\varepsilon}_{yy}^2+\dot{\varepsilon}_{zz}^2+2\dot{\varepsilon}_{xy}^2+2\dot{\varepsilon}_{xz}^2+2\dot{\varepsilon}_{yz}^2\right)$$

the energy density of the second focus

$$\ddot{w} = \frac{2\mu\nu}{1-2\nu}\left(\ddot{\varepsilon}_{xx}+\ddot{\varepsilon}_{yy}+\ddot{\varepsilon}_{zz}\right)^2 + \mu\left(\ddot{\varepsilon}_{xx}^2+\ddot{\varepsilon}_{yy}^2+\ddot{\varepsilon}_{zz}^2+2\ddot{\varepsilon}_{xy}^2+2\ddot{\varepsilon}_{xz}^2+2\ddot{\varepsilon}_{yz}^2\right)$$

and the density of the interaction energy of the focuses

$$w_I = \frac{4\mu\nu}{1-2\nu}\left(\dot{\varepsilon}_{xx}\ddot{\varepsilon}_{xx}+\dot{\varepsilon}_{yy}\ddot{\varepsilon}_{yy}+\dot{\varepsilon}_{zz}\ddot{\varepsilon}_{zz}+\dot{\varepsilon}_{xx}\ddot{\varepsilon}_{yy}+\dot{\varepsilon}_{xx}\ddot{\varepsilon}_{zz}+\dot{\varepsilon}_{yy}\ddot{\varepsilon}_{zz}\right)$$
$$+2\mu\left(\dot{\varepsilon}_{xx}\ddot{\varepsilon}_{xx}+\dot{\varepsilon}_{yy}\ddot{\varepsilon}_{yy}+\dot{\varepsilon}_{zz}\ddot{\varepsilon}_{zz}+2\dot{\varepsilon}_{xy}\ddot{\varepsilon}_{xy}+2\dot{\varepsilon}_{xz}\ddot{\varepsilon}_{xz}+2\dot{\varepsilon}_{yz}\ddot{\varepsilon}_{yz}\right).$$

Integrating $\dot{w}$ and $\ddot{w}$ we obtain the focus energies in the form (3.19). The interaction energy of the focuses is

$$W_I = \iiint_V w_I \, dV$$

(4.7)

For integrating turn to the bipolar coordinates in (4.7). The elastic energy density is a scalar, and the change to the bipolar coordinates is performed by replacing the Cartesian coordinates by the bipolar ones (4.7) using the formulas [8] :

$$x = \frac{a\sinh\tau}{\cosh\tau-\cos\sigma}, \quad y = \frac{a\sin\sigma\cos\varphi}{\cosh\tau-\cos\sigma}, \quad z = \frac{a\sin\sigma\sin\varphi}{\cosh\tau-\cos\sigma}, \quad Y = \frac{a^3\sin\sigma}{(\cosh\tau-\cos\sigma)^3}$$

(4.8)

The coordinate surface for $\tau = \mathbf{const}$ is the radius sphere

$$R = \frac{a}{\sinh \tau} \tag{4.9}$$

with the center on the abscissa at the point $a\cosh\tau$. At $\tau > 1$ the value of $\cosh\tau$ is close to unity and the center of the coordinate sphere, in fact, coincides with that of the IEF. For instance, already for $\tau = 1.5$ the value of $\cosh\tau = 1.10$. Therefore, one can expect with sufficiently great accuracy that the centers of the focuses and the centers of the coordinate spheres coincide. For this reason, the limits of integration over τ, in accordance with (4.9), are as follows

$$\tau_1 = \operatorname{arcsinh}\frac{a}{\dot{R}}, \qquad \tau_2 = \operatorname{arcsinh}\frac{a}{\ddot{R}}.$$

The elastic interaction energy (4.7) in the bipolar coordinates is written as

$$W_I = 2\mu\int_{\tau_1}^{\tau_2} \mathrm{d}\tau\int_0^{\pi} \mathrm{d}\sigma\int_0^{2\pi} w_I Y \mathrm{d}\varphi$$
$$= -\frac{3\pi\mu\dot{A}\ddot{A}}{8a^3}\eta\sin\frac{\xi}{2}\left[2\sqrt{1+\ddot{\zeta}^2} - \frac{1}{\ddot{\zeta}^3}\ln\left(\ddot{\zeta}+\sqrt{1+\ddot{\zeta}^2}\right) - 2\sqrt{1+\dot{\zeta}^2} + \frac{1}{\dot{\zeta}^3}\ln\left(\dot{\zeta}+\sqrt{1+\dot{\zeta}^2}\right)\right]. \tag{4.10}$$

Here it is designated

$$\eta = \cos\dot{\vartheta}\cos\ddot{\vartheta} - \cos(\dot{\varphi}-\ddot{\varphi}) + 3\cos(\dot{\varphi}+\ddot{\varphi}+2\psi)\sin\dot{\vartheta}\sin\ddot{\vartheta}, \quad \dot{\zeta} = \frac{\dot{R}}{a}, \quad \ddot{\zeta} = \frac{\ddot{R}}{a} \tag{4.11}$$

Since the first TR is an artificial source of the moment of force, the distance to the focus will always be much larger than its size $\dot{R}$, therefore,

$$\dot{\zeta} \ll 1$$

hence

$$2\sqrt{1+\dot{\zeta}^2} - \frac{1}{\dot{\zeta}^3}\ln\left(\dot{\zeta}+\sqrt{1+\dot{\zeta}^2}\right) \approx -\frac{1}{\dot{\zeta}^2}$$

For the second TR the following relation can be true $\ddot{\zeta} \sim 1$, however, in both cases, we deal with the equality

$$\left|2\sqrt{1+\ddot{\zeta}^2} - \frac{1}{\ddot{\zeta}^3}\ln\left(\ddot{\zeta}+\sqrt{1+\ddot{\zeta}^2}\right)\right| \ll \frac{1}{\dot{\zeta}^2}$$

As a result, we derive from (4.10)

$$W_I = \frac{\dot{K}\Omega\ddot{R}^4\eta}{8\pi a\dot{R}^2}\sin\frac{\xi}{2}\sqrt{\frac{\rho}{15\mu}}$$

The total energy will be written as

$$W = \dot{W} + W_I + \ddot{W} = \dot{W} + \frac{\dot{K}\Omega\ddot{R}^4\eta}{8\pi a\dot{R}^2}\sin\frac{\xi}{2}\sqrt{\frac{\rho}{15\mu}} + \ddot{W} \tag{4.12}$$

Here, instead of a, the value of $s = 2a$ is used, which is the distance between the centers of the focuses. Differentiating the energy W with respect to s we obtain a generalized force acting between the focuses in the form

$$f = \frac{\mathrm{d}W}{\mathrm{d}s} = -\frac{\dot{K}\Omega\ddot{R}^4\eta}{4\pi\dot{R}^2 s^2}\sin\frac{\xi}{2}\sqrt{\frac{\rho}{15\mu}} \tag{4.13}$$

If $f < 0$, then the focuses are attracted to each other, and if $f > 0$, then they repel. The sign depends on the coefficient η from (4.11) The following unknowns are included here: $\ddot{R}$ is the IEF size, $\sin\frac{\xi}{2}$, the distance between the focuses s, and $|\eta| \le 5$. The values $\dot{R}$, ρ, Ω and μ are known, they are specified during an experiment. After measuring the value f for various orientations of the vector $\dot{K}$, we determine the orientation factor η, that is the IEF orientation. The above method does not allow defining the value $\ddot{R}$ separately from $\sin(\xi/2)$. However, we can suppose that the angle ξ will not be very small, otherwise the energy of the focus of a starting earthquake will not be very small too, so it may be that $\sin(\xi/2)$ will be of the order of ~ 1, then the size of the focus of a starting earthquake $\ddot{R}$ and its energy are evaluated according to (3.24) and (4.13). Making measurements for various orientations one can define the unknown parameters $\ddot{\vartheta}$, $\ddot{\varphi}$ and ψ and hence the IEF orientation.

The force f acting on the first focus can be considered as the concentrated force. The displacement field of the concentrated force can be found from the following considerations. As the value of the concentrated force does not depend on the radius of the integration sphere in (3.3), so the stress field must decrease with the distance from the point of its application as $1/r^2$, and then the displacement field will decrease as $1/r$. Consider that the concentrated force is directed along the axis z, therefore, let us take the displacement vector such as

$$\mathbf{u} = \left\{0, 0, \frac{B}{\sqrt{x^2+y^2+z^2}}\right\}$$

Here B is the arbitrary constant. The above vector does not satisfy the equation of elastic equilibrium (3.1), whereas a check proves that the following equality is satisfied

$$\nabla(\nabla\cdot\mathbf{u})=\nabla\times\nabla\times\mathbf{u}$$

We have

$$\nabla\cdot\mathbf{u}=-\frac{Bz}{\left(x^2+y^2+z^2\right)^{3/2}}$$

The potential Ψ satisfying the equation

$$\nabla\cdot\nabla\Psi=-\frac{Bz}{\left(x^2+y^2+z^2\right)^{3/2}}$$

is

$$\Psi=\frac{Bz}{2\sqrt{x^2+y^2+z^2}}$$

The vector

$$\tilde{\mathbf{u}}=\mathbf{u}-\frac{1}{2(1-\nu)}\nabla\Psi=\frac{B}{8(1-\nu)r}\{\cos\varphi\sin2\vartheta,\sin\varphi\sin2\vartheta,7-8\nu+\cos2\vartheta\}$$

is the solution of the Equation (3.1). Here the vector $\tilde{u}$ is written in the Cartesian system of coordinates. The force (3.3) caused by the field $\tilde{u}$ does not depend on the size of the integration sphere S and is

$$\tilde{f}=4\pi\mu B \qquad (4.14)$$

This force is directed along the line connecting the IEF centers. Equating the forces (4.13) and (4.14) we obtain the value of the arbitrary constant B

$$B=-\frac{\dot{K}\Omega\ddot{R}^4\eta}{16\pi^2\dot{R}^2s^2}\sqrt{\frac{\rho}{15\mu^3}}\sin\frac{\xi}{2}.$$

The displacement vector component $\tilde{u}_z$ is directed along the line connecting the IEF centers and can be written as

$$\tilde{u}_z=\frac{7-8\nu}{8(1-\nu)r}B+\frac{\cos2\vartheta}{8(1-\nu)r}B.$$

It consists of two terms. The first term is the displacement of the radius sphere r as a whole without deformation. The second term is the deformation component. Substituting the coefficient B with the change of $r=\dot{R}$ into the first term, we get the displacement of the first TR in the form

$$\gamma=-\frac{7-8\nu}{8(1-\nu)}\frac{\dot{K}\Omega\ddot{R}^4\eta}{16\pi^2\dot{R}^3s^2}\sin\frac{\xi}{2}\sqrt{\frac{\rho}{15\mu^3}}.$$

Measuring γ for various orientations and at several points we obtain a set of equations to determine the angles ϑ, φ, ψ the distance s and the size of the focus $\ddot{R}$. In this way we determine the focus location, its orientation and energy. For instance, for the following parameters: $\Omega=0.0000729$ rad/s, $\dot{R}=0.1$ m, a = 40000 m, $\eta=1$, $\rho=3000$ kg/m^3, $\mu=10^9$ N/m^2, $\nu=0.3$, $\ddot{R}=20000$ m (such a value of $\ddot{R}$ corresponds to the focus energy of the order of $10^{16}-10^{17}$ J), we obtain $\gamma\sim17$ mm. The direction of the displacement vector γ is always oriented to the focus of a starting earthquake. Depending on the sign of the orientation factor η the displacement will be either toward IEF or from it. Measuring γ at several points one can define the IEF coordinates and evaluate the energy level by the value γ. According to (3.24) and (3.25), even for the most intensive earthquakes of an energy of the order of 10^{22} J, when $\ddot{R}=200000$ m, the maximum stress will be $\sim4\times10^7$ N/m^2. In the IEF there is no stress concentration.

To locate an IEF, it is necessary to determine the displacement of an artificial TR with an accuracy up to a millimeter and better. In order to create the moment of force, one can use precession. Between the moment of force $\mathbf{K}$ normal to the moment of momentum $\mathbf{M}$ and the angular velocity of precession ω there exists the relation [5]

$$\mathbf{K}=[\boldsymbol{\omega}\times\mathbf{M}]$$

The above method can be the simplest one. It allows achieving an arbitrary orientation of the moment of force, which is necessary to perform measurements.

CONCLUSIONS

In the present work a method of locating the focus of a starting earthquake is suggested. The existence of the IEF is the effect of the Earth's rotation. Its analogue is the Coriolis force. Cyclones and anticyclones are caused by the Earth's rotation. In the lithosphere they are analogous to the IEF. Undoubtedly, the IEF evolution takes much longer than a cyclone, and the rates of gas flows in the atmosphere are by orders of magnitude higher than tectonic rates. However, it should not lead to denying or ignoring inertial effects in the lithosphere.

The existence of tectonic plates with localization of elastic stresses is the main hypothesis of earthquake focuses. The reasons for stress localization can be different as well as their space dependence. If there exist two regions with a similar space distribution of stresses, then between them there is the energy of elastic interaction. Using the above energy it is possible to locate the focus of a starting earthquake creating artificially a region with a specified stress profile. Mathematical calculation of other models of earthquake focuses will be somewhat different from the one discussed here, which is not a problem.

REFERENCES

1. Lothe, J. and Hirth, J.P. (1967) Theory of Dislocations. McGraw-Hill Book Company, New York. [Paper reference:1]
2. Vikulin, A.V. and Ivanchin, A.G. (1997) Seismic Process Model. Computing Technologies, 2, 35-54. [Paper reference:1]
3. Vikulin, A.V. and Ivanchin, A.G. (1998) Rotational model of Seismic Process. Pacific Geology, 17, 18-29.
4. Vikulin, A.V., Ivanchin, A.G., et al. (2012) Data-Processing and a Computing System of Simulation of

Seismic and Volcanic Processes as a Basis for Studying Wave Geodynamic Phenomena. Computer Technologies, 3, 34-54. [Paper reference:1]

5. Landau, L.D. and Lifshitz, E.M. (1960) Mechanics. Pergamon Press Ltd., Oxford/London [Paper reference:2]
6. Landau, L.D. and Lifshits, E.M. (1986) Theory of Elasticity. 3rd Edition, Elsevier Butterworth-Heimenann, Oxford. [Paper reference:2]
7. Ivanchin, A. (2010) Potential. Solution of Poisson's Equation, Equation of Continuity and Elasticity. http://arXiv.org/abs/1011.4723 [Paper reference:2]
8. Korn, T.M. and Korn, G.M. (1968) Mathematical Handbook for Scientists and Engineers. McGraw-Hill Book Company, New York. [Paper reference:1]
9. Ivanchin, A. (2008) Nonpotential Solution of the Electron Problem.http://arXiv.org/abs/0902.1286 [Paper reference:1]

Appendix. Nonpotential Solution

As is known, along with the potential solution there can also be a nonpotential one [7] [9] . A general solution will be their linear combination, whose coefficients are determined from the condition of the total energy minimum.

To derive the nonpotential solution take the vector

$$\mathbf{U}^+ = \left\{0, 0, \frac{z}{r^3}\right\}$$

The rotor is

$$\nabla \times \mathbf{U}^+ = \frac{3z}{r^5}\{-y, x, 0\} \tag{6.1}$$

One must find the solenoidal vector $\mathbf{G}$, such that

$$\nabla \times \nabla \times \mathbf{G} = -\nabla \times \mathbf{U}^+ = \frac{3z}{r^5}\{y, -x, 0\},$$

then the vector

$$\tilde{\mathbf{U}} = \mathbf{U}^+ - \frac{1}{1-2\nu}\nabla \times \mathbf{G} \tag{6.2}$$

is the solution (3.1). For simplification, let us integrate (6.1) over z :

$$\Delta \boldsymbol{\Psi} = \int \nabla \times \mathbf{U}^+ dz = \frac{1}{r^3}\{y, -x, 0\} \tag{6.3}$$

Here it is designated

$$\boldsymbol{\Psi} = \{\Psi_x, \Psi_y, 0\} = \int \mathbf{G} dz.$$

Since $\mathbf{G}$ is the solenoidal vector, that is $\nabla \cdot \mathbf{G} \equiv 0$, the following equality takes place $\nabla \times \nabla \times \mathbf{G} = -\Delta \mathbf{G}$. Proceeding from (6.3) it is necessary to solve two equations

$$\Delta \Psi_x = \frac{y}{r^3} \tag{6.4}$$

$$\Delta \Psi_y = -\frac{x}{r^3} \tag{6.5}$$

Integrate (6.4) over and (6.5) over x . As a result, we will derive, practically, the same equation

$$\Delta \Phi_1 = -\frac{1}{r}, \quad \Delta \Phi_2 = \frac{1}{r}$$

Here it is designated

$$\Phi_1 = \int \Psi_x dy, \quad \Phi_2 = \int \Psi_y dx$$

In the spherical coordinates we have

$$\frac{1}{r^2}\frac{\partial}{\partial r}\left(r^2 \frac{\partial \Phi_1}{\partial r}\right) = -\frac{1}{r}, \quad \frac{1}{r^2}\frac{\partial}{\partial r}\left(r^2 \frac{\partial \Phi_2}{\partial r}\right) = \frac{1}{r}$$

Its solution is

$$\Phi_1 = -\frac{1}{2}r, \quad \Phi_2 = \frac{1}{2}r$$

Differentiating we obtain the components of the vector $\boldsymbol{\Psi}$ in the form

$$\Psi_x = \frac{\partial \Phi_1}{\partial y} = -\frac{y}{2r}, \quad \Psi_y = \frac{\partial \Phi_2}{\partial x} = \frac{x}{2r}, \quad \Psi_z = 0 \tag{6.6}$$

The vector divergence is

$$\nabla \cdot \boldsymbol{\Psi} \equiv \mathbf{0}.$$

which means that the vector $\boldsymbol{\Psi}$ is solenoidal. Differentiating (6.6) over z we find

$$\mathbf{G} = \frac{z}{2r^3}\{y, -x, 0\},$$

hence

$$\nabla \times \mathbf{G} = \frac{r^2 - 3z^2}{2r^5}\{x, y, z\}.$$

The nonpotential solution (6.2) is written in Cartesian and spherical coordinates as

$$\tilde{\mathbf{U}} = \mathbf{U}^+ - \frac{1}{1-2\nu}\nabla \times \mathbf{G} = -\frac{z}{2(1-2\nu)r^5}\left\{x\left(r^2 - 3z^2\right), y\left(r^2 - 3z^2\right), z\left[r^2(4\nu - 1) - 3z^2\right]\right\}$$

$$= -\frac{1}{2r^2}\left\{\frac{4\nu - 3 + (4\nu - 5)\cos 2\vartheta}{2(1-2\nu)}, \sin 2\vartheta, 0\right\}.$$

In the deformation tensor for the displacement $\tilde{\mathbf{U}}$ the component $\tilde{\varepsilon}_{r\varphi} = 0$ and in the deformation tensor (3.17) only this component is other than zero. It means that the interaction energy of the potential and nonpotential solutions is zero. Hence, the elastic energy minimum is achieved using the potential solution (3.9).

CHAPTER 6

THE RESEARCH ON EARTHQUAKE RADON ANOMALIES

Liangquan Ge, Jiankun Zhao and Yaoyao Luo

Keywords: The Earthquake Radon, The Water Radon Anomalies, Earthquake Prediction

INTRODUCTION

To make the relationship between earthquake and radon anomaly clear, the Chinese seismologists had done a lot of related research in radon and thoron exhalations during fracture in 1970s. Luo Guangwei proposed that radon concentration increased significantly under the uniaxial pressure (Luo, 1980). Based on the data of the AMRITASR Earthquake Monitoring Station of India, Han Xuehui of TYUT present some neodoxy in 2000 as followed: after a new round of tectonic stress action, geologic body will be to experience "compaction stage" firstly, and at this stage the tectonic stress impact on geologic body increase gradually, the primary fracture will decrease with the increase of stress until to be closed, which shut down the original radon exhalation channel. With all of these reasons, the radon concentration will present downward obviously at this stage (Han, 2000). After the Wenchuan Ms 8.0 earthquake, Gu Yi of Chengdu University of Technology (CDUT) evaluated the earthquake fault activity by measuring the soil radon concentration on the activity fault section, and the results show that the method is effective and reliable (Gu, 2009).

THE FORMING MECHANISM OF EARTHQUAKE RADON ANOMALIES

In the process of earthquake preparation, the underground rock in seismogenic zone also undergoes the experiences above the stages. When the rock cannot expand any more, the tiny fractures will cluster to the main rupture leading to the earthquake swarm which is regarded as the precursor for megaseism (QIN, 2011). In the initial accumulation stage, the connectivity between inner pores become worse and pore size gets smaller. It not only reduces the effective space of the storage for radon, but also blocks the migration of free radon in pores and fractures. Therefore, in the process of stress accumulation, the free radon concentration in the pores or fractures will be reduced which is confirmed in the experiments. When the rock under continuous stress or greater stress, the porosity increases significantly, so does the pore connectivity. This not only increases the free radon concentration in the pore, but also drives the migration of radon easily which is conducive to the detection. When the rock undergoes a tensile stress for longtime, the crack of rock will expand. Obviously, rock porosity and pore connectivity will continue to increase, which is conducive to increase the free radon concentration in rock pore and to easily migrate. As a result, no matter compression, stretching, or deformation, the free radon concentration in rock pores will change and an Earthquake Radon Anomaly (ERA) will form in seismogenic zone. Based on the ERA, we not only may predict the changes of rock stress at the monitoring plots, but also indicate the types of stress loaded on the rocks, such as compression is tension stress.

THE DISCUSSION OF ERA MORPHOLOGY

In the process of seismogenic, the ERA morphology is related to many factors including the monitoring area, measuring objects and the sensitivity of instruments. At present, the object of long-term monitoring radon is groundwater. Since phreatic fractures and pores are fully filled of water, the radon emits from rocks and dissolves in underground water immediately. Groundwater becomes the carrier of radon migration. So, the ERA is

Citation: Ge, L. Q., Zhao, J. K., & Luo, Y. Y. (2014). The Research on Earthquake Radon Anomalies. Journal of Geoscience and Environment Protection, 2, 38-40. http://dx.doi.org/10.4236/gep.2014.25006.

closely relative to underground water movement. According to the behaviors of radon in the process of seismogenic, the morphology of ERA can be expressed as "slope style", "hook style" and "pulse style", shown in Figure 1. The slope type of ERA (Figure 1(a)) is forming in seismogenic stretch area. In this area, tensile stress forces on rocks and the tiny fractures in rocks may increase and expand gradually, which is in favor of the storage and space for radon migration. The radon concentration in groundwater will gradually increase. In the process of rack sustained compression, the "slope style" anomaly may develop in multistep. The hook style of ERA (Figure 1(b)) may be observed in the seismogenic compressive area. The ERA consists of descent stage and ascent stage. The descent stage represents the initial compaction and elastic deformation. In this stage, the fractures are compressed and the path of radon migration may be closed, which leads to reduce radon concentration in groundwater gradually. The ascent stage of ERA shows that cracks grow crazily and rocks break. In this stage, porosity and pore connectivity increase fast, so does the concentration. The rate of ascent is significantly faster than that of descent. The pulse style (Figure 1(c)) describes the radon concentration in underground water increases and decrease rapidly. Its intensity is strong, also can be superimposed on the "slope style" or "hook style". This kind of

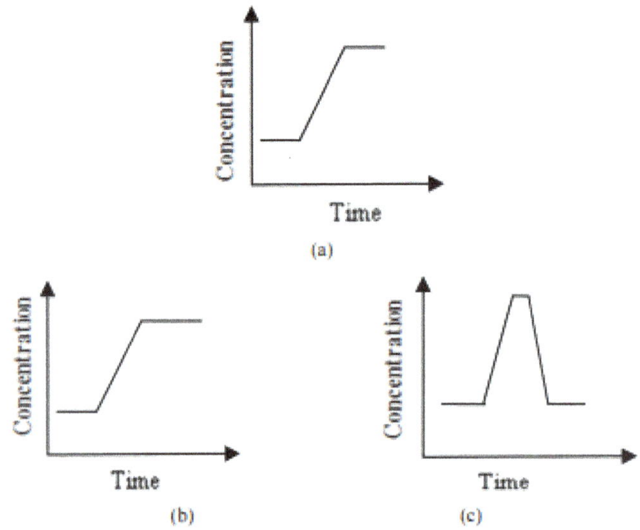

Figure 1. The patterns of ERA (a) slope-type; (b) hook-type; (c) pulse-type.

anomaly may be caused by a alternant deformation to expand the porosity or the radon releases from the rock fitfully.

THE DISCUSSION ABOUT ERA IN TIME-SPACE DOMAIN

As the half-life of 222Rn is only 3.825 days, the diffusion length of radon in water is just a few centimeters. For this reason, the range of a radon anomaly in underground water is strongly depended on the movement of under groundwater. Since the radioactive decay equilibrium between 222Rn and226Ra in a closed water-rock system takes 38 days that is about 10 times half-time of 222Rn, the range of ERA should cover the distance of underground water movement in 38 days. The time span of ERA is the whole time of rock stress accumulating. To the short-time "slope type" pattern caused by single compression change, it takes 38 days to balance the radium-radon decay. For the long time tensile stress accumulating, the "slope style" anomaly develops in multistep. It will impact the whole process of sustained compression. The descent stage of "hook style" anomaly may decline in multistep; the ascent stage presents rock broken superimposed many times. The "pulse style" anomaly caused by radon released fitfully takes very short time to be formed. According to the monitoring data to free radon released from soil, the pulse style is measured in units of hours or days. While it caused by a rock alternant deformation, the anomaly will sustain for longer time. In addition, the more "pulse style" anomalies appear the more tremendously stress change will happen near the monitoring station.

CONCLUSION

In the process of seismogenic, no matter rocks endure compression, stretching, or deformation, the free radon concentration in rock pores will change and an Earthquake Radon Anomaly (ERA) will form in seismogenic zone. The morphology of ERA can be described as "slope-type", "hook-type" and "pulse-type". The slope type represents the underground radon anomaly in the seismogenic stretch area. The ERA can cover where the underground water with radon moves in 38 days and go through the whole time of rock sustained compression. The ERA can reflect the states the rock in such as stretch, compression and broken. So the monitoring station should be located not only tensile area but also compressive zone. By analyzing the regional ERA comprehensively, we provide important reference information for seismology and earthquake prediction.

ACKNOWLEDGEMENTS

The authors gratefully acknowledge the financial support of: (1) National Natural Science Foundation of China through Grant No. 41074093 entitled "Research on mechanism of earthquake radon anomaly". (2) The National High-tech Research and Development Program of China through Grant 2012AA061803 entitled "Development of high precision nuclear radiation spectrometry".

REFERENCES

1. Luo, G. W., & Shi, X. Z. (1980). Experimental Results of Radon and Thorium Emanations from Rock Specimen under Pressure. Acta Seismologica Sinica, 2, 198-204.
2. Han, X. H. (2000). The Ideal Curve Model on Earthquake Prediction by Radon Monitoring. Journal of Taiyuan University of Technology, 31, 101-103.
3. Gu, Y (2009).Analysis and Evaluation of Faults Activities in Chengdu Region with Radon Concentration Measurements after Wenchuan Earthquake. Journal of Engineering Geology, 17, 296-300.
4. Qin, S. Q. (2011). Seismogenic Law and Mode of Strong Earthquakes. Journal of Earth Sciences and Environment, 33, 311- 316.

CHAPTER 7

SEISMIC SEQUENCE STRUCTURE AND EARTHQUAKES TRIGGERING PATTERNS

Giulio Riga and Paolo Balocchi

Keywords: Foreshock, Classification, Sequence, Cycle, Pattern, Trigger, Earthquake

INTRODUCTION

The analysis of the seismic sequence structures underlines (points out) that magnitude values do not unpredictably change, but follow some trends that can be identified in a chart.

Starting from this principle, we can speculate that some behaviors cyclically recur, making it possible, if identified on time, for the formulation of hypothesis about the future evolution of the seismic sequence.

The checks we have performed on several big earthquakes occurred in the last 40 years, have highlighted that it is possible to identify some earthquakes triggering patterns in the short and medium term, whose space-time complexity is strictly linked to the following items: a) the seismic region heterogeneity [1] ; b) the stress state inside and around the source of a big earthquake [2] [3] ; c) the long-range interactions among different seismic and heterogenic regions.

The identification, in advance, of the warning pattern within the seismic sequence, provides several useful insights to understand what is happening in the source region area, which can be used to improve earthquakes predictions.

Below, we will explain the known seismic sequences features, which include warning patterns we have classified and discussed in relation to the magnitude values' trend.

We performed the analysis of the seismic sequence using: a) spatial coverage of datasets ranging from portions of a region to several countries (from a few tens of kilometers to thousands of kilometers); b) magnitude range of 2.0 - 10 M; c) depths range of 1 - 50 km.

METHOD OF ANALYSIS

We analyzed and classified the time series of magnitude values using the results of some researches carried out in the field of financial markets studies.

The analysis provides some characteristics of the values' trend in a time series which are useful for assessing whether the phenomenon occurs according to the predicted scheme.

In particular, for the study of the movements of the seismic cycle we used the Elliott's wave theory [4] - [6] , where the concept of pattern is one of the most significant aspects.

The methodological approach adopted for the classification of earthquake triggering patterns and the assessment of the main earthquake magnitude is based on the technical analysis [7] [8] , which consists in a methodology for the evaluation of financial markets used to identify trends and assess future trends possibility. Similarly, it may be used as a safe tool to analyze the most significant magnitude values' oscillations and to predict likely future events, mainly through graphical methods and statistics.

STRUCTURE OF SEISMIC SEQUENCE

Types of Seismic Sequences

Earthquakes are generally part of a sequence with variable and sharp features. Mogi [9] divided these sequences in three types: 1) mainshock-aftershock (Figure 1(a)); 2) foreshock-mainshock-aftershock (Figure 1(b)); 3) swarm (Figure 1(c)). The foreshocks and aftershocks sequences are strictly

Citation: Giulio Riga, Paolo Balocchi (2016) Seismic Sequence Structure and Earthquakes Triggering Patterns. Open Journal of Earthquake Research,05,20-34. doi: 10.4236/ojer.2016.51003.

associated to a bigger event called mainshock, while the swarm is a sequence of earthquakes and is not associated with a dominant earthquake. The subsequent events of mainshock-aftershock sequences of similar dimension (Figure 1(d)) are defined as swarm, but they clearly differ from an ordinary swarm (Figure 1(c)).

Moreover, numerical results [9] show that different homogeneity index will produce different patterns of seismic sequences. We have three types of sequences: a) swarm; b) foreshock-mainshock-aftershock; c) mainshock-aftershock that happen specifically in these cases: a) extremely heterogeneous; b) mildly heterogeneous and c) homogeneous [1]

Isolated earthquakes occur in presence of a great pre-stress and/or low stress at break. Foreshocks-main- shock-aftershock seismic sequences happen in presence of a mild pre-stress and stress at break. Swarms happen in case of low pre-stress and/or high resistance to break. Very low values of pre-stress and/or very high break strength are the best conditions for the aseismic creep event [10] to occur. Other studies underline that in areas where the time variation of the strain-rate (measured by geological surveys GPS) is low, higher magnitude earthquakes are expected compared with areas with high strain-rate [11] .

Simulations [10] suggest that an isolated earthquake can be interpreted as the extension of a break in a

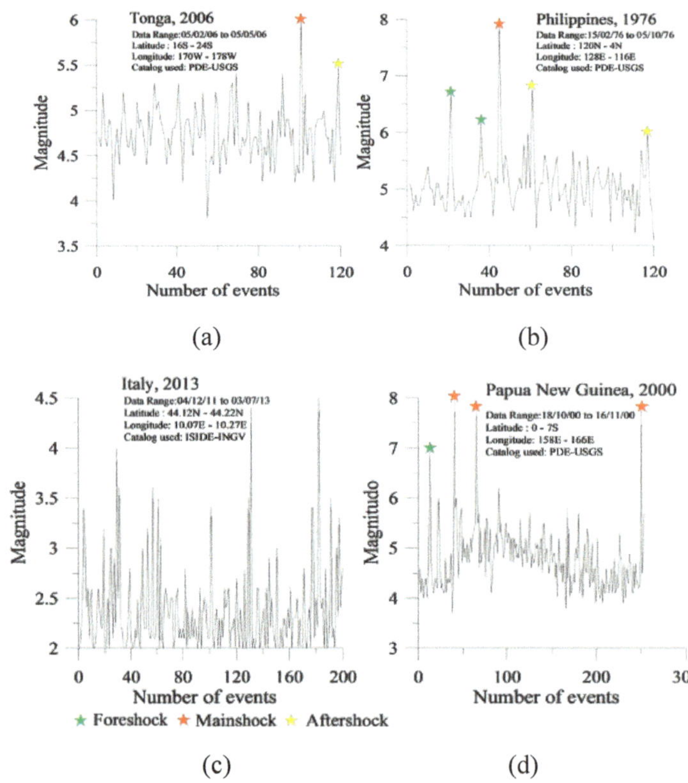

Figure 1. Examples of earthquake sequences: (a) mainshock-aftershock sequence; (b) foreshock-mainshock-after- shock sequence; (c) earthquake swarm; (d) subsequent occurrence of mainshock-aftershock sequence.

non-uniform field, while a foreshocks-mainshock-aftershocks sequence results from a break extension to a non- uniform field having modest fluctuations and swarms are a consequence of a break extension to a very inhomogeneous sector.

Swarms are a sequence of earthquakes that often start and finish gradually, where there is no evidence of any earthquake shock with higher magnitude [12] . However, shocks that make up swarms are actually included in seismic cycles in which the main shock, usually of mild magnitude, occurs always at the end of the nucleation process.

Swarms are divided in two typologies. The first one is the ordinary swarm in which the activity changes fairly irregularly (Figure 1(c)), and has a duration between some hours and little more than a year. Most of the swarms happening in volcanic zones show this model, which reflects the magmatic activity in depth. The second typology features a succession of main shocks with similar magnitude (Figure 1(d)) whose duration is between several days and months.

Seismic Cycle's Basic Structure

The long-term seismic sequence consists of several cycles that are repeated according to the representation scale and whose duration varies from a few minutes to several years.

The basic seismic cycle is a collection of processes that develop before, during and after an earthquake, which are highlighted by the variations in the magnitude values that develop in short and medium term time windows. The term "cycle" suggests that big earthquakes can be either periodic or regular events over time. The analysis of the succession of the seismic cycles included in a seismic sequence shows that the periodicity can be prevalent among big earthquakes recorded in areas even with different extension.

According to the magnitude values fluctuations, seismic cycles can be divided in four stages: 1) trigger; 2) energy release; 3) breaking; 4) adjustment (Figure 2).

Trigger stage. In this phase, the magnitude values vary within a very close range in which a small shock, after some fluctuations, triggers the next one.

Energy release stage. This step follows the trigger phase, where the magnitude values fluctuations produce increasing maximum values along with an increase in the number of shocks. In type 1 sequences, (mainshock- aftershock) these characteristics are often not found.

Breaking stage. This is the energy release phase's top. It corresponds to the main breaking point i.e. when the mainshock occurs, generally marked by the highest number of recorded shocks. If no additional main breaking (multiple events) follows, the magnitude values fall quickly under the peak magnitude value recorded.

Adjustment stage. The magnitude values and the number of shocks decrease, until it gets back to the trigger phase once again.

The transition from the energy release phase to the adjustment one can be confirmed by the values decrease before and after the mainshock, called "transition line" (Figure 3(a)).

We should bear in mind, however, that usually after being dropped below the "transition line" (point 3), the magnitude values temporarily return to the same line level or even slightly above, before definitely starting to decrease up to the next minimum value.

In case of "transition line" crossing during the increase phase, a second mainshock may occur (subsequent occurrence of mainshock-aftershock sequences).

In the case of minimum "1" and "2" descending values, the step from energy release to the adjustment one is always confirmed by the minimum "2" value formation (Figure 3(b)).

The Seismic Cycle Movements

In many long-term seismic sequences, the analysis of magnitude values trends during the energy release phase shows a cyclical fluctuation, which is characterized by one or more waves (foreshocks-mainshock). In the adjustment phase, we observe several aftershocks that partially rebalance all the impulsive movement in the previous step.

Based on Elliot's theory [4] - [6] , a seismic sequence can be considered as a succession of cyclic variations of magnitude values whose duration may vary over time (Figure 4 and Table 1), where the most energetic seismic

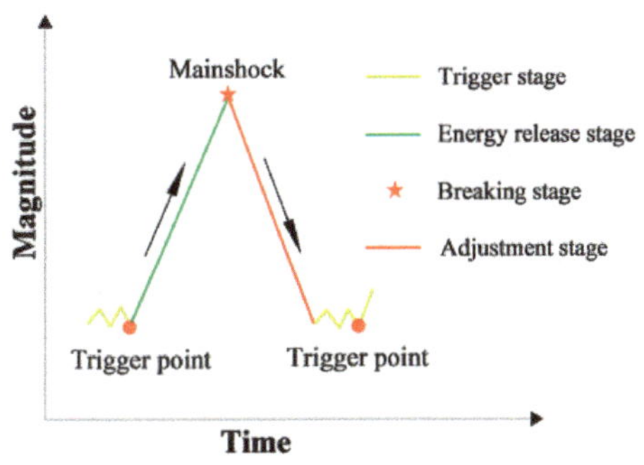

Figure 2. Schematic representation of the seismic cycle's four stages.

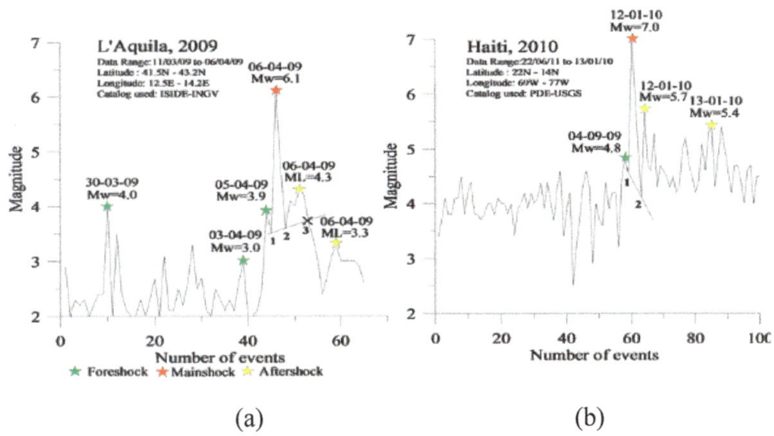

Figure 3. Sequence of earthquake imminent period: (a) L'Aquila earthquake, 06-04-2009 (Italy); (b) Haiti earthquake, 12-01-2010.

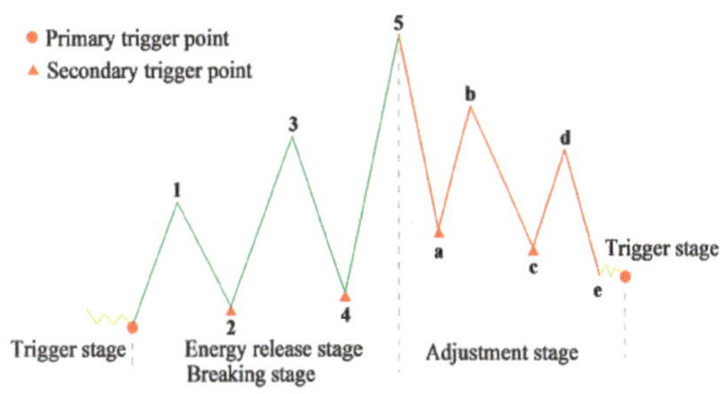

Figure 4. Schematic representation the seismic cycle's waves.

Table 1. Seismic cycle-energy release and adjustment phases' waves.

Point	Waves	Description
1	Energy release	Formation of a first foreshock
2	Adjustment	Formation of a relative minimum
3	Energy release	Formation of a second foreshock
4	Adjustment	Formation of a relative minimum
5	Energy release	Formation of a mainshock
a	Adjustment	Formation of a relative minimum
b	Energy release	Formation of a first aftershock
c	Adjustment	Formation of a relative minimum
d	Energy release	Formation of a second aftershock
e	Adjustment	Formation of a relative minimum

events in the energy release phase occur according to a pattern consisting of several waves. Most frequently the pattern consists of five waves (1, 2, 3, 4, 5) whose values in 1-3-5 show an increasing magnitude trend (foreshocks-mainshock), while in the adjustment stage it consists of five waves (a, b, c, d, e) whose magnitude values in "b" and "d" are increasing compared to the values in "a" and "c" (significant aftershocks).

The energy release stage starts after the formation of the primary trigger point, while the transition from a decreasing movement to an increasing one and vice versa is marked by a secondary trigger point.

Table 1 describes the fluctuations of the magnitude values over time and reports the description of the maximum values that are formed at the end of each wave is given as well.

The basic chart is characterized by a series of several seismic cycles, showing a nonlinear structure with an up and down progression (Figure 5) in which the different phases of the seismic process before a big earthquake are clearly visible. Typically, a big earthquake may be preceded by two types of energy release phase: 1) progressive earthquakes and 2) flash earthquakes.

In the progressive earthquakes-type energy release phase, a big earthquake is preceded by one or more foreshocks of various kinds, while in the flash earthquakes type, a big earthquake is not preceded by any foreshock at all.

The adjustment phase consists of two or more primary aftershocks with a lower magnitude compared to the mainshock. They usually occur days, months, sometimes even years after the main event to rebalance the elastic deformation caused by the mainshock.

Figure 6(a) shows the chart of the magnitude values recorded in Japan from 1973 to 2015. Here, we can observe two similar medium-term seismic cycles. The first consists of three energy release phases and one adjustment phase. The first ended on May 8th, 1974 with a 6.7 Mw foreshock, the second on June 12th, 1978 with a 7.7 Mw foreshock and the third on May 26th, 1983 with the 7.8 Mw mainshock, followed by the adjustment phase which ended on November 11th, 1989 with the aftershock.

The second cycle, not completed yet, consists of the energy release phases, represented by the foreshocks occurred on 28 December 1984 (7.8 Mw) and the 25 September 2003 (8.3 Mw), while the last one, with the high energy earthquake occurred on 11 March 2011 which has not been followed by a significant aftershock yet.

Figure 6(b) shows the chart of the magnitude values recorded in Greece in 2008: here, we can observe a traditional cycle consisting of one energy release phase with two short-term foreshocks, the mainshock and one adjustment phase with two main aftershocks.

Progressive Earthquakes Type

Typically, the energy release phase that precedes a large earthquake consists of a series of warning shocks of

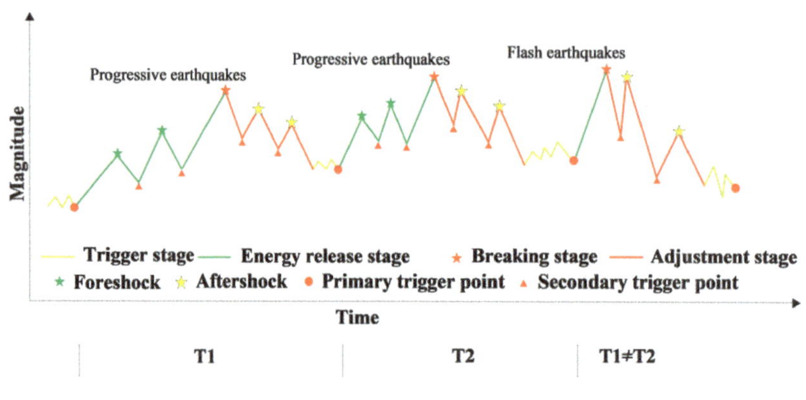

Figure 5. Schematic representation of the seismics cycle. The green, red and yellow-colored stars indicate the foreshocks, the mainshocks and the aftershocks, respectively. The red-colored circle indicates the energy release phase's trigger point.

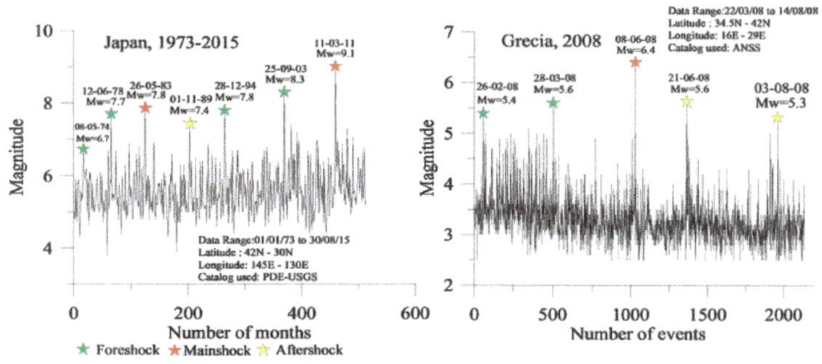

(a) (b)

Figure 6. Examples of seismic cycles: (a) long-term sequence; (b) short-term sequence.

different order (foreshocks) featuring an increasing magnitude characterizing the breaking process that ends with a big earthquake.

It begins to develop at the end of an adjustment phase with an opposite motion compared to the preceding one.

Based on the magnitude values' fluctuations that characterize the energy release phase, it becomes clear that the greater the amplitude between two foreshocks, the greater the magnitude of the mainshock expected.

The form varies according to the temporal position and the foreshocks' magnitude values that precede the mainshock.

Depending on the foreshocks position, the pattern may consist of:

a) One or more first-order or medium-short-term foreshocks and manshock (Figure 7(a));

b) One second-order or impending term foreshock and mainshock (Figure 7(b));

c) One or more first-order or medium-short-term foreshocks, one second order or impending-term foreshock and mainshock (Figure 7(c));

d) One or more various order foreshocks and several mainshocks (Figure 7(d)).

The adjustment phase consists of two or more main aftershocks with a lower magnitude compared to the mainshock. Usually the aftershocks occur days, months, sometimes even years after the main event to rebalance the elastic deformation the mainshock caused.

Figure 8(a) shows the pattern that anticipated the earthquake occurred in Nicaragua on September 2nd, 1992. This pattern consists of two first-order foreshocks and one second-order foreshock (M 5.8) that occurred shortly before the mainshock on 2 September 1992 in the proximity of its hypocenter.

The second-order foreshock recorded on 10 August 1992 with 5.8 M, most likely indicates the beginning of the breaking phase that ended with the mainshock.

Figure 8(b) shows the pattern of the earthquake occurred in Nicaragua on September 2nd, 1992, which was processed by using a seismic area extended to the state of Guatemala. We can note how the pattern retains the same foreshocks, mainshocks and aftershocks distribution of Figure 13(a), but includes seismic events that occurred in different times and different areas.

Figure 9(a) shows the pattern that preceded the earthquake that took place in Vanuatu in 2009. Usually, only one mainshock occurs, but here the strongest earthquake was preceded on the same day (October 7th, 2009) by a high magnitude event classifiable as mainshock.

The second-order foreshock on 10 August 2009, with 6.6 Mw magnitude, marks the beginning of the breaking phase which ended with the three strong mainshocks.

Figure 9(b) shows the pattern which preceded the earthquake in Loma Prieta (California, USA). We can see the two first-order foreshocks that preceded the mainshock on 18 October 1989 and the two main aftershocks that followed it.

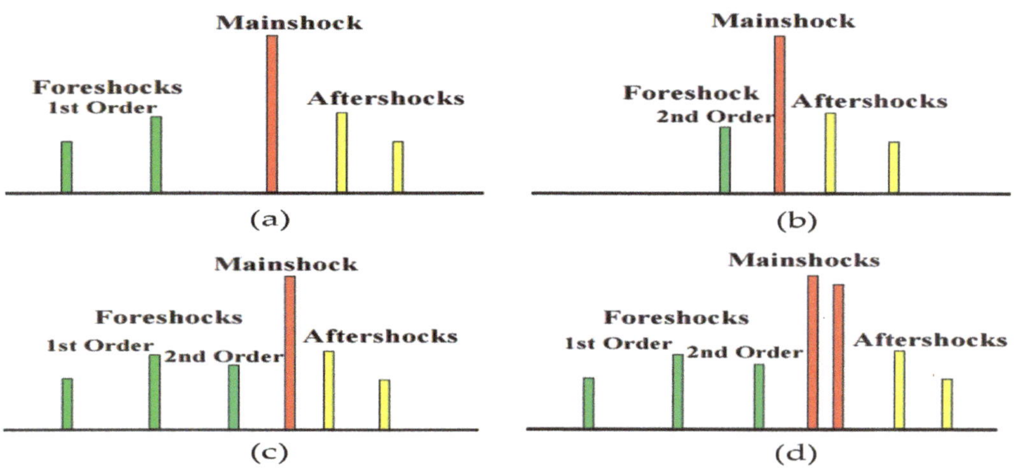

Figure 7. Schematic representation of progressive earthquakes patterns.

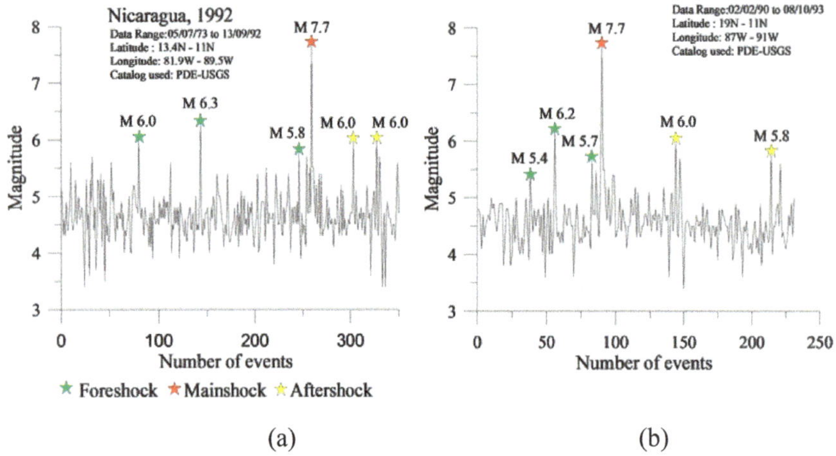

Figure 8. Progressive earthquakes patterns of Nicaragua earthquake, 1992: (a) restricted area; (b) wide area.

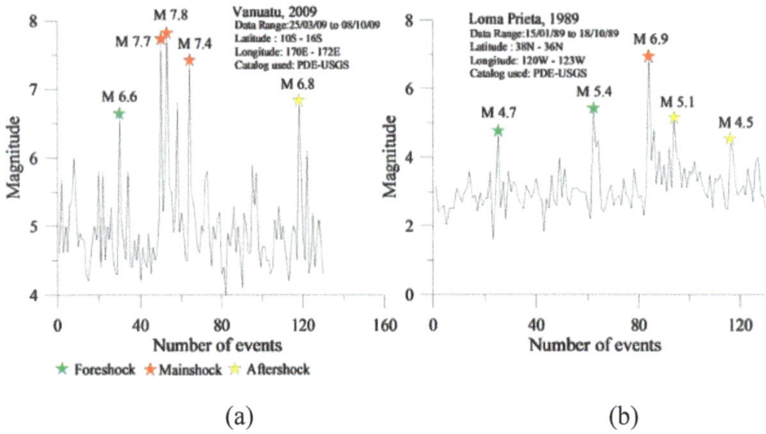

Figure 9. Progressive earthquakes patterns: (a) Vanuatu earthquake, 2009; (b) Loma Prieta earthquake (USA).

Flash-Type Earthquake

This type of energy release phase is quite uncommon. Here, the big earthquake develops suddenly (with no foreshocks) after a long adjustment phase following the breaking stage and consists of aftershocks whose magnitude slowly decreases over time.

The magnitude of the expected mainshock depends on the adjustment stage time amplitude.

Based on the adjustment stage duration, it is clear that the more extended over time, the stronger the mainshock.

Depending on the aftershocks magnitude the pattern may consist of:

a) One or more significant aftershocks with decreasing magnitude (Figure 10(a));

b) One or more significant aftershocks with a similar magnitude (Figure 10(b)).

Figure 11(a) shows the pattern that preceded the earthquake in Taiwan on 20 September 1999. The maishock was followed by a first main aftershock whose magnitude was slightly lower than the mainshock's, while Figure 11(b) shows the pattern that preceded the earthquake in Tonga on 3 May 2006. Here we observe how a high magnitude mainshock was preceded by a sequence, whose magnitude values are consistent with the rest of the sequence in the post-mainshock period.

EARTHQUAKE-TRIGGERING PATTERNS

The sequence of aftershocks that develops after a mainshock or a foreshock event may have a significant impact on the next earthquake triggering pattern that anticipates another mainshock formation.

The effect is more apparent in those with the longest duration that consist of aftershocks associated with a large earthquake. Some earthquakes triggering patterns demonstrate how a sequence of earthquakes can trigger other earthquakes or how small earthquakes significantly contribute to the triggering of a big earthquake.

Statistical studies performed on seismic sequences worldwide showed that pre-earthquake seismic patterns, in relation to the evolution usually occurring over time, can be broadly divided in two types:

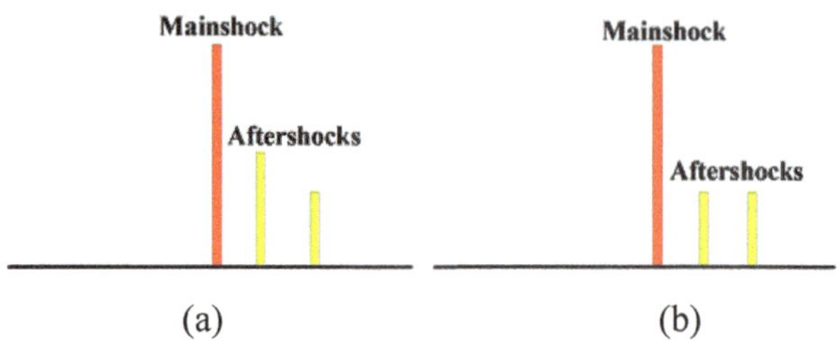

Figure 10. Flash type earthquakes' pattern illustration.

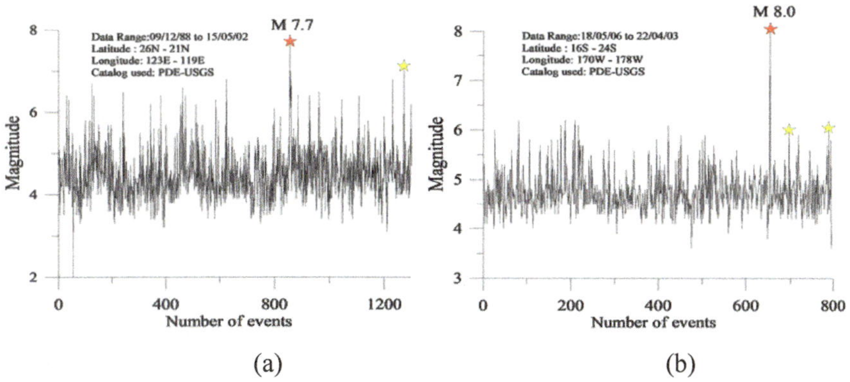

Figure 11. Examples of flash earthquales types: (a) Taiwan earthquake, 1999; (b) Tonga earthquake, 2006.

- Flash earthquakes patterns;
- Progressive earthquakes patterns.

In flash earthquakes patterns, the magnitude values vary within two trend lines - inclined or horizontal - until the pattern reaches a critical state (breaking conditions) imposed by the rocks' strength properties and by the fault surface, where the small shock triggers the energy release phase that occurs all of sudden.

The upper trend line crossing determines the beginning of the breaking stage (breaking stage).

A small scale decrease in the crustal heterogeneity along with an increase in its depth, suggests that an increase in the normal stress inhibits the foreshocks occurrence [13] .

Usually, the mainshock is followed by aftershocks that are smaller compared to the mainshock, although occasionally an aftershock can be strong enough to assume the appearance of a subsequent mainshock.

Overall, the adjustment stage appears to be extended due to the effect of a large number of events that occur after the mainshock, but first and foremost it is characterized by events with decreasing magnitude.

In particular, the number of shocks recorded from the trigger point (first minimum value) up to the mainshock is 50% smaller compared to the total number of shocks recorded up to the second minimum value formed after the first significant aftershock (Figure 12(a)).

As to the chart, the final minimum value will be placed on a lower level compared to the trigger point: this makes the pattern take an apparently sinusoidal shape with maximum decreasing values (mainshock and aftershocks).

The time this pattern needs to complete varies from a few weeks up to a number of years.

In progressive earthquakes patterns, the magnitude values oscillate within two trend lines which diverge until the main shock.

The mainshock, therefore, is preceded by foreshocks that may occur at different times both inside and outside of the source region.

The number of shocks usually used from the trigger point up to the mainshock is greater than 50% of the total number of registered shocks to reach the second minimum value that is formed after the mainshock (Figure 12(b)).

This means that, until the mainshock, the energy release phase usually is more extended than the adjustment one due to a higher number of events with increasing magnitude.

As regards the chart, the second minimum value will be higher than the Trigger point's: this makes the cycle take the apparently classical sinusoidal shape with increasing maximum and minimum values that tend to return to the initial level.

Earthquake Triggering Patterns Classification

The proposed classification is the result of a thorough graphical analysis of seismic sequences, which is the basis for the first part of the short-term, deterministic forecasting model currently in the testing phase.

The characteristics taken into consideration in the classification process are: a) the fluctuations of the magnitude minimum and maximum values; b) the foreshocks observation before the mainshock.

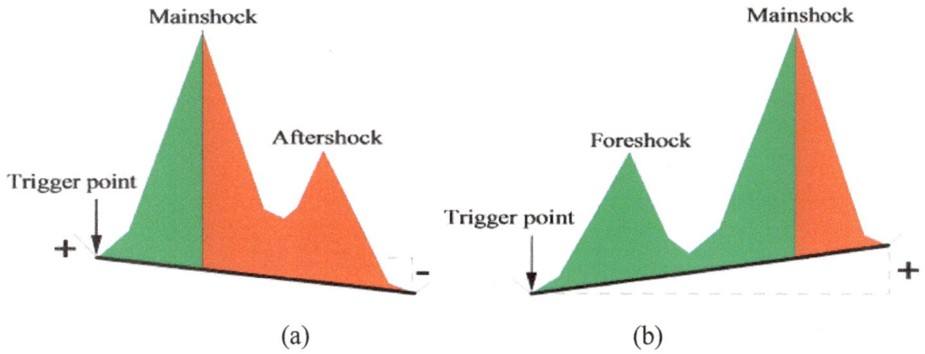

Figure 12. Schematic representation of pre-earthquake seismic types (a) flash earthquakes patterns; (b) progressive earthquakes parterns.

The patterns identified and classified using the technical graphic analysis [7] [8] , reflect the energy release and breaking phases and attempt to provide useful information to create connections between the pattern and the preparation process for a big earthquake.

(A) Symmetrical triangle pattern

The preparation process for a big earthquake starts with a first medium-strong shock, a short-medium term foreshock, followed by minor aftershocks, which are increasingly weaker and less frequent, included in a phase in which the deformation of the rock is probably minimal.

Observing this diagram in details (Figure 13), we note how the magnitude values oscillate between two trend lines that define maximum decreasing and minimum increasing values, which converge towards a triangle intersection point (Apex). The fluctuations' amplitude, which is formed inside the triangle, gradually decreases and the earthquake occurs when the magnitude values are nearing the apex. Therefore, the more extended over time the magnitude and amplitude values fluctuations, the stronger the earthquake associated with the pattern.

Usually, the critical phase begins when the difference in magnitude between points 4 and 5 is 50% less compared to the one between points 2 and 3 (triangle base). The crossing of the upper trend line currently occurs between 50% and 75% of its length.

The energy release phase may consist of the mainshock preceded by an impending term foreshock (progressive earthquakes), or by the mainshock preceded by no foreshock (flash earthquakes).

The pattern is considered as an "energy adjustment or accumulation", because the magnitude maximum values and the number of shocks decrease during the development, while increase upon the pattern breaking.

The symmetrical triangle pattern is typical of areas with little fractured, homogeneous rocks and it is most likely associated with planes of fault featuring little ruggedness.

The magnitude target (TM) of this pattern, once completed, is obtained by projecting from the breaking point (point 5), the width of the triangle base.

(B) Falling wedge pattern

The wedge is a pattern very similar to the triangle's, being the magnitude values limited by two converging trend lines, but differs from it because both sides have a different downward inclination and often the magnitude values remain locked up in the pattern up to the apex (Figure 14).

The upper trend line crossing usually occurs between 50% and 75% of the base length.

Usually, the energy release phase consists of the mainshock preceded by an impending term foreschock (progressive earthquakes), or by the mainshock preceded by no foreshock (flash earthquakes).

Just like in the triangle pattern, the number of shock decreases during its formation.

The TM of this pattern is calculated by projecting the width of the wedge base from the breaking point (point 3).

(C) Rectangle pattern

The rectangular pattern (Figure 15 and Figure 16) is one of the easiest patterns to identify because it is characterized by fluctuations in magnitude values within a more or less extensive strip. Most likely, it represents a stage of energy accumulation before the medium-strong event.

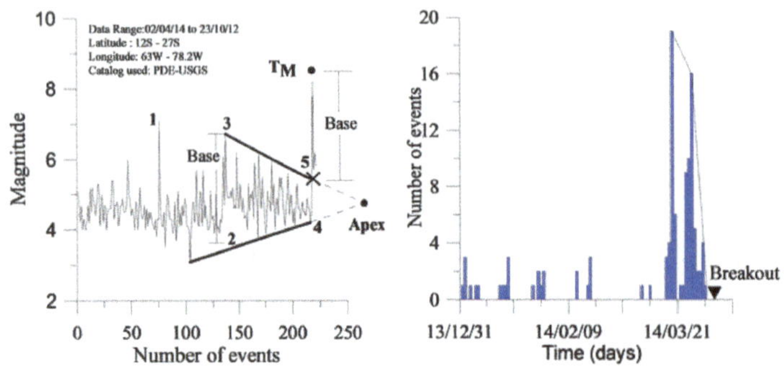

Figure 13. Symmetrical triangle pattern (Chile earthquake, 2014).

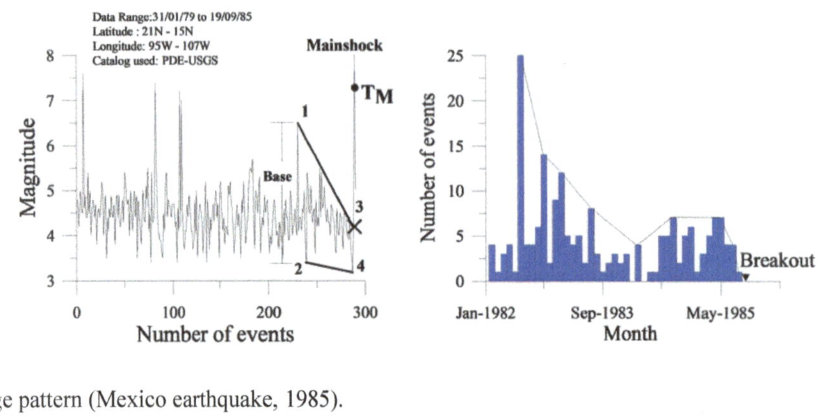

Figure 14. Falling wedge pattern (Mexico earthquake, 1985).

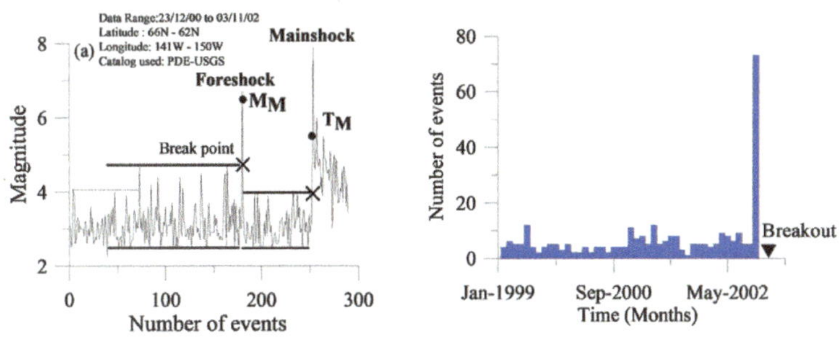

Figure 15.Horizontal rectangle pattern (Denali earthquake , 2002).

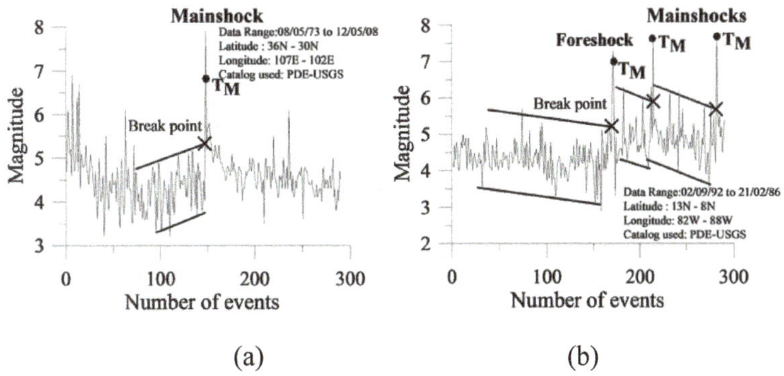

Figure 16. Rectangle Pattern: (a) ascending (China earthquake, 2008); (b) descending (Costa Rica-Nicaragua earth- quakes, 1990-1992).

This pattern does not allow us to early identify the exact time when the upper trend line will be crossed.

The most frequent energy release phase consists of the mainshock preceded by no foreshock (flash earthquakes).

Sometimes it is characterized by an increase in the number of shocks before the upper trend line crossing, as it happened, for example, during the earthquake in Denali (Alaska) coinciding with the foreshock on 23 October 2002 that preceded the earthquake on November 3rd, 2002 whose magnitude amounted to 7.9 Mw (Figure 15).

It includes three types: a) horizontal, b) ascending, c) descending, according to the arrangement of the minimum and the maximum values that are formed.

The TM of this pattern is calculated by projecting the rectangle height from the breaking point.

(D) Expansion triangle pattern

This is an uncommon pattern consisting of a series of oscillations of the magnitude minimum and maximum values that expand over time (Figure 17). The completion of the model does not coincide with the upper trend line crossing.

This pattern can be interpreted as alternate small cycles of energy accumulation and gradual release that continue until they reaches the critical stress level and, therefore, the mainshock.

The number of shock decreases during the first part of the pattern development and increases when it gets close to the breaking phase.

Pattern's special features considered, it is possible to calculate the TM at the intersection point of the line joining the maximum points with the vertical line passing from the last recorded event plus one (n + 1).

CORRELATION BETWEEN THE SEISMIC PATTERN AND THE FOCAL MECHANISM

The seismic patterns observed in many sequences can be correlated with the major earthquakes' focal mechanisms [14] . Table 2 shows some patterns that we could identify within the short-medium term time windows. As we see, the flash earthquakes patterns are generally associated with reverse or strike-slip faults, while progressive earthquakes patterns are related to normal faults.

Reverse faults tend to feature a higher percentage of big earthquakes not preceded by foreshocks compared to normal faults, while we have an average percentage with the strike-slip ones.

Previous studies on sequences with foreshocks showed that it would be probably very difficult to observe foreshocks before a big earthquake associated with reverse faults because normal stress would increase significantly from reverse faulting to normal [15] .

Some patterns with foreshocks recorded in an environment with reverse faults seem to be associated with a greater degree of small-scale heterogeneity in crust [16] .

CONCLUSIONS

With this study, we wanted to define the earthquakes triggering patterns that develop in seismic sequences before an energy earthquake that contains a series of information including the seismic history of the area under investigation.

Following a thorough graphical analysis performed over several seismic sequences, we classified the earthquakes triggering patterns according to the fluctuations of the minimum and maximum magnitude values, which in some cases may be considered as true earthquakes precursors.

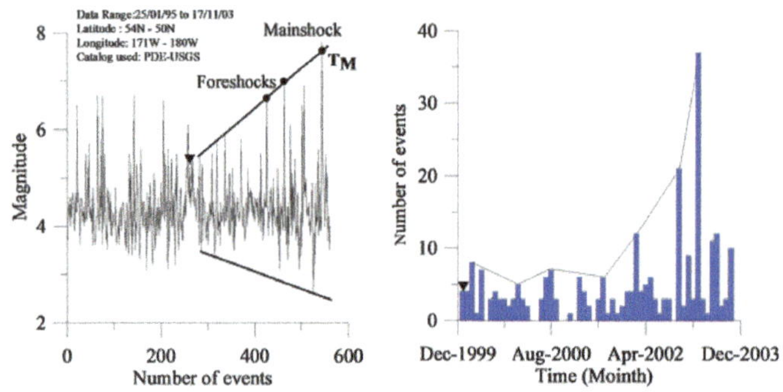

Figure 17. Expansion triangle pattern (Aleutian Islands Earthquake of 17 November 2003).

Table 2. Earthquake pattern type and focal mechanism correlations.

Earthquake	Date	Earthquake triggering patterns	Focal mechanism
L'aquila-Italy	06/04/2009	Progressive earthquakes pattern	Normal
Colfiorito-Italy	26/09/1997	Progressive earthquakes pattern	Normal
Irpinia-Italy	23/11/1980	Progressive earthquakes pattern	Normal
Emilia-Italy	20/05/2012	Progressive earthquakes pattern	Thrust
Loma Prieta-CA, USA	19/10/1989	Progressive earthquakes pattern	Transpression
Nicaragua	02/09/1992	Progressive earthquakes pattern	Thrust
Vanuatu	07/10/2009	Progressive earthquakes pattern	Thrust
Aleutian Islands	17/11/2003	Progressive earthquakes pattern	Thrust
Haiti	12/01/2010	Progressive earthquakes pattern	Srike-slip
Turkey	23/10/2011	Progressive earthquakes pattern	Reverse strike-slip
Taiwan	14/11/1986	Progressive earthquakes pattern	Reverse
Chile	01/04/2014	Progressive earthquakes pattern	Low-angle reverse strike-slip
Japan	11/03/2011	Progressive earthquakes pattern	Thrust
New Zealand	03/09/2010	Flash earthquakes pattern	Dextral strike-slip
Landers-CA, USA	28/06/1992	Flash earthquakes pattern	Srike-slip
China	14/04/2010	Flash earthquakes pattern	Srike-slip
Mexico	19/09/1985	Flash earthquakes pattern	Reverse
Pakistan	18/11/2011	Flash earthquakes pattern	Normal strike-slip
Pakistan	24/09/2013	Flash earthquakes pattern	Normal strike-slip
Nicaragua	02/09/1992	Flash earthquakes pattern	Reverse
Nicaragua	04/02/1976	Flash earthquakes pattern	Srike-slip
Uzbekistan	19/03/1984	Flash earthquakes pattern	Reverse
Uzbekistan	17/05/1976	Flash earthquakes pattern	Reverse
Japan	26/05/1983	Flash earthquakes pattern	Reverse
Alaska Gulf	30/11/1987	Flash earthquakes pattern	Srike-slip
Philippines	16/08/1976	Flash earthquakes pattern	Reverse
Australia	22/01/1988	Flash earthquakes pattern	Reverse
Colombia	15/11/2004	Flash earthquakes pattern	Reverse
Colombia	18/10/1992	Flash earthquakes pattern	Reverse strike-slip
Colombia	13/07/1974	Flash earthquakes pattern	Low-angle reverse
Indonesia	11/04/2012	Flash earthquakes pattern	Low-angle reverse
Indonesia	06/04/2010	Flash earthquakes pattern	Srike-slip
Sumatra	26/12/2004	Flash earthquakes pattern	Thrust
Nepal	25/04/2015	Flash earthquakes pattern	Reverse
Emilia-Italy	29/05/2012	Flash earthquakes pattern	Reverse

Friuli-Italy	06/05/1976	Flash earthquakes pattern	Reverse strike-slip
Denali (Alaska)	03/11/2002	Flash earthquakes pattern	Reverse strike-slip
Tonga	03/05/2006	Flash earthquakes pattern	Srike-slip (transform)
Russia	05/12/1997	Flash earthquakes pattern	Srike-slip (transform)
Australia	22/01/1988	Flash earthquakes pattern	Reverse
Uganda	05/12/2005	Flash earthquakes pattern	Transtensive
Taiwan	20/09/1999	Flash earthquakes pattern	Reverse

The statistical studies performed on seismic sequences worldwide demonstrated that the earthquakes triggering patterns, in relation to the evolution usually shown over time, tend to group into two types: 1) flash earthquakes patterns; 2) progressive earthquakes patterns.

By considering the focal mainshock mechanisms, we saw how the earthquakes triggering patterns, associated with reverse and strike-slip faults tend to have a higher percentage of big earthquakes of flash earthquakes patterns type compared to normal faults that develop earthquakes of progressive earthquakes patterns type.

Some patterns featuring foreshocks recorded in an environment with reverse faults seem to be associated with a greater degree of small scale crustal heterogeneity and maybe even to a reduced normal stress on the fault plane.

The aforementioned approach unquestionably has some advantages in terms of prediction. In fact, the early detection of the earthquakes triggering patterns, allows us not only to predict the likely future changes in magnitude values but also to graphically estimate the magnitude of the strongest event.

REFERENCES

1. Jiao, M., Tang, C., Zhang, G., Shi, Y. and Hou, W. (2003) A Numerical Test on Influence of Mesoscopic Heterogeneity on Macroscopic Behavior of Rock Failure and Seismic Sequance Types. Chinese Journal of Geophysics, 46, 943-953.http://dx.doi.org/10.1002/cjg2.414 [Paper reference:2]
2. Matsu'ura, R.S. (1986) Precursory Quiescence and Recovery of Aftershock Activities before Some Large Aftershocks. Bulletin of the Earthquake Research Institute, 61, 1-65. [Paper reference:1]
3. Matsu'ura, R.S. (1991) Case 1: Precursory Quiescence and Recovery of Aftershock Activity before Some Large Aftershocks. In: Wyss, M., Ed., Evaluation of Proposed Earthquake Precursors, American Geophysical Union, Washington DC, 8-11.http://dx.doi.org/10.1029/SP032p0008 [Paper reference:1]
4. Phetking, C., Sap, M.N.M. and Selamat, A. (2008) Time Series Representation for Elliott Wave Identification in Stock Market Analysis.http://comp.utm.my/pars/files/2013/04/Time-Series-Representation-for-Elliott-Wave-Identification-in-Stock-Market-Analysis.pdf [Paper reference:2]
5. Frost, A.J. and Prechter, R.R. (1985) Elliott Wave Principle. New Classics Library, Georgia.
6. Prechter Jr., R.R. (1995-2004) The Basics of the Elliott Wave Principle. New Classics Library, Gainesville.http://www.yinyangforex.com/download/trading_books_and_documents/EW%20Basics%20Booklet.pdf [Paper reference:2]
7. Bulkowski, T.N. (2002) Trading Classic Chart Patterns. John Wiley & Sons, Inc., Hoboken. http://fx-arabia.com/vb/uploaded/178_01275676445.pdf [Paper reference:2]
8. Matras, K. (2010) Chart Patterns Trader Supplement Understanding and Trading Classic Chart Patterns. Zacks Investment Research, Inc.http://www.sju.edu/int/academics/hsb/finance/wstr/pdf/CPGuide1.pdf [Paper reference:2]
9. Mogi, K. (1963) Some Discusions on Aftershocks, Foreshocks and Earthquake Swarms: The Fracture of a Semi-Infinite Body Caused by Inner Stress Origin and Its Relation to the Earthquake Phenomena (3rd Paper). Bulletin of the Earthquake Research Institute, 41, 615-658. [Paper reference:2]
10. Chen, Y.T. and Knopoff, L. (1987) Simulation of Earthquake Sequences. Geophysical Journal International, 91, 693-709. http://dx.doi.org/10.1111/j.1365-246X.1987.tb01665.x [Paper reference:2]
11. Riguzzi, F., Crespi, M., Devoti, R., Doglioni, C., Pietrantonio, G. and Pisani, A.R. (2012) Geodetic Strain Rate and Eartquake Size: New Clues for Seismic Hazard Studies. Physics of the Earth and Planetary Interiors, 206-207, 67-75.http://dx.doi.org/10.1016/j.pepi.2012.07.005 [Paper reference:1]
12. Scholz, C.H. (2002) The Mechanics of Earthquakes and Faulting. 2nd Edition, Cambridge University Press, Cambridge, UK.http://dx.doi.org/10.1017/cbo9780511818516 [Paper reference:1]
13. Abercrombie, R.E. and Mori, J. (1996) Occurrence Patterns of Foreshocks to Large Earthquakes in the Western United States. Nature, 381, 303-307.http://dx.doi.org/10.1038/381303a0 [Paper reference:1]
14. Global CMT Catalog Search. http://www.globalcmt.org/CMTsearch.html [Paper reference:1]
15. Sibson, R.H. (1974) Frictional Constraints on Thrust, Wrench and Normal Faults. Nature, 249, 542-544. http://dx.doi.org/10.1038/249542a0 [Paper reference:1]
16. Lin, C.-H. (2004) Repeated Foreshock Sequences in the Thrust Faulting Environment of Eastern Taiwan. Geophysical Research Letters, 31, L13601.http://dx.doi.org/10.1029/2004GL019833 [Paper reference:1]

CHAPTER 8

PREDICTING EARTHQUAKES WITH MICROSEQUENCES AND REVERSED PHASE REPETITIVE PATTERNS

Giulio Riga and Paolo Balocchi

Keywords: Microsequences, Foreshocks, Pattern, Triggers, Earthquake, Mainshock

INTRODUCTION

In any earthquake prediction study, it is essential to identify the precursory phenomena within the seismic data to be analyzed [1]. To this end, it is necessary to analyze a solid scientific database to identify and classify the premonitory models that precede big seismic events, whose nature and shape, variable from one event to another, are probably controlled by the tectonic environment (geometries of the fault, strain rate) and the fault plan heterogeneity [2]. The seismic sequence of a given area includes all the information needed to understand its evolution in space and time [3]; through an in-depth analysis of its inner structure it is possible to draw conclusions related to strong events prediction. In fact, the long-term sequence includes smaller-scale, medium-term sequences, which, in turn, contain short- and imminent-term sequences that influence the sequence evolution in relation to development speed and magnitude (space-time evolution). The analysis performed on several seismic sequences has identified in short and imminent-term time windows, particular groupings of shocks (microsequences and repetitive patterns), which anticipate a phase inversion [4]. Usually, micro-sequences and repetitive patterns are formed before the beginning/end of the energy release energy, since the distribution of shocks over time is not random but follows rules that allow us to understand the development level achieved by the sequence and to make more reliable predictions. The analysis of the seismic sequence structure is performed using: a) a spatial coverage of variable datasets ranging from regional scale to international scale (or global, which covers an area ranging from few tens of km2 to thousands of km2); b) a range of magnitude values of 2.0 - 10 M; (c) a range of depths of 1 - 50 km.

METHODOLOGY

Microsequences TT-7S and DB-3SE

We all are aware of the relative importance of the small and large earthquakes to change static stresses and trigger earthquakes. The small shocks observed before strong earthquakes, suggest that they have specific properties that can be used as of earthquakes precursors. Even though big earthquakes are unquestionably more important compared to smaller ones in relation to the energy release, small shocks collectively have the same influence [5]. Microsequences TT-7S (Triple Maximum—Seven Shocks) and DB-3SE (Double Minimum—Three small shocks) are graphic patterns that develop inside the seismic sequence and allow us to locate the point of closure/beginning of an energy release phase. In microsequence TT-7S (Figure 1), we identify a peculiar trend, which involves the magnitude values fluctuations that is completed during the energy release phase: it develops a first maximum (point 2) followed by a minimum value (point 3). Hereinafter, the magnitude values begin to rise again (point 4) reaching higher values than the previous maximum, and a second minimum (point 5), which generally has greater value compared to the previous one. The next step is characterized by a third peak (point 6) lower than the second maximum, followed by the last minimum (point 7) that is placed below the "transition line" [6] joining the minimum points 3 and 5, respectively. This completes the microsequence TT-7S [7]. Overall, the microsequenceTT-7S appears as a strong fluctuation of the

Citation: Riga G. and Balocchi P., Predict Earthquakes with Micro-sequences and Reversed Phase Repetitive Patterns, http://dx.doi.org/10.4236/ojer.2016.53012.

magnitude values that occur and are completed during the energy release phase. In Figure 2 the green circles indicate the fourth shock in the microsequence whose magnitude M(4) must comply with the following condition:

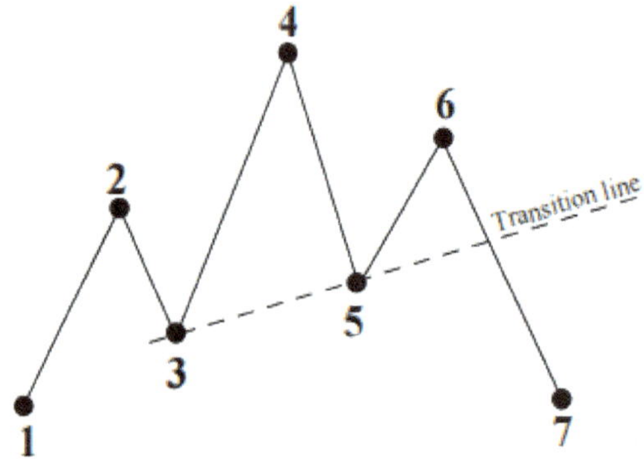

Figure 1. Microsequence TT-7S pattern.

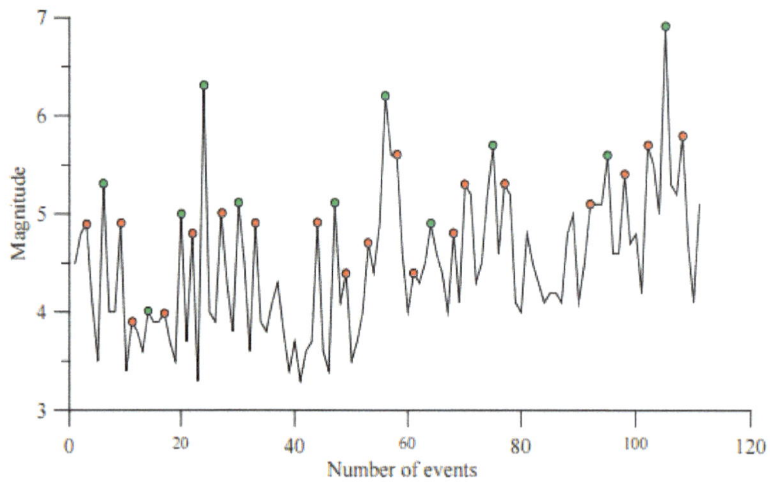

Figure 2. Microsequences TT-7S fluctuations.

$$M(2) \leq M(4) \geq M(6) \quad (1)$$

While the red circles indicate the points 2 and 6, respectively. The microsequence BD-3SE [4] show more reliable features for imminent-term predictions. It is characterized by three small shocks fluctuations within two shocks of greater magnitude (Figure 3), where their completion establishes both the reversal of the energy accumulation phase and a critical condition, in which the successive small shocks trigger either average magnitude or catastrophic events. The microsequence's characteristic lies in the presence of a double minimum (points 1 and 3), and can be either symmetric, if the two minimums have the same magnitude value (Figure 3(a)) or asymmetrical, if the two minimums values are different (Figures 3(b)-(e)). The microsequence BD-3SE is identifiable on charts concerning both the daily/monthly magnitude values trends or the progressive number of shocks, and generally they are completed after a very short period. Conversely, the release phase is active for one to eight days/months or eight events (in relation to the sequence's development time) after the trigger point, which is represented by the second minimum of the microsequence DB-3SE (red triangle). After the trigger point, it is possible that the magnitude values progressively increase up to the final target, thus generating, a progressive earthquake- or flash-earthquake type energy release phase [4]. The procedure for identifying the microsequence DB-3SE is the following: 1) Identify the absolute maximum and relative values P(n), which represent the extreme points (Figure 4); 2) Identify the extreme points $P_1(n)$ (second shock in the microsequence DB-3SE) whose magnitude $M_1(n)$ must comply with the following conditions:

$$M_1(n-1) \geq M_1(n) < M_1(n+1)$$
$$M_1(n-1) > M_1(n) \leq M_1(n+1) \quad (2), (3)$$

where $M_1(n-1)$, $M_1(n+1)$ are the magnitude values of the extreme point that precedes and follows the point $P_1(n)$, respectively (Figure 5). 3) Draw the line joining the second shock in the microsequence DB-3SE (blue line) (Figure 6); 4) Draw the line joining the fourth shock in the microsequence TT-7S (black line). During a seismic sequence development it is possible to identify the beginning of an energy release phase by assuming the following condition:

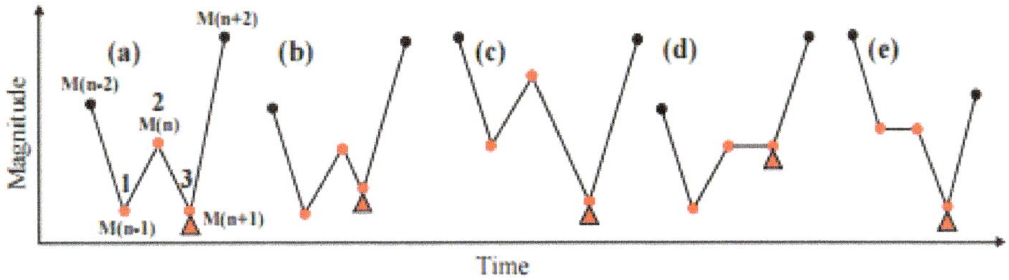

Figure 3. Schematic representation of microsequence DB-3SE. The red circles indicate the microsequence, while the red triangle indicates the trigger point of the energy release phase. The black circles represent the microsequence's starting and completion shocks.

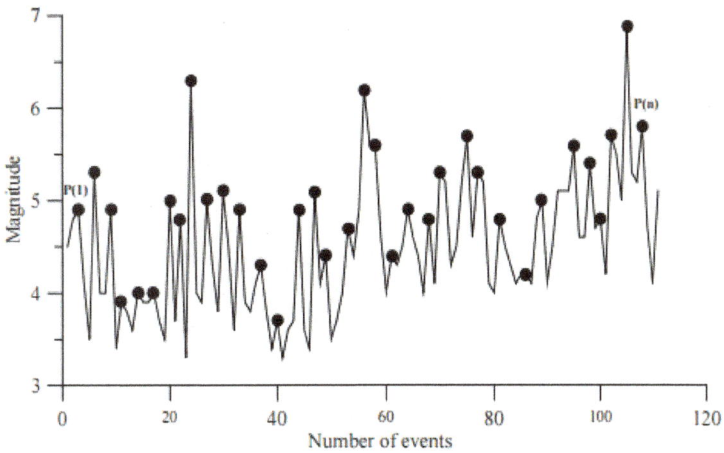

Figure 4. Identification of absolute and relative maximum values (black circles).

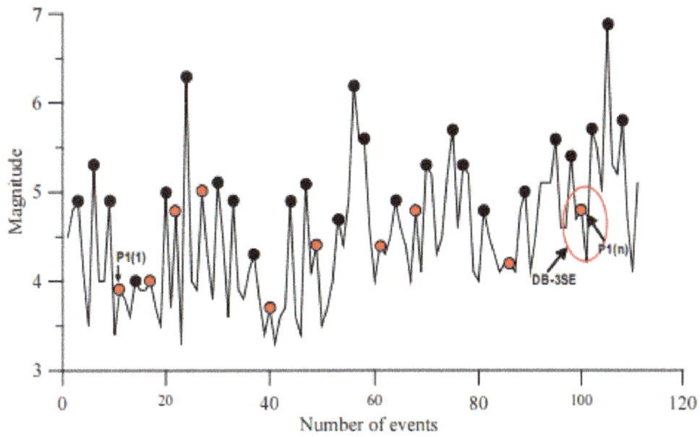

Figure 5. Procedure apt to identify the microsequences DB-3SE. In the red ellipse an example of microsequence DB-3SE is shown. The red circle indicates the second shock in the microsequence. The black circles indicate the extreme points (absolute and relative maximum values)—Earthquake occurred in China on 13 April 2010.

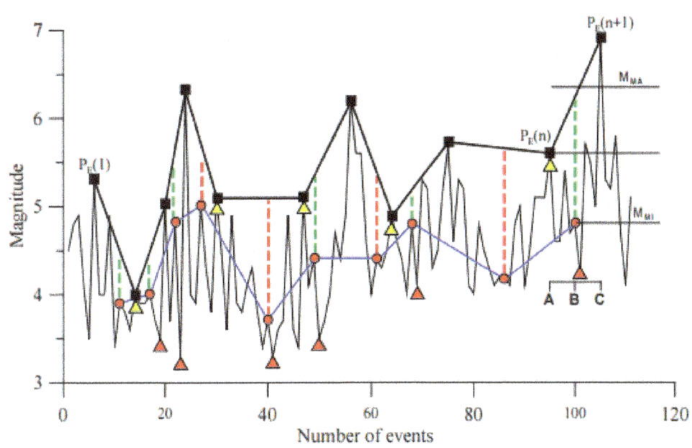

Figure 6. Definition of energy accumulation and release phases. The yellow triangles represent the first-order warning signals, while the red triangles represent the second and third order ones; graphical identification of minimum (M_{MI}) and maximum magnitude (M_{MA}).

If $M_E(n) \geq M_E(n-1)$ is the beginning of an energy release phase (yellow triangles). Where $M_E(n)$ is the magnitude of the fourth shock $P_E(n)$ in the microsequence TT-7S that must be confirmed by the following condition as well:

$$AB < BC$$

where AB is the temporal distance or number of shocks between the point $P_E(n)$ (yellow triangle) and the second shock in the microsequence DB-3SE (red circle), while the segment BC represents the temporal distance or the number of shocks between the microsequence DB-3SE and the last recorded event during the current phase. The AB > BC condition indicates an ongoing energy accumulation phase, while the AB = BC condition is regarded as the "watershed" between the energy accumulation and release phases. In general, the magnitude value of the second shock in the micorsequence DB-3SE corresponds to the minimum expected magnitude (M_{MI}) during the energy release phase, while an indicative value of the maximum expected magnitude (M_{MA}) shall be calculated by adding the difference between the magnitude of the latter point and the minimum expected magnitude value (M_{MI}) to the magnitude value of the fourth shock in the microsequence TT-7S (point $P_E(n)$). In order to obtain a more accurate method, however, in some cases it is necessary to use other types of signals to confirm the ongoing phase. For example, a clear signal of phase inversion is given by the fluctuations of the red line (Figure 7) that joins the extreme points of the seismic sequence to the second shock in the microsequence DB-3SE. In the lower part of the graph, the amplitudes of the imminent-short term energy accumulation and release phases are indicated by red and green horizontal bars. From the graph it is possible to infer the following information: 1) The microsequence DB-3SE indicates the end of the energy accumulation phase and the beginning of the energy release phase (point A). 2) Point B represents an imminent-short term foreshock and the first maximum in microsequence TT-7S. 3) The point C represents the mainshock or a stronger shock and the second maximum in microsequence TT-7S. 4) The point D represents the first shock of the energy accumulation (aftershock) and also the third maximum in microsequence TT-7S. The seismic sequence development takes place according to an evolution that follows this pattern: after the

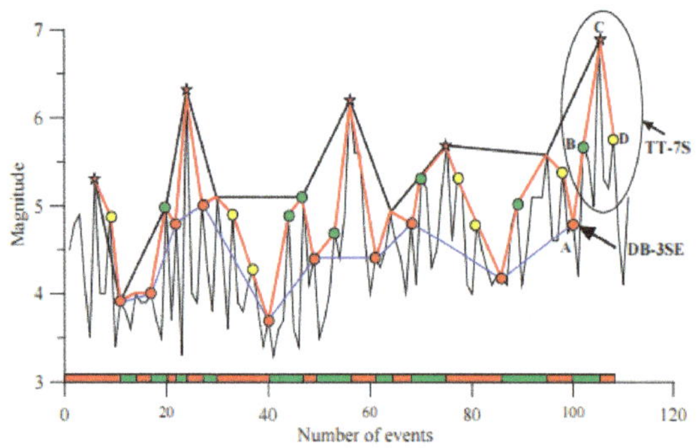

Figure 7. Alternating accumulation and release of energy phases.

microsequence DB-3SE formation (point A), a single shock of greater magnitude than the second shock in microsequence DB-3SE or more shocks whose magnitude increases over time (points B and C) may occur. In point C the closure of the energy release p occurs if the third maximum in the microstructure TT-7S is formed (point D), which may coincide with a second mainshock or with an imminent-term aftershock (yellow circle), while the imminent-term energy accumulation phase ends with the formation of a microsequence DB-3SE

Warning Signs

A traditional use of the microsequence DB-3SE involves the triggering of a first alert in correspondence of the occurrence of the third shock it consists of. However, this method does not allow to know the extension of the energy release phase. A more reliable method is to associate alerts in increasing order with the microsequence DB-3SE and with the shocks that follow (Figure 8). For example, the third shock in microsequence DB-3SE that precedes an extreme point PE(n) of magnitude $M_E(n) \leq M_E(n-1)$ is associated with a first-order alert (green triangle), while the extreme point PE(n) is associated with second order alert (yellow color). The imminent-term, higher order signals (red triangles) are associated with microsequences DB-3SE that forms after the second order signal. The graph clearly shows how the warning signs occur following a repetitive 1-2-3 or 1-2-3-4 order that allows to follow the sequence development during the energy release phase and to establish the levels of danger that grow over time. It is possible to use a different system to define the various order warning signs, using the time/number distance of shocks between the extreme point PE(n) of magnitude $M_E(n) \leq M_E(n-1)$ and the second shock in microsequence DB-3SE. The extreme point PE(n) constitutes a first order warning sign, the third shock in microsequence DB-3SE is a second-order warning sign, while the point C, placed at a distance from the point B equal to AB, represents a third order alert (Figure 9).

Repetitive Patterns

A strong earthquake is preceded by shocks of different magnitude organized in repetitive patterns that anticipate the energy release phase, whose identification and understanding constitute a sound method to make short and medium term predictions.

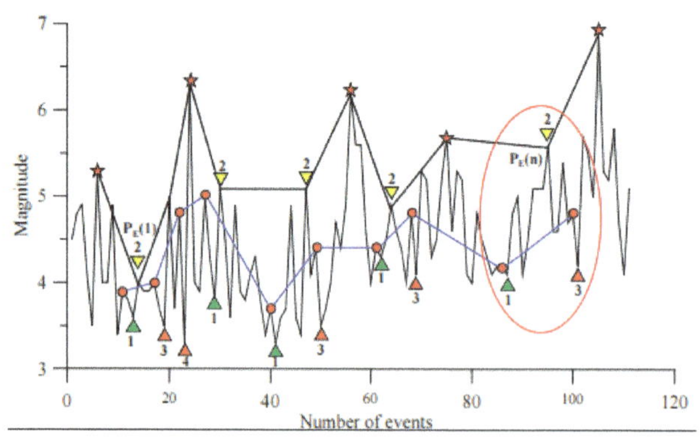

Figure 8. Various orders warning signals. The red star indicate the strongest earthquakes.

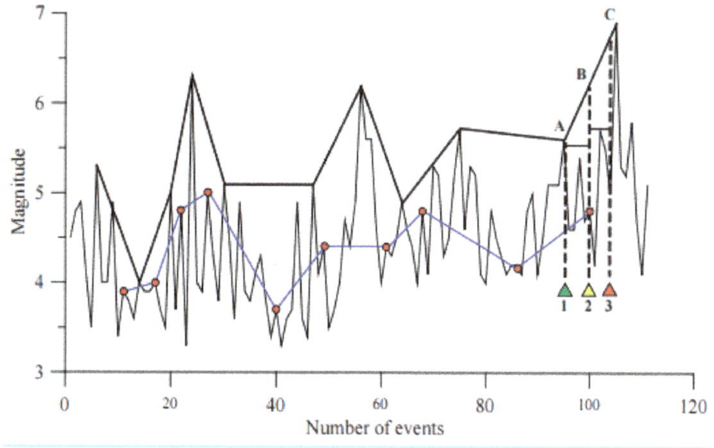

Figure 9. Various order warning signs.

The method for identifying the repetitive patterns consists in drawing the line that joins the fourth shock in microsequence TT-7S (magenta-colored upper line) and the line joining the low extreme points (magenta-colored bottom line) whose magnitude values must comply with the following conditions (Figure 10):

$$M_I(n-1) \geq M_I(n) < M_I(n+1) \tag{4}$$

$$M_I(n-1) > M_I(n) \leq M_I(n+1) \tag{5}$$

where M_I (n − 1) and M_I (n + 1) are the magnitude values of the low extreme points that precede and follow the point $P_I(n)$ of magnitudo $M_I(n)$. The two lines constitute the fluctuation band of the extreme points values.

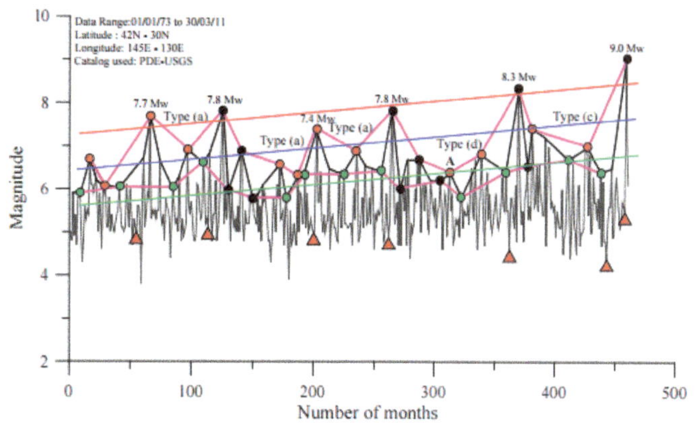

Figure 10. Procedure to identify the repetitive patterns. We observe some structures of repetitive patterns based on the type shown in Figure 11. The red triangles indicate the third-order warning signs generated by the microsequence DB-3SE.

The distinctive feature of the band is that the spaces between the top and the bottom line will vary as a function of the ongoing phase. Basically, the fluctuation band provides information on the ongoing phase in the seismic sequence: if the sequence is currently in an energy release phase then the band undergoes a progressive widening, while if it is in an accumulation phase, the band undergoes a shrinkage. In particular, during the activation step of the energy release phase the band shrinks further to contain the small shock. The flattening of the fluctuation band usually means that a trigger point of the energy release phase is about to form. If the magnitude values begin to exceed the interpolation line of the extreme points values (solid blue line), it means that the motion will continue usually upwards (with increasing magnitude values), while a downwards crossing will indicate the presence of an energy accumulation phase. By default, after a strong earthquake the magnitude values always tend to return below the straight interpolation line. The band upper line and the straight interpolation line movements provide some information on the magnitude values of the strongest earthquakes. The method adopted involves the drawing of the parallel line to the interpolation straight line (green line) from the minimum point of the oscillation band upper line (point A) and the projection of a second parallel (line red) on it, whose amplitude is equal to that between the interpolation straight line and the lower parallel line. The magnitude values supplied by the second parallel line over time may be assumed as the strongest earthquakes' expected target. Figure 11 shows the five repetitive patterns consisting of four shocks (two maximum and two minimum) generated by the oscillation band of the extreme points. Models a-b-c will form at the end of a energy accumulation cycle, while models d-e during the energy release cycle (they are the most important, because the shock S3 is a foreshock). In addition, the pattern are followed by warning signals generated by the microsequence DB-3SE that anticipate the strongest shocks.

RESULTS

In Figure 12 shows the results obtained from the application of the methodology explained above to the seismic sequence of the earthquake occurred in 1989 in Loma Prieta.

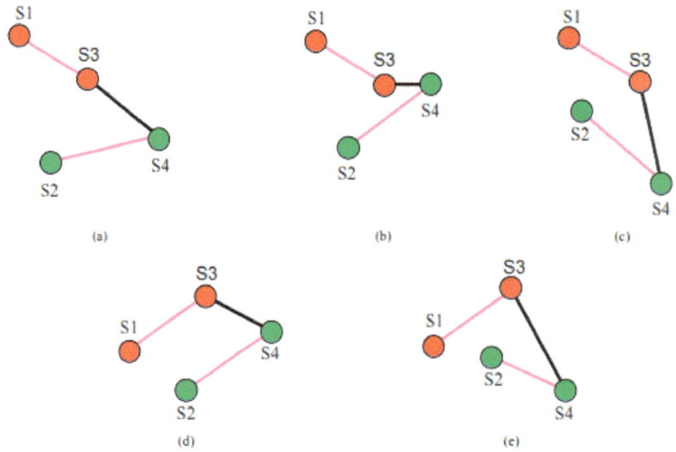

Figure 11. The repetitive phase inversion patterns.

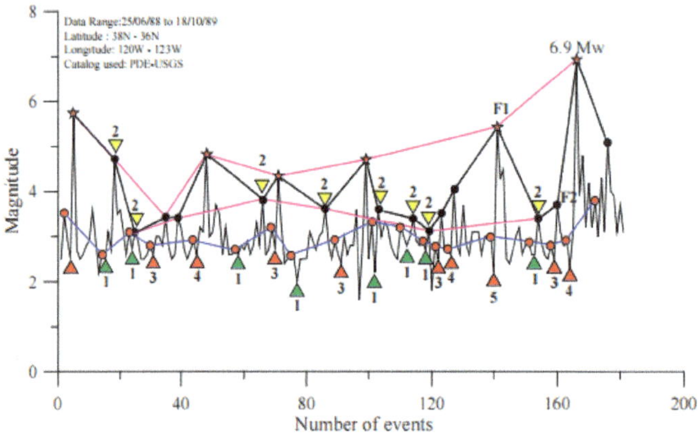

Figure 12. Warning signs before the earthquake in Loma Prieta, 1989.

The main shock occurred on 18 October 1989 (6.9 Mw) was preceded by a progressive earthquake-type energy release phase [6] characterized by a short-term foreshock of magnitude 5.4 ML recorded on 08 August 1989 (point F_1) and preceded by a fifth-order imminent-term warning sign generated by the microsequence DB-3SE. The subsequent evolution of the seismic recordings over time shows the development of an impending-term accumulation phase consisting of shocks of magnitude less than 3.5 M, some of which are organized in microsequences DB-3SE that have generated various-order warning signs.

In particular, we note the formation of a first-order signal (point 1) generated by the third shock in microsequence DB-3SE, which has formed after the first foreshock (point F1), followed by the second-order signal generated by the extreme point (point 2) and by subsequent points (points 3 and 4). The latter preceded respectively the second foreshock (point F2) and the mainshock with magnitude of 6.9 Mw, generated by the third shock in microsequence DB-3SE. In addition, prior to the energy release phase, the oscillation band shrinks, while after the first foreshock it expands. Even in this case, the oscillation band formed by the extreme points lines (black line) and by the second shocks in microsequence DB-3SE (blue line) can be used to identify phase inversion patterns (Figure 13). In fact, this oscillation band is an excellent oscillator to understand when the seismic sequence reaches a critical stage where small shocks can trigger shocks of greater magnitude. Essentially, when the band narrows, we reach the terminal step of the energy accumulation phase: this may suggests that in the short term the magnitude values could begin to grow. The band provides also significant warning signs (you should always bear in mind that this is a predictive analysis, i.e. it tries to predict future events and therefore not yet occurred) generated by the intersection between the descending sections extensions of the line joining the extreme points and the ascending sections extensions of the line joining the second shock in the microsequence DB-3SE (red triangle). The magnitude minimum value (M_{MI}) expected in the energy release phase and associated with the warning sign, corresponds to the magnitude value of the extreme point that precedes the signal (green circle). In order to further reduce the errors related to the interpretation of the signals which inevitably the procedure can generate, we recommend the adoption of one or more filters as the simplified Aroon oscillator [4], with the purpose of identifying the energy accumulation areas in relation to the data provided by the oscillation band. Starting from the previous example, in the graph shown in Figure 14 various order warning signs are reported (green, yellow and red triangles), which were identified through the distance between the extreme point (black circle) and the second shock in microsequence DB-3SE (red circle). The first order warning sign coincides with the extreme point. The main shock has been triggered by an asymmetric microsequence DB-3SE (second order signal), while the third-order signal, placed at a distance BC = AB (point C) has formed a few seconds before the mainshock. The graph displayed in Figure 15 shows some details of the microsequences TT-7S that have formed during the energy release phase and after the most hazardous seismic events. In particular, it is possible to observe two imminent-term microsequences TT-7S in which the second maximum

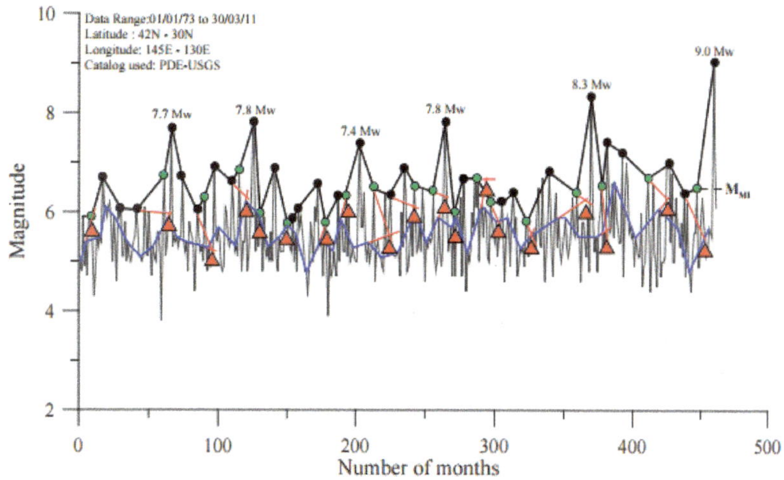

Figure 13. Warning signs (red triangles generated by the oscillation band of the extreme points-microsequence DB-3SE.

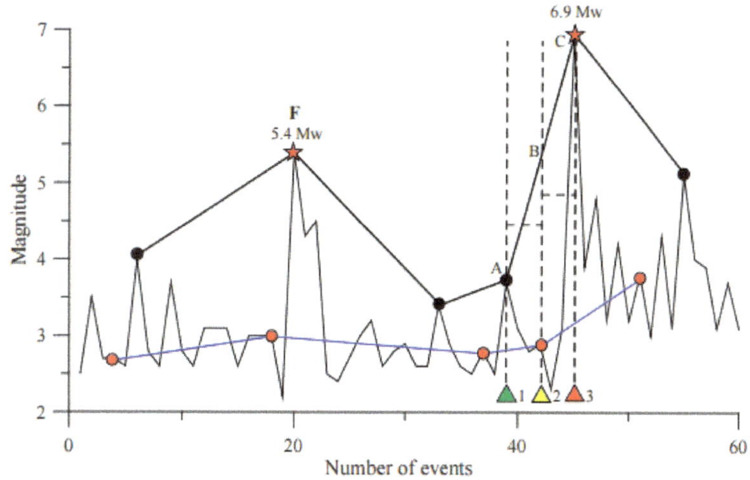

Figure 14. Warning signs before the earthquake in Loma Prieta, 1989. The red star indicates the mainshock, the letter "F" the foreshock.

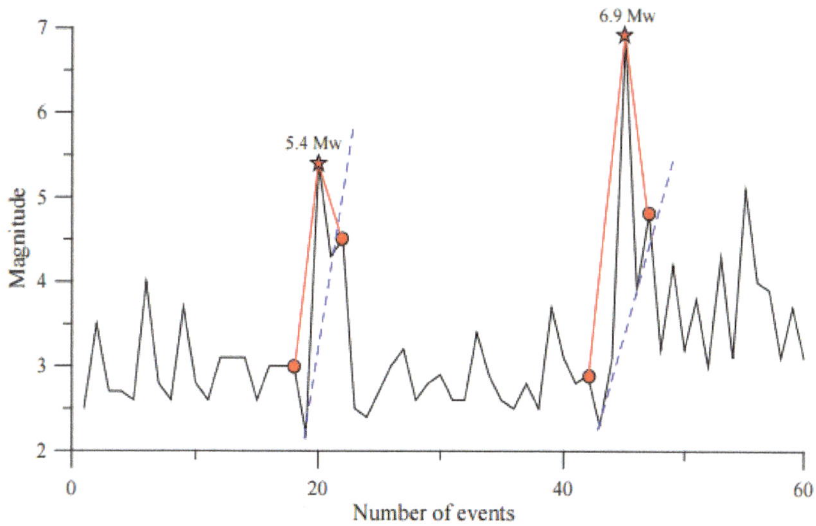

Figure 15. Microsequence TT-7S in energy release phase. The red symbols indicate the three maximum values in the microsequence.

coincides with the most energetic shocks (foreshock and mainshock), while the third maximum, which closes the energy release phase, is represented by an aftershock. The transition from the energy release to the energy accumulation phase is also confirmed by the magnitude values decrease below the "transition line" (dashed blue line) [6].

CONCLUSIONS

A seismic sequence that occurs in a given area consists of microsequences and repetitive patterns formed by a grouping of seismic events with particular features. Their analysis allows studying the short-term evolution of a seismic sequence by identifying the warning signals that precede a strong earthquake. In particular, the microsequence TT-7S allows us to locate the closure point of an energy release phase and identify a second order warning sign. The microsequence DB-3SE, identifies the beginning of the energy release phase and explains how from low energy earthquakes moderate to high magnitude events can result, whose occurrence time can be known with sufficient advance. The band allows us to understand when the seismic sequence is located within an energy release or accumulation phase; therefore if it reaches a critical stage where small shocks can trigger shocks of greater magnitude. The aforementioned method, based on the microsequence analysis and on repetitive patterns can be easily applied to the study concerning earthquakes forecasts both at regional and global level. The method is based on a simple graphical method that can be used through the data of different seismicity catalogues.

REFERENCES

1. Wakita, H., Nakamura, Y. and Sano, Y. (2013) Case 3: Short-Term and Intermediate-Term Geochemical Precursors. In: Wyss, M., Ed., Evaluation of Proposed Earthquake Precursors, American Geophysical Union, 15-20. http://onlinelibrary.wiley.com/doi/10.1029/SP032p0015/summary
2. Kanamori, H. (1981) The Nature of Seismicity Patterns before Large Earthquakes. In: Simpson, D.W. and Richards, P.G., Eds., Earthquake Prediction, American Geophysical Union, Washington, 1-19. http://onlinelibrary.wiley.com/doi/10.1029/ME004p0001/summary http://dx.doi.org/10.1029/me004p0001
3. Mantovani, E., Babbucci, D., Tamburelli, C. and Viti, M. (2009) A Review on the Driving Mechanism of the Tyrrhenian-Apennines System: Implications for the Present Seismotectonic Setting in the Central-Northern Apennines. Tectonophysics, 476, 22-40. http://dx.doi.org/10.1016/j.tecto.2008.10.032
4. Riga, G. and Balocchi, P. (2016) Short-Term Earthquake Forecast with the Seismic Sequence Hierarchization Method. Open Journal of Earthquake Research, 5, 79-96. http://www.scirp.org/Journal/PaperInformation.aspx?PaperID=66332 http://dx.doi.org/10.4236/ojer.2016.52006
5. Helmstetter, A., Kagan, Y.Y. and Jackson, D.D. (2005) Importance of Small Earthquakes for Stress Transfers and Earthquake Triggering. Journal of Geophysical Research: Solid Earth. http://onlinelibrary.wiley.com/doi/10.1029/2004JB003286/full
6. Riga, G. and Balocchi, P. (2016) Seismic Sequence Structure and Earthquakes Triggering Patterns. Open Journal of Earthquake Research, 5, 20-34. http://www.scirp.org/Journal/PaperInformation.aspx?PaperID=63538 http://dx.doi.org/10.4236/ojer.2016.51003
7. Bulkowski, T. N. (2002) Trading Classic Chart Patterns. John Wiley & Sons, Inc. http://fx-arabia.com/vb/uploaded/178_01275676445.pdf

CHAPTER 9

APPLICATION OF COMMENSURABILITY IN EARTHQUAKE PREDICTION

Youjin Su and Hui Hu

Keywords: Commensurability, Orders in the Natural World, Earthquake Prediction

INTRODUCTION

Earthquake (EQ) research, especially EQ prediction, is accepted as a worldwide difficult scientific problem [1]. People have made arduous and unremitting efforts for searching for the effective methods of EQ prediction for a very long time. However, most of the basic concepts and systems of methods do not break away from the system of quantitative analysis of the inertial system so that the problem of prediction of serious natural disasters like EQs has not been solved. A Chinese famous geophysicist, Weng Wenbo (1912-1994) studies the Titius-Bode law [2] and firstly pointed out that the occurrence time of EQ has the commensurability and applied the principle of the commensurability which was firstly proposed in astronomy to the prediction of major natural disasters such as earthquakes, droughts and water-loggings, etc., thereby developed it into the theory of prediction which have been widely used in the prediction of disasters [3] - [7]. Based on this theory, the occurrence of a few EQs, droughts and water-loggings, and other natural disasters were successfully predicted.

COMMENSURABILITY AND ITS EXPANSION

Commensurability of the word was first proposed by the German famous astronomer Titius who found and proposed it in the study of the average distance between the planets within the solar system to the sun. Later, another famous German astronomer Bode made further studies. This is the famous Titius-Bode Law. From Titius- Bode law the distance of the planet n from the sun can be expressed as:

$$a_n = 0.4 + 0.32 \times 2^{n-2}, \qquad (1)$$

It also can be written as the following form:

$$\beta = \frac{a_{n+1}}{a_n}, \qquad (2)$$

where a_n is the distance of the planet n to the sun, reckoned in astronomical unit, and n is the number of the planets away from the sun to the far side. For Mercury, the number n is not 1, instead it is taken as $-\infty$. β is the commensurable value for the planets in the solar system [8].

Just as the famous geophysicist Weng Wenbo [9] pointed out that the commensurability is one of the orders in the natural world. The Equation (2) itself brings to light the distribution law of the matter in a space region, and for time domain the commensurability can be expressed as:

$$\Delta X = \frac{X_{i+\Delta i} - X_i}{K} \qquad (3)$$

Here k is an integer. If the above relation is tenable, then the data set $\{X_i\}$ is commensurable. ΔX is the commensurable value of the data set $\{X_i\}$ and X_i and $X_{i+\Delta i} \in \{X_i\}$. If k is equal to 1, then ΔX is the period of $\{X_i\}$ [10].

PREDICTION PRACTICE

Forecast by Weng Wenbo on the EQ Ms7.4 in California of United States on June 28, 1992

Before 1992 Weng Wenbo made a few forecasts of EQs occurring in United States, most of these forecasts corresponded quite well with the later occurred actual events

Citation: Su Y. and Hu H. (2015) Application of Commensurability in Earthquake Prediction. International Journal of Geosciences, 6, 619-624. doi: 10.4236/ijg.2015.66049.

[11] . As example, on January 17, 1992, according to a request from Prof. Cecil H. Green, Director of American Geophysical Unit, and upon obtaining ratification from senior authorities, Weng Wenbo presented his forecast report to Prof. Green [12] ;

Time: June 19 1992

Magnitude: 6.8

Place: Broad region of San Francisco

The predicted EQ occurred on June 28, 1992. Its magnitude Ms7.4, was only Ms0.6 greater in magnitude, and the occurrence date was only 9 days later than the date stated by the forecast report of Weng Wenbo [13] .

Prediction on the Wenchuan EQ 8.0 and Lushan EQ 7.0 in China and the Iquique EQ 8.2 in Chile of 2014

The Wenchuan EQ Ms8.0 occurred in Sichuan province of China on May 12, 2008. It caused greatest heavy losses of the life and property in China in recent history. Before the EQ, Long Xiaoxia et al. based on the commensurability using the following commensurable equations with five elements rather precisely predicted that a strong EQ might occur in 2008 [14] :

$$X_{23} + X_{15} + X_1 - X_3 - X_7 = 2008$$
$$X_{23} + X_{20} + X_{12} - X_{18} - X_7 = 2008$$
$$X_{24} + X_{21} + X_8 - X_{18} - X_6 = 2008$$
$$X_{25} + X_{15} + X_7 - X_{19} - X_3 = 2008$$
$$X_{22} + X_{15} + X_5 - X_{13} - X_3 = 2008$$
$$X_{21} + X_{11} + X_9 - X_{13} - X_1 = 2008 \quad (4)$$

The EQ catalogue used by Long Xiaoxia et al. is given in Table 1. From Table 1 the commensurable value of the EQs occurrence is equal to 2.44 years. Taking k = 5 we obtain:

$$2008\text{-}05\text{-}12 = 1996.09 + 2.44 \times 5 = 2008^y 4^m 15^d + 27 \text{ days} \quad (5)$$

Its error which is the opposite of the occurrence date of the earthquake is 27 days. Its relative error is 0.03.

An EQ 7.0 occurred in Lushan of Sichuan province, China on 2013-04-20. Similar, on the basis of the Table 1 we obtain:

$$2013\text{-}04\text{-}20 = 2008.36 + 2.44 \times 2 = 2013^y 3^m 29^d + 22 \text{ days} \quad (6)$$

It occurred just at the commensurable point equal to 2 times of its time axis. Its absolute error is 22 days. Its relative error also is 0.03.

An EQ 8.2 in Iquique, northern Chile, occurred on 2014-04-01. By the research of Engdahl and Villasenor [15] the catalogues of EQs of $M \geq 7.0$ have been complete and reliable since 1900. Therefore we have analyzed the EQs of $M \geq 7.0$ in Chile since 1900.0, and found its commensurable value is 0.59 years (see Table 2). So,

$$2004\text{-}04\text{-}01 = 2010.15 + 0.59 \times 7 = 2014^y 04^m 12^d - 11 \text{ days} \quad (7)$$

It occurred just at the commensurable point equal to 7 times of its time axis. Its absolute error is 11 days. Its relative error is 0.05.

On the basis of many years' research for commensurability according to the commensurable principle, we compiled a Fortran program for commensurable analyses of the data. By means of the Fortran program our analyzed results have been all given in the tables. The two tables all are directly output results by computer.

Table 1. Commensurability of earthquakes in Sichuan-Yunnan region since 1900.0.

Earthquakes date $X_i - X_{i-1}$				$K\Delta X$ $X_i - K\Delta X$		
No.	YMD	(year)	(year)	K	(year)	(year)
1	19131221	1913.97				
2	19170731	1917.57	3.60	1	2.44	1.16
3	19230324	1923.22	5.65	2	4.88	0.77
4	19250316	1925.20	1.98	1	2.44	−0.46
5	19330825	1933.64	8.44	3	7.32	1.12
6	19360427	1936.32	2.68	1	2.44	0.24
7	19410516	1941.37	5.05	2	4.88	0.17
8	19420201	1942.08	0.71	0	0.00	0.71
9	19480525	1948.40	6.32	3	7.32	−1.00
10	19500203	1950.09	1.69	1	2.44	−0.75
11	19520930	1952.75	2.66	1	2.44	0.22
12	19550414	1955.28	2.53	1	2.44	0.09
13	19601109	1960.85	5.57	2	4.88	0.69
14	19670830	1967.66	6.81	3	7.32	−0.51

15	19700105	1970.01	2.35	1	2.44	−0.09
16	19710428	1971.32	1.31	1	2.44	−1.13
17	19730206	1973.09	1.77	1	2.44	−0.67
18	19740511	1974.35	1.26	1	2.44	−1.18
19	19760823	1976.64	2.29	1	2.44	−0.15
20	19790315	1979.20	2.56	1	2.44	0.12
21	19810124	1981.06	1.86	1	2.44	−0.58
22	19881106	1988.84	7.78	3	7.32	0.46
23	19890425	1989.31	0.47	0	0.00	0.47
24	19950712	1995.52	6.21	3	7.32	−1.11
25	19960203	1996.09	0.57	0	0.00	0.57
26	20080512	2008.36	12.27	5	12.20	0.07
Commensurable value			2.44			
Mean			−0.031			
Standard deviation (σ_{n-1})			0.717			

Table 2. Commensurability of EQs in Chile since 1900.0.

Earthquakes date $X_i - X_{i-1}$				KΔX X_i − KΔX		
No.	YMD	(year)	(year)	K	(year)	(year)
1	19040319	1904.21				
2	19060817	1906.62	2.41	4.	2.36	0.05
3	19060830	1906.66	0.04	0.	0.00	0.04
4	19061226	1906.98	0.32	1.	0.59	−0.27
5	19090608	1909.43	2.45	4.	2.36	0.09
6	19100906	1910.67	1.24	2.	1.18	0.06
7	19110915	1911.70	1.03	2.	1.18	−0.15
8	19140130	1914.07	2.37	4.	2.36	0.01
9	19180520	1918.38	4.31	7.	4.13	0.18
10	19181204	1918.92	0.54	1.	0.59	−0.05
11	19221107	1922.85	3.93	7.	4.13	−0.20
12	19221111	1922.86	0.01	0.	0.00	0.01
13	19250515	1925.36	2.50	4.	2.36	0.14
14	19270414	1927.28	1.92	3.	1.77	0.15
15	19281201	1928.92	1.64	3.	1.77	−0.13
16	19310318	1931.20	2.28	4.	2.36	−0.08
17	19330223	1933.14	1.94	3.	1.77	0.17
18	19360713	1936.53	3.39	6.	3.54	−0.15
19	19390125	1939.06	2.53	4.	2.36	0.17
20	19390418	1939.29	0.23	0.	0.00	0.23
21	19401004	1940.76	1.4	2.	1.18	0.29
22	19420708	1942.51	1.75	3.	1.77	−0.02
23	19430314	1943.20	0.69	1.	0.59	0.10
24	19430406	1943.26	0.06	0.	0.00	0.06
25	19450913	1945.70	2.44	4.	2.36	0.08
26	19460802	1946.58	0.88	1.	0.59	0.29
27	19490420	1949.29	2.71	5.	2.95	−0.24
28	19490425	1949.31	0.02	0.	0.00	0.02
29	19530506	1953.34	4.03	7.	4.13	−0.10

30	19600521	1960.38	7.04	12.	7.08	−0.04
31	19600522	1960.39	0.01	0.	0.00	0.01
32	19600522	1960.39	0.00	0.	0.00	0.00
33	19600620	1960.46	0.07	0.	0.00	0.07
34	19601202	1960.92	0.46	1.	0.59	−0.13
35	19620214	1962.12	1.20	2.	1.18	0.02
36	19650328	1965.23	3.11	5.	2.95	0.16
37	19661228	1966.98	1.75	3.	1.77	−0.02
38	19671221	1967.97	0.99	2.	1.18	−0.19
39	19710709	1971.51	3.54	6.	3.54	0.00
40	19740818	1974.62	3.11	5.	2.95	0.16
41	19750510	1975.35	0.73	1.	0.59	0.14
42	19811016	1981.79	6.44	11.	6.49	−0.05
43	19811107	1981.85	0.06	0.	0.00	0.06
44	19831004	1983.75	1.90	3.	1.77	0.13
45	19850303	1985.16	1.41	2.	1.18	0.23
46	19850304	1985.17	0.01	0.	0.00	0.01
47	19850409	1985.26	0.09	0.	0.00	0.09
48	19870305	1987.17	1.91	3.	1.77	0.14
49	19870305	1987.17	0.00	0.	0.00	0.00
50	19870808	1987.60	0.43	1.	0.59	−0.16
51	19880119	1988.05	0.45	1.	0.59	−0.14
52	19880205	1988.09	0.04	0.	0.00	0.04
53	19950730	1995.57	7.48	13.	7.67	−0.19
54	19971015	1997.78	2.21	4.	2.36	−0.15
55	19980130	1998.07	0.29	0.	0.00	0.29
56	20050613	2005.44	7.37	12.	7.08	0.29
57	20100227	2010.15	4.71	8.	4.72	−0.01
Commensurable value			0.590			
Mean			−0.027			
Standard deviation (σ_{n-1})			0.142			

From Table 1 and Table 2 the results show that the EQs basically all occur at the commensurable point of its time axis, respectively. It also shows that the EQs occurrence is not accidental, and there is its pattern and inevitability, only the commensurable value is different for EQs occurred in different areas.

In the Tables, ΔX is the commensurable value of the studied region. K means that the EQ lies on K times of the commensurable value ΔX behind the latest recent EQ occurred in the area in its time axis. In the Tables, $X_i - X_{i-1}$ is the time interval between major EQs occurring in the same area. For example, EQ No. 3 in Table 1, X_3 = 1923.22, X_2 = 1917.57, the time interval between the 3rd and 2nd EQ is 5.65 year (unit: years). The corresponding K value under EQ No. 3 is $K_3 = 2$, i.e. the 3rd EQ occurred on the second commensurable point in its time axis after the 2nd EQ. According to K_3 and the commensurable value, the predicting point equals to 1917.57 adding the product of 2 and the commensurable value (i.e. 1917.57 + 2ΔX = 1917.57 + 4.88). Its prediction error (i.e. the difference between the predicted point and actual time of occurrence of the third EQ) is 0.77 years.

DISCUSSION AND CONCLUSION

1) The commensurability revealed by Titius-Bode itself brings to light the distribution law of the matter in a space region, and the expanding commensurable theory proposed by Weng Wenbo reveals the time law of the occurrence of the events in a specified space region. It can be seen that the commensurability is present in various natural phenomena and has universality. It is helpful to study the complicated relationships among various matters, and thus merits further in-depth research.

2) The occurrence of the events seems to be a random accident. In fact, that is not the case. It is in the accident that the necessity resides. Therefore, the commensurability can provide a scientific basis for the prediction of events which may occur.

3) After commensurable value can be determined, K values should be used in order of K = 1, 2, 3, · · · , i.e. when K = 1 is used but earthquake does not occur, then use K = 2, · · · . The predicted point extrapolated in the time axis by the commensurable value is only a

necessary condition, and therefore certain false forecasts are also inevitable, because the exact occurrence time is determined by multiple complex factors. In order to obtain precise prediction, this method must be used in collaboration with other relevant methods, taking the approach of comprehensive analysis.

ACKNOWLEDGEMENTS

We extend heartfelt thanks to the referees for their valuable comments. Dr. Chen I-wan, British (ancestor Chinese), Former Advisor of Committee of Natural Hazard Prediction of China Geophysics Society gave us valuable discussion and help. We express our heartfelt thanks to him.

Youjin Su, Hui Hu (2012BAK19B01).

REFERENCES

1. Geller, R.J., Jackson, D.D., Kagan, Y.Y., et al. (1997) Earthquakes Cannot Be Predicted. Science, 275, 1616-1617. http://dx.doi.org/10.1126/science.275.5306.1616
2. Nieto, M.M. (1972) The Titius-Bode Law of Planetary Distances, Its History and Theory. Pergamon Press, New York.
3. Dong, G.S. (1987) Preliminary Study on Commensurability of Strong Earthquakes (M > 8) within the Continent of China. Journal of Seismological Research, 10, 441-446.
4. Men, K.P. (2003) Ordered Network Feature of M > 7 Earthquakes and Its Prediction in Northern Part of Qinghai- Xizang Plateau. Progress in Geophysics, 18, 765-771.
5. Hu, H. and Han, Y.B. (2005) Prediction of the Hualian Earthquakes in Taiwan and an Extended Discussion on the Method of Commensurability. Applied Geophysics, 2, 194-196. http://dx.doi.org/10.1007/s11770-005-0023-7
6. Fan, C.R. and Li, X.B. (2008) Commensurable Network Structural Drawing of Major Drought and Flood in Three Gorges of Changjiang and Bengbu of Huaihe. China Flood & Drought Management, 2, 41-45.
7. Lin, H., Su, Y.J., Han, Y.B. and Hu, H. (2009) Commensurability of the Earthquakes of Kunlun Area in Xinjiang. Journal of Natural disasters, 18, 189-191.
8. Zhang, Y.Z., et al. (1980) Encyclopaedia of China: Astronomy. Encyclopaedia of China Press, Beijing, 375,
9. Weng, W.B. (1981) Commensurability. Acta Geophysical Sinica, 24, 151-154.
10. Weng, W.B. (1984) Fundmentals of Forecasting Theory. Petrochemical Industry Press, Beijing.
11. Xu, D.Y. (1992) The Prediction of California Earthquakes in USA in 1992—A Lecture Given by Prof. Weng Wen-Bo. Acta Geophysica Sinica, 36, 801. (In Chinese)
12. Weng, W.B. (1992) Letter of Green Giving to Weng Wenbo (Personal Communicate of Letter). In: Wang, M.T., et al., Eds., Academician Weng Wenbo and His Forecast for Natural Disaster, Petrochemical Industry Press, Beijing, 235.
13. Green, C.H. (1992) Letter of Weng Wenbo Giving to Green (Personal Communicate of Letter). In: Wang, M.T., et al., Eds., Academician Weng Wenbo and His Forecast for Natural Disaster, Petrochemical Industry Press, Beijing, 233.
14. Long, X.X., Yan, J.P., Sun, H. and Wang, Z.Z. (2006) Study on Earthquake Tendency in Sichuan-Yunnan Region Based on Commensurability. Journal of Atastrophology, 21, 81-84.
15. Engdahl, E.R. and Villasenor, A. (2002) Global Seismicity: 1900-1999. In: Lee, H.K., Kanamori, H., Jennings, P.C., et al., Eds., International Handbook of Earthquake and Engineering Seismology, Part A, Academic Press, Amsterdam.

CHAPTER 10

THE CO-PLANARITY AND SYMMETRY PRINCIPLE OF EARTHQUAKE OCCURRENCE

Zhe Yin

Keywords: Earthquake, Co-planarity, Symmetry

INTRODUCTION

(Heading 1) Research on earthquakes is one of the most important elements about disaster prevention study, because of its devastating and destructive. Especially in recent years, with the rapid development of modern economic activity, destruction to the Earth was continuous. The frequency tends to rise and there are more large earthquakes. Therefore, how to analyze the reason of the earthquake correctly, and predict earthquakes effectively becomes a hot topic in nowadays. Under the assumption that the earth movement tendency is a sphere and mountains fall to the surface of the earth; estrogens and basin rise to the surface of the Earth. This article gives out the co-planarity and symmetry theory of earthquake occurrence according to research about the earthquakes occurrence between 2001 and 2010 which over 6 scales of magnitude.

Questions and Relevant Definitions

The Earth, like a very sensitive being, according to the laws of nature, it has the desire to achieve a balance of features. In this process, it will show a particular trend. As shown in Figure 1, we give some relevant definitions here:

Ecliptic axis: an axis through the center of the Earth which is a vertical line to the ecliptic plane.

Equatorial coordinate system: a celestial coordinate system of the equatorial plane and the axis of rotation. Ecliptic system of coordinates: a celestial coordinate system of the ecliptic plane and ecliptic axis.

Citation: Zhe Yin, The Co-planarity and Symmetry Principle of Earthquake Occurrence, doi:10.4236/ijg.2010.11005.

Tangent plane parallel to the equator: the Earth's equatorial plane parallel to the tangent plane (Tangent plane parallel the equator), shown as TPPTE.

Ecliptic line tangent plane: parallel to the ecliptic of the earth tangent plane (Ecliptic line tangent plane), shown as ELTP.

Equatorial vertical plane: after the Earth's surface

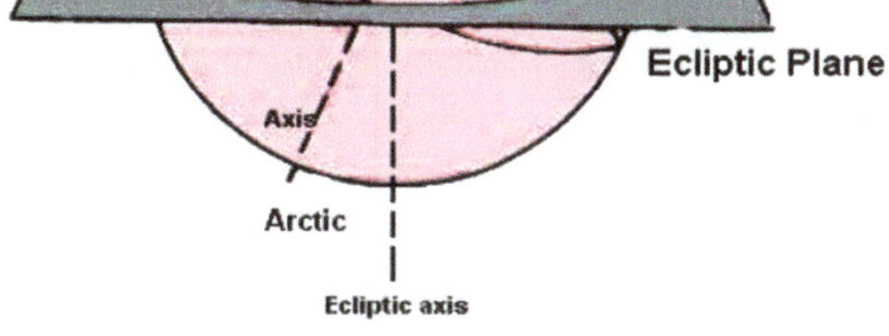

Figure 1. Plane figure.

shock point and the spin axis of the tangent plane (Equatorial vertical plane), shown as EVP.

Longitudinal ecliptic plane: after a shock point and the yellow surface of the Earth axis tangent plane (Ecliptic plane vertical), shown as EPV.

Circular symmetry point: Circular , shown as CSP.

Similar point: Similar Point, shown as SP.

Earthquake effects: measure the Earth's surface; the intensity level of shock is unstable, the stability of the boundary tangent plane level. Plane is the ball away from the more deviation, the more poor results, and the more unstable and more likely point of this earthquake.

BASIC ASSUMPTIONS AND THE CO-PLANARITY AND SYMMETRY PRINCIPLE OF EARTHQUAKE

Basic Assumptions

- The Earth became positive sphere with earthquakes which rise or low the ground. That is, because of the ecliptic
- (23.26 degrees), the Ecliptic Plane TPPTE line tangent planes ELTP have tended to be round;
- The Earth has tended to balance, and because the earthquake, to re-balance;
- Crustal thickness tends to vary. Earth's core and mantle movement space tends to a ball space;
- With the earthquake, the earth rotation axis offset.

When there is a earthquake which makes the low rise, the earth rotation axis moves to the seismic vertical, those makes the high low is opposite.

Co-planarity and Symmetry Principle

After the observation of earthquakes on Earth N (154 earthquakes over 6 scales of magnitude in this article) times, it is occur time H_K, H_{K+1} of adjacent earthquakes M_K, M_{K+1} measure with UTC.

A) The previous earthquake is a cause of the after. The tangent plane balances was destroyed, then tend to a new balance, which Ecliptic line tangent plane generate new state plane all the times.

B) Seismic vertical M_K and seismic vertical are on the same plane at the time or H_K or H_{K+1}. That is parallel to the cut-plane in the equatorial plane or Ecliptic line cut the ecliptic plane, or longitudinal vertical plane or equatorial plane.

C) If the tangent of the plane where seismic vertical M_{K+1} and seismic vertical are tends to round, or fall-rise movement in the earthquake tends to stable, M_K and M_{K+1} tends to symmetry in the circle; otherwise, coplanar circle from the difference of the most unstable point will have an earthquake.

D) After earthquake on M_K occurs, if the most unstable point (poor seismic effect) is still M_K, earthquake M_{K+1} will take place near the same point.

RESEARCH BASIS AND CALCULATION METHOD

Assumptions (1), (2), (3) are prerequisite of conclusion (A), (B), (C) and (D); symmetry points including the spin axis symmetry, yellow axis symmetry and geocentric axial symmetry. With the existence of the ecliptic obliquity (ecliptic plane and the equatorial plane angle 23.26 degrees) and the Ecliptic plane tangent line, parallel to the equatorial plane tangent tends to a circle, the shape of the earth tends to a sphere becomes possible. This can be concluded with (C) and (D). Assumptions (4) decided the longitudinal movement of the after. Assumptions (1) and (3) show the possibility that the earthquake and the earthquake potential basis for the order. Be noted that symmetrical points at any time change the fact that, in the premise revolution, a startled different points of the Ecliptic OK UTC timing of your cutting surface is subject to change, and cut parallel to the equatorial plane remain unchanged. MK+1 MK Earthquake time: the time the earthquake is very important, when the earthquake occurred; it generates and identifies Ecliptic line tangent plane, tangent plane parallel to the equator, the ecliptic plane and the equatorial vertical longitudinal cut tangent plane, as well as the corresponding point of the plane of symmetry at the same time.

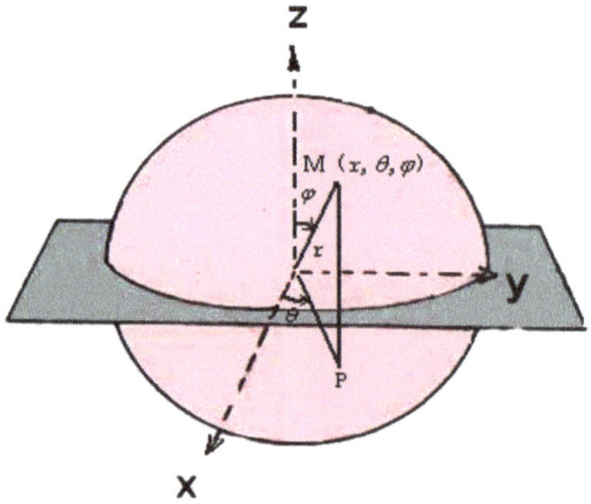

Figure 2. Caculations.

Shock point set M of lat = W, long = J, time (UTC) = a: b, then $\varphi_0 = 90 - W$, $\theta_0 = J + (a + b/60) \times 15$

Step2:

$$\begin{cases} x = \sin\varphi_0 \cos\theta_0 \\ y = \sin\varphi_0 \sin\theta_0 \\ z = \cos\varphi_0 \end{cases}$$

$$0^\circ \le \theta_0 \le 360^\circ, 0^\circ \le \varphi_0 \le 180^\circ$$

Step3: To the counterclockwise rotation 23°26', then

$$\begin{cases} x' = x\cos 23^\circ 26' + z\sin 23^\circ 26' \\ y' = y \\ z' = x\sin 23^\circ 26' - z\cos 23^\circ 26' \end{cases}$$

M point symmetry point assignment

Step4: Clockwise rotation to 23°26', then

$$\begin{cases} x'' = x'\cos 23^\circ 26' - z'\sin 23^\circ 26' \\ y'' = y' \\ z'' = x'\sin 23^\circ 26' + z'\cos 23^\circ 26' \end{cases}$$

Step5

$$\begin{cases} x'' = \sin\varphi_1 \cos\theta_1 \\ y'' = \sin\varphi_1 \sin\theta_1 \\ z'' = \cos\varphi_1 \end{cases}$$

Step6: symmetry point, lat = $W' = 90 - \varphi_1$, long = $\theta_1 - (a + b/60) \times 15$

RESULTS

According to the information of 154 earthquakes over 6 scales of magnitude over the world from the USA Earthquake Observation Council, the causal relationship common area, symmetry of them can be concluded as following (Table 1) [1]: in the Comments, the next item TPPTE cut parallel to the earthquake in the equatorial plane, ELTP that the next seismic line cutting in the Ecliptic plane, EVP that the next earthquake in the vertical plane perpendicular to the equator, EPV that the next earthquake in the vertical longitudinal ecliptic plane. A total of circular symmetry point, then increasing the CSP, such as TPPTECSP; similar point with the SP said. Two adjacent seismic M_K and M_{K+1}, present, earthquake moment H_K, when the total surface, then added 1; when the time after the earthquake , when the total surface, then added 2.

Table 1. Original data.

2001

Date	Time (UTC)	Place	Lat.	Long.	Magnitude	Comments
Jan 13	17:33	Salvador	13.04	-88.66	7.7	ELTPCSP1
Jan 26	03:16	India	23.39	70.23	7.7	TPPTECSP1
Feb 13	14:22	Salvador	13.67	-88.94	6.6	ELTP1
Feb 28	18:54	USA	47.11	-122.6	6.8	EPV2
Jun 23	20:33	Peru	-16.3	-73.55	8.4	ELTPCSP1
Nov 14	09:26	China	36.2	90.9	7.8	ELTP1

2002

Date	Time (UTC)	Place	Lat.	Long.	Magnitude	Comments
Feb 3	07:11	Turkey	38.573	31.271	6.5	TPPTE1
Mar 3	12:08	Afghani-stan	36.543	70.424	7.4	SP1
Mar 25	14:56	Afghani-stan	36.06	69.32	6.1	TPPTE1
Jun 22	02:58	Iran	35.63	49.05	6.5	TPPTE1
Oct 31	10:32	Italy	41.79	14.87	5.9	ELTPCSP1
Nov 3	22:12	Alaska, USA	63.52	-147.4	7.9	EPV1

2003

Date	Time (UTC)	Place	Lat.	Long.	Magnitude	Comments
Jan 22	02:06	Mexico	18.8	-104.1	7.6	TPPTECSP1
Feb 24	02:03	China	39.61	77.23	6.3	TPPTE1
May 1	00:27	Turkey	39.01	40.46	6.4	TPPTE1
May 21	18:44	Algeria	36.96	3.63	6.8	TPPTE1
Sep 25	19:50	Japan	41.82	143.91	8.3	ELTP2
Nov 17	06:43	USA	51.15	178.65	7.8	
Dec 22	19:15	USA	35.71	-121.1	6.6	ELTP2
Dec 26	01:56	Iran	29.00	58.31	6.6	ELTP1

2004

Date	Time (UTC)	Place	Lat.	Long.	Magnitude	Comments
Feb 5	21:05	Indonesia	-3.61	135.53	7.0	ELTP1
Feb 24	02:27	Gibraltar	35.14	-3.99	6.4	TPPTE1
May 28	12:38	Iran	36.29	51.61	6.3	TPPTE1
Sep 5	10:07	Japan	33.07	136.62	7.2	SP1
Sep 5	14:57	Japan	33.14	137.07	7.4	ELTP1
Oct 9	21:26	Nicaragua	12	-86	7.0	ELTPCSP1
Oct 23	08:56	Japan	37.3	138.8	6.6	EVP1
Nov 11	21:26	East Timor	-8.15	124.87	7.5	ELTPCSP1
Nov 15	09:06	Colombia	4.7	-77.51	7.2	ELTP1
Nov 20	08:07	Costa Rica	9.6	-84.17	6.4	TPPTE1
Nov 21	11:31	Islands	15.68	-61.71	6.3	TPPTECSP1
Nov 26	02:25	Indonesia	-3.60	135.40	7.1	EPV1
Nov 28	18:32	Japan	42.94	145.28	7.0	EVP1
Dec 23	14:59	New Zealand	-50.24	160.13	8.1	ELTP1
Dec 26	00:58	Indonesia	3.30	95.87	9.1	ELTP1

2005

Date	Time (UTC)	Place	Lat.	Long.	Magnitude	Comments
Feb 22	02:25	Iran	30.726	56.817	6.4	TPPTE1
Mar 20	01:53	Japan	33.54	130.12	6.6	EPV1
Mar 28	16:09	Indonesia	2.08	97.11	8.6	ELTPCSP1
Jun 13	22:44	Chile	-19.89	-69.12	7.8	EPV1
Jun 15	02:50	California	41.284	-125.9	7.2	TPPTE1
Aug 16	02:46	Japan	38.259	148.98	7.2	EPV1
Sep 9	07:26	New Guinea	-4.539	153.47	7.6	TPPTE1
Sep 26	01:56	Peru	-5.68	-76.4	7.5	EPVD2
Oct 8	03:50	Pakistan	34.43	73.54	7.6	TPPTE1
Nov 26	00:49	China	29.7	115.7	5.2	TPPTE1
Nov 27	10:22	Iran	26.77	55.86	5.9	EPV1
Dec 5	12:19	Tanzania	-6.212	29.599	6.8	EPV1

2006

Date	Time (UTC)	Place	Lat.	Long.	Magnitude	Comments
Jan 8	11:34	Greece	36.250	23.498	6.7	EPV1
Feb 22	22:19	Mozambique	-21.32	33.58	7.0	EPV1
Apr 20	23:25	Russia	61.075	167.08	7.6	EPV1
May 3	15:26	Tonga	-20.13	-174.1	8.0	EVP1
May 16	10:39	New Zealand	-31.55	-179.2	7.4	ELTP2
May 27	22:54	Indonesia	7.977	110.31	6.3	EPV1
Jul 17	08:19	Indonesia	-9.334	107.26	7.7	EPV1
Aug 20	03:41	Scotia Sea	-61.02	-34.37	7.0	EPV1
Sep 10	14:56	Mexico	26.339	-86.56	5.9	TPPTE1
Oct 15	17:07	USA	19.801	-156.0	6.7	EPV1
Nov 15	11:14	Russia	46.616	153.22	8.3	ELTP1
Dec 26	12:26	Taiwan	21.818	120.53	7.1	ELTP2

2007

Date	Time (UTC)	Place	Lat.	Long.	Magnitude	Comments
Jan 13	04:23	Russia	46.288	154.44	8.1	ELTP1
Jan 21	11:27	Molucca Sea	1.207	126.29	7.5	EPV1
Feb 12	11:27	Portugal	36.09	10.26	6.0	ELTP2
Mar 6	05:49	Indonesia	-0.490	100.52	6.4	ELTP2
Mar 25	00:40	Vanuatu	-20.59	169.41	7.1	EPV1
Mar 25	00:42	Japan	37.537	136.43	6.7	EPV1
Apr 1	20:39	Islands	-8.474	156.95	8.1	ELTP1
May 16	08:56	Laos	20.470	100.70	6.3	ELTP1
Jun 2	21:34	China	23.013	101.05	6.1	ELTPCSP1
Jun 13	19:29	Guatemala	13.628	-90.73	6.7	ELTPCSP1
Jul 16	01:13	Japan	37.574	138.44	6.6	ELTP2
Jul 18	17:30	Tanzania	-2.586	36.281	5.9	ELTP1
Aug 1	17:08	Vanuatu	-15.67	167.60	7.2	EPV2
Aug 2	02:37	Russia	47.259	141.75	6.2	EPV1
Aug 8	17:04	Indonesia	-5.968	107.65	7.5	ELTPCSP2
Aug 15	23:40	Peru	-13.32	-76.50	8.0	ELTPCSP1
Sep 12	11:10	Indonesia	-4.517	101.38	8.5	ELTP1
Sep 28	13:38	USA	21.980	142.68	7.5	EPV1
Sep 30	05:23	New Zealand	-49.39	163.84	7.4	SP1
Oct 15	12:29	New Zealand	-44.68	167.21	6.8	EPV1
Oct 24	21:02	Indonesia	-3.909	101.06	6.8	ELTP2
Oct 31	03:04	USA	37.432	-121.7	5.6	ELTP1
Oct 31	03:30	USA	18.854	145.31	7.2	ELTP2
Nov 14	15:40	Chile	-22.18	-69.84	7.7	ELTPCSP1
Nov 25	16:02	Indonesia	-8.294	118.36	6.5	EPVD1
Nov 29	19:00	France	14.951	-61.24	7.4	EPV1
Dec 9	07:28	Fiji	-26.15	-177.4	7.8	EVP1
Dec 19	09:30	Alaska, USA	51.495	-179.4	7.1	ELTP1

2008

Date	Time (UTC)	Place	Lat.	Long.	Mag-nitude	Comments
Jan 5	11:01	Canada	51.299	-130.7	6.6	ELTPCSP1
Feb 3	07:34	Congo	-2.314	28.896	5.9	EVP1
Feb 14	10:09	Greece	36.646	21.833	6.9	ELTP2
Feb 20	08:08	Indonesia	2.751	95.966	7.4	EPV1
Feb 21	15:26	(Norway)	77.41	14.48	6.1	ELTP1
Feb 21	14:16	USA	41.076	114.77	6.0	TPPTE1
Apr 18	09:37	USA	38.450	87.890	5.2	ELTP1
May 7	16:45	Japan	36.141	141.54	6.9	ELTP2
May 12	06:28	China	31.099	103.27	7.9	ELTP2
May 24	19:20	Colombia	4.447	-73.67	5.9	EPV2
May 29	15:46	Iceland	63.992	-21.01	6.3	ELTP2
Jun 8	12:25	Greece	38.029	21.464	6.4	TPPTE1
Jun 13	23:43	Japan	39.122	140.67	6.9	ELTP2
Jul 15	03:26	Greece	35.983	27.785	6.4	ELTP1
Jul 19	02:39	Japan	37.627	142.11	7.0	SP1
Jul 23	15:26	Japan	39.807	141.46	6.8	ELTP2
Jul 29	18:42	California	33.955	-117.7	5.5	ELTP2
Aug 21	12:24	China	25.066	97.737	6.0	EPV2
Aug 25	13:21	China	30.893	83.614	6.7	ELTP2
Aug 30	08:30	China	26.277	101.91	6.0	EPVD2
Sep 8	18:52	Vanuatu	-13.51	166.96	6.9	ELTP1
Sep 10	11:00	Iran	26.823	55.825	6.1	ELTP1
Sep 11	00:20	Japan	41.979	143.62	6.8	EPV1
Sep 29	15:19	New Zealand	-29.87	-177.6	7.0	
Oct 5	15:52	Kyrgyzstan	39.515	73.768	6.6	ELTP2
Oct 6	08:30	China	29.759	90.302	6.3	ELTP1
Oct 11	09:06	Russia	43.271	46.262	5.8	ELTP2
Oct 16	19:41	Mexico	14.443	-92.42	6.7	ELTP2
Oct 19	05:10	Tonga	-21.85	-173.8	6.9	ELTP2
Oct 28	23:09	Pakistan	30.653	67.323	6.4	ELTP1
Nov 16	17:02	Indonesia	1.290	122.10	7.3	EPV1
Nov 24	09:02	Okhotsk	54.194	154.31	7.3	EPV2

2009

Date	Time (UTC)	Place	Lat.	Long.	Mag-nitude	Comments
Jan 3	19:43	Indonesia	-0.510	132.78	7.6	ELTPCSP1
Jan 8	19:21	Costa Rica	10.197	-84.15	6.1	ELTPCSP2
Jan 15	17:49	Russia	46.862	155.15	7.4	EPV1
Feb 11	17:34	Indonesia	3.902	126.40	7.2	ELTP1
Mar 19	18:17	Tonga	-23.05	-174.6	7.6	ELTPCSP2
Apr 6	01:32	Italy	42.334	13.334	6.3	ELTPCSP2
Apr 7	04:23	Islands	46.088	151.49	6.9	ELTPCSP1
Apr 16	21:27	Afghanistan	34.197	70.065	5.4	ELTPCSP1

May 28	08:24	Honduras	16.730	-86.20	7.3	TPPTECSP1
Jul 9	11:49	China	25.619	101.08	5.7	EPV2
Jul 15	09:22	New Zealand	-45.75	166.58	7.8	EPV1
Aug 9	10:55	Japan	33.144	138.04	7.1	ELTP1
Aug 10	19:55	Andaman	14.013	92.923	7.5	EPV2
Sep 2	07:55	Indonesia	-78.09	107.25	7.0	TPPTE1
Sep 12	20:06	Venezuela	10.70	-67.92	6.3	ELTP1
Sep 29	17:48	Samoa	15.509	-172.0	8.1	ELTP1
Sep 30	10:16	Indonesia	-0.725	99.856	7.6	ELTP2
	22:03		-13.05	166.18	7.6	SP1
Oct 7	22:18	Vanuatu	-12.55	166.32	7.8	SP1
	23:13		-13.14	166.29	7.3	ELTP1
Dec 19	13:02	Taiwan	23.763	121.68	6.4	ELTP2

2010

Date	Time (UTC)	Place	Lat.	Long.	Mag-nitude	Comments
Jan 3	22:36	Solomon	-8.912	157.30	7.2	ELTP2
Jan 10	00:27	California	40.645	-124.7	6.5	EPV1
Jan 12	21:53	Haiti	18.451	-72.44	7.0	ELTPCSP1
Feb 26	20:31	Japan	23.472	123.71	7.0	ELTPCSP1
Feb 27	03:34	Chile	-35.84	-72.71	8.8	SP1
Feb 27	15:45	Argentina			6.1	ELTPCSP1
Mar 4	0:18	Taiwan			6.4	EPV1
Mar 5	23:29	Sumatra			6.5	ELTP1
Mar 8	4:32	Turkey			6.0	ELTP1
Mar 11	11:39	Chile	-34.25	-71.88	6.9	ELTPCSP1
Mar 14	9:57	Indonesia			6.4	EPV1
Mar 14	8:08	Japan			6.6	ELTPCSP1
Mar 16	2:21	Chile			6.7	EPV1
Apr 4	22:40	Mexico	2.128	115.3	7.2	ELTP1

ANALYSIS

According to the information of 154 earthquakes over 6 scales of magnitude over the world from the USA Earthquake Observation Council, the causal relationship common area, symmetry of them we can get the following results, as shown in Table 2:

Table 2. Results.

Type	Equatorial level Line plane	Which Symmetry	Ecliptic Line plane	Among Symmetry	Vertical Flat	Similar point	Other Not coplanar	Total
Quantity	22	(Four)	77	(23)	45	7	3	154
Percentage of total	14%	(3%)	50%	(15%)	29%	5%	2%	100%

CONCLUSIONS

This study allowed existence of the errors, but those errors do not interfere the co-planarity and the symmetry principle. Errors may come from the inaccuracy of observation time, perhaps the errors accelerate the process of the Earth becoming to sphere. In the next step of our research, we plan to study on earthquake prediction and the relationship between earthquake and economic development.

REFERENCES

1. Http://earthquake.usgs.gov/recenteqsww/quakes/quakes_ all.html

CHAPTER 11

A NUMERICAL INVESTIGATION OF EARTHQUAKE SHIELDING WITH SEISMIC CRYSTALS

Baris Baykant Alagoz and Serkan Alagoz

Keywords: Earthquake Shielding, Seismic Metamaterials, Seismic Wave Attenuation, Band Gap Analysis

INTRODUCTION AND PRELIMINARIES

In recent research about photonic, phononic and sonic crystals, band gaps were observed in the band structure characteristics of crystals with various lattice geometries [1-20]. As crystal structures exhibit very low transmission in the frequencies within band gaps, they were also referred to as "stop bands." Band gaps appear in the wave transmission spectrum when a homogenous wave propagation medium turns into an inhomogeneous medium with periodic defects. Those periodical defects in the medium disturb the wave propagation at certain frequency groups, which make up the so-called band gap. In general, the periodical defects in a medium are commonly called scatterers and the medium in which the wave propagates is called the host material of the crystal structure. In order to form such a periodically inhomogeneous medium, photonic crystals, used for electromagnetic waves, are composed of periodically modulated dielectric materials [9-12]; sonic crystals for acoustic waves are mostly composed of periodic arrays of solid rods in the air [1,3-5] and phononic crystal are made of solid or liquid host material with a periodic inhomogenization [15-17]. For a seismic crystal implementation, in our study, we preferred periodic holes in the ground as the scatterers and the ground itself as the host material. A representation of a 2D seismic crystal structure and its relevant design parameters is given in Figure 1. The hole scatterer-solid host configurations and their full band gap effect were demonstrated for surface acoustic waves on two-dimensional phononic crystal in the sub-meter scales, such as a piezoelectric phononic crystals [15-17]. As a result of scalability features of two dimensional crystal lattice with respect to wavelength (a/l is constant), the similar band gap effects can be obtained for the seismic waves via a larger crystal implementation in the scales of kilometers [14]. This idea was our basic motivation in this theoretical work. In general notation, the distance between the centers of the cylindrical scattering materials is referred to as the lattice constant and denoted by a , and radius of the cylindrical scattering material is denoted by r. The parameter h is the height of the holes and λ represents the wavelength. Band gap properties of crystal structures have been used practically in the application of wave isolation in a selected frequency band [5,6,8,18,19]. In a similar fashion, we numerically investigate crystal structures built on the ground in order to attenuate the longitudinal ground vibrations propagating on the surface of the earth by means of band gaps. We employed the hole-ground type crystal structures to be a sort of seismic crystal and applied them towards isolating a region from seismic waves, i.e., to earthquake shielding. Another technique to isolation, recently made progress, is to cloaking method that can isolate a region by steering wave envelopes around to isolated region instead of attenuating their amplitudes [20,21]. However, a wide-band cloaking technique [21] should be developed for the efficient applications.

Citation: B. Alagoz and S. Alagoz, "A Numerical Investigation of Earthquake Shielding with Seismic Crystals," Open Journal of Acoustics, Vol. 1 No. 3, 2011, pp. 63-69. doi: 10.4236/oja.2011.13008.

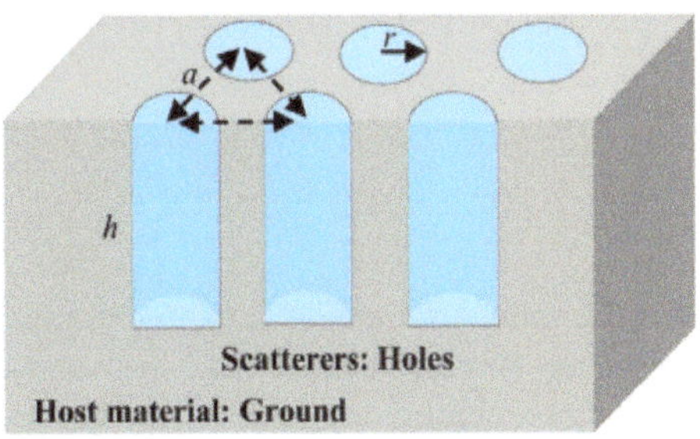

Figure 1: Representation of 2D seismic crystal.

In order to theoretically demonstrate the wave attenuation effect in the proposed crystal structures, we applied the Finite Difference Time Domain (FDTD) method, which is a very common numerical analysis technique used in acoustic wave simulations in inhomogeneous materials [1,3]. In this simulation method, seismic waves in the ground were assumed to travel as compression and decompression fields in a granular structure via displacements of granules. In order to model this type of the wave propagation numerically, a linear wave equation system based on parameters of granule (particle) velocity and pressure field were solved for a two-dimensional plane by using the FDTD method. In our earthquake simulations, a narrow band vibration signal containing very low sampled frequency components up to 10 Hz was generated in the pressure field. By using spatiatemporal solution of wave propagation, the Richter scaled map of the ground vibration, which is composed of the maximum instant granule velocities, was obtained to show the efficiency of the earthquake shielding. Additionally, wavelength responses of some basic seismic crystal configurations were analyzed to give a comparison of several crystal configuration types.

METHOD

Theoretical Background

To do a numerical analysis of wave propagation on the surface layer of the Earth, the ground is assumed to be composed of dynamic tiny granular structures as illustrated in Figure 2. In this system, the pressure in a unit volume of the ground is dependent on the density of the volume and it can be expressed as a function of the density, P=P(r). Thickness of surface layer is assumed to be narrow enough to ignore variation in density r depending on the gravity.

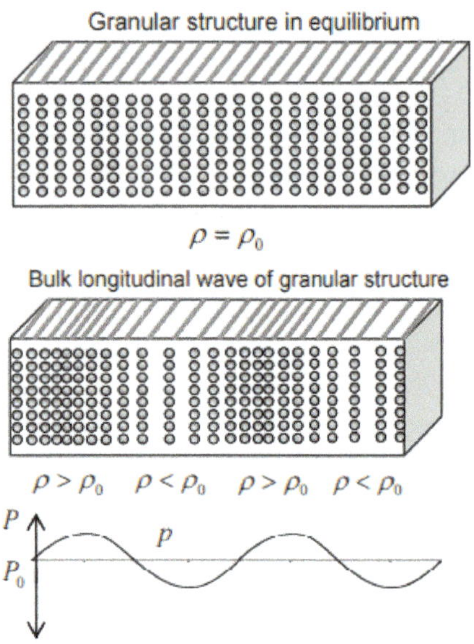

Figure 2: Granule displacements and corresponding pressure field in one dimension.

When the granules forming the ground are allowed to a short directional displacement around their resting location, the density will exhibit a small variation around an equilibrium density ρ_0. In this case, by applying a linear approximation, the pressure function P(r) can be written as,

$$P \cong P_0 + \left(\frac{\partial P}{\partial \rho}\right)_{\rho_0} (\rho - \rho_0) \tag{1}$$

Therefore the deviation in pressure from the equilibrium pressure of the ground result in the forces that move granular structures in ground, it will be meaningful to consider the pressure deviation, p P-P_0 in the analysis of seismic waves. Hence, considering the Equation (1), the pressure field of the seismic wave can be reorganized as

$$p = K_o s \tag{2}$$

where the parameter K_0 is bulk modulus of media and expressed as $K_0 = \rho_0 \left(\partial P/\partial \rho\right)_{\rho_0}$ and s denotes squeezing rate and defined as $s = (\rho - \rho_0)/\rho_0$.

In this system, the granule displacements even in short distances lead to alternations in local pressure field variations and result in compression (high pressure) and decompression (low pressure) fields inside the ground as represented in Figure 2. Meanwhile, the gradient in the pressure fields causes new granule motions. This mutual interaction between the pressure and the granule displacements results in wave propagation through the medium. Since this motion of granules is the small in magnitudes, the Euler equation, which is also known as the linear inviscid force equation, can be used to model this phenomenon [25].

$$\nabla p = -\rho_0 \frac{\partial V}{\partial t} \tag{3}$$

where the vector V is the velocity of the granules. On the other hand, the motion of granules in a unit volume of ground must comply with the mass continuity equation and it was expressed as [25]

$$\frac{\partial \rho}{\partial t} + \nabla \cdot (\rho V) = 0 \tag{4}$$

Solid granules have quite low viscosity, so the ground can be assumed to be a inviscid media. $(s \ll 1)$ In that case, the mass continuity equation for linear mediums can be obtained as,

$$\left(\frac{\partial (p/K_0)}{\partial t}\right) + \nabla \cdot V = 0 \tag{5}$$

Here, the r density of medium is considered as $\rho = \rho_0 (1 + s)$ and s is written as $s = p/K_0$ according to Equation (2). If the Equations (3) and (5) are reorganized for a medium, where variation in bulk modulas in temporal domain are negligibly small, we obtains the following linear wave equation system,

$$\frac{\partial V}{\partial t} = -(1/\rho_o)\nabla p \quad \text{and} \quad \frac{\partial p}{\partial t} = -K_0 \nabla \cdot V \, , \tag{6}$$

The Equation (6) was normalized by the material parameters of a simple uniform medium which will be the ground in our case. The following normalized linear wave equation system is obtained for the wave motion simulation in an in homogeneous media:

$$\frac{\partial \boldsymbol{v}}{\partial u} = -\rho \nabla p \text{ and } \frac{\partial p}{\partial u} = -K\boldsymbol{\nabla}\cdot\boldsymbol{v}, \tag{7}$$

where denotes the normalized granule velocity and is the pressure. The material parameters of inhomogeneous structures are described by the normalized medium density, $\rho = \rho_0/\rho_s$, and the normalized bulk modulus, $K = K_s/K_o$. Here, ρ_0 is the density of the host material and K_0 is the bulk modulus of the host material. ρ_s and K_s are the corresponding parameters for the scattering material. To obtain an FDTD based numerical solution of (7), finite difference equations are derived for the spatial plane seen in Figure 3(a). The calculation of granule velocity vectors, defined as $v = [v_x \ v_y]$ is done by using discrete finite difference formulas:

$$v_x^{n+\frac{1}{2}}\left(i+\frac{1}{2},j\right) = v_x^{n-\frac{1}{2}}\left(i+\frac{1}{2},j\right) - \rho\left(i+\frac{1}{2},j\right)R_x\left[p^n(i+1,j)-p^n(i,j)\right] \tag{8}$$

$$v_y^{n+\frac{1}{2}}\left(i,j+\frac{1}{2}\right) = v_y^{n-\frac{1}{2}}\left(i,j+\frac{1}{2}\right) - \rho\left(i,j+\frac{1}{2}\right)R_y\left[p^n(i,j+1)-p^n(i,j)\right] \tag{9}$$

The calculation of the pressure field is made with the following discrete finite difference formula:

$$p^{n+1}(i,j) = p^n(i,j) - K(i,j)R_x \cdot \left[v_x^{n+\frac{1}{2}}\left(i+\frac{1}{2},j\right) - v_x^{n+\frac{1}{2}}\left(i-\frac{1}{2},j\right)\right] - K(i,j)R_y\left[v_y^{n+\frac{1}{2}}\left(i,j+\frac{1}{2}\right) - v_y^{n+\frac{1}{2}}\left(i,j-\frac{1}{2}\right)\right] \tag{10}$$

For the stability of the wave propagation in FDTD simulation, R_x and R_y parameters should satisfy the Courant condition [1] and expressed as,

$$R_x = \frac{\Delta u}{\Delta x} \text{ and } R_y = \frac{\Delta u}{\Delta y}. \tag{11}$$

The Δ_x and Δ_y represent unit distances in the spatial domain and Δ_u is the normalized unit time that is defined as $\Delta_u = c\Delta t$. Here, Δ_t and c is the unit time increment and wave velocity, respectively. Swells on the surface level of the ground accompany compressed and decompressed fields in the ground. For the sake of simplicity, we used a linear modeling for the ground swelling for a cubic unit volume on the surface layer of ground seen in Figure 3(b). The rise in upper surface of the unit volume (Δh) is considered to be proportional to deviation of density from its equilibrium value, that is, $\Delta h = \xi(\rho - \rho_0)$. The parameter x , expressed as $\partial h\ \partial r$, is the swelling to density rate of the unit volume. The Equation (2) can be written in the form of $p = c^2(\rho - \rho_0)$ wning to $c^2 = K_0/\rho_0$. In that case, the change in the surface level of a unit volume can be estimated by

$$\Delta h = \alpha p \tag{12}$$

Here, $\alpha = \xi/c^2$ is the swelling coefficient. Equations (7) were previously applied in the simulations of acoustic wave propagation in sonic crystals [1, 3,4]. In our work, we see that seismic waves can be supposed as the earthquakes sounds propagating through the ground and many properties observed in acoustic wave propagation in a solid media, such as the reflection, the refraction and the attenuation of waves, are also valid in the seismic wave propagation on the ground. In the following sections, we will numerically analyze frequencyselective attenuation effects of periodically defected grounds on the seismic wave propagation.

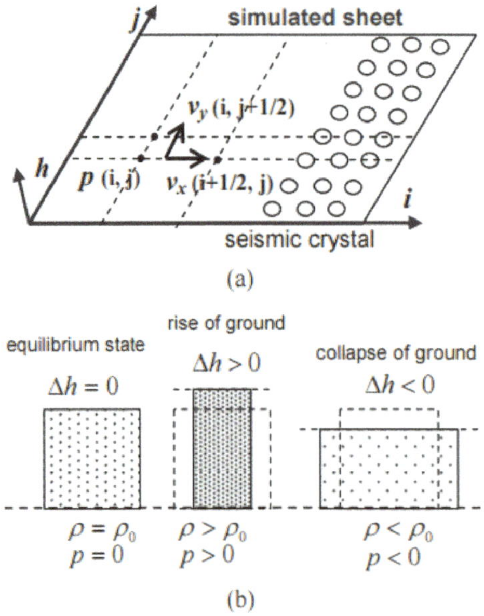

Figure 3: (a) Simulation plane and corresponding parameters used in discrete solution points grid; (b) Ground swelling by the pressure.

Earthquake Simulation Results In this section, FDTD based simulation results of earthquake shielding are presented in the Richter scale, which is the logarithm of the maximum displacement in microns ($M = \log \Delta x_{micron}$). In the simulation, a rectangular sheet representing the host material was stimulated by a signal of a pressure field (P) containing multi-frequency components as illustrated in Figure 4. This signal was generated by the following function,

$$Ps(n) = \sin(\pi \cdot n) + \sin(2\pi \cdot n) + \sin(4\pi \cdot n) + \sin(6\pi \cdot n) + \sin(8\pi \cdot n) + \sin(10\pi \cdot n) + \sin(20\pi \cdot n) \tag{13}$$

where n is the simulation time index. The simulation time can be expressed as $u = n \cdot \Delta u / c$. The basic steps of the algorithm are presented in Figure 5. In the proposed FDTD based wave simulations, the wave velocity in the host material (c) is to 3000 m/sn. For the scattering material, the normalized medium density ρ was set to 10 and the normalized bulk modulus K to 0.1. The Δ_x and Δ_y unit distances in spatial domain were set to 40 m and Δ_u normalized unit time was set to 4.66× 10^{-3} second. By considering Equation (12), the pressure wave pattern (p) also forms a map of the ground swelling (Δh) for a =1. Vibration maps in Richter scale are composed of the norm of the maximum instant velocity vector parallel to ground. It is calculated by using the following formula,

$$V_{\max}(i,j) = \max\left(\sqrt{\left(v_x^{n-1/2}(i,j)\right)^2 + \left(v_y^{n-1/2}(i,j)\right)^2}\right) \quad \text{for} \quad n \in [0, n_{sim}] \tag{14}$$

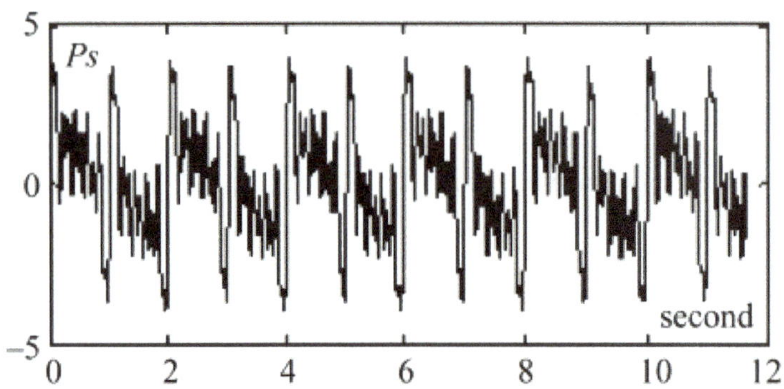

Figure 4: Multi-frequency pressure signal produced at the earthquake center in the simulation.

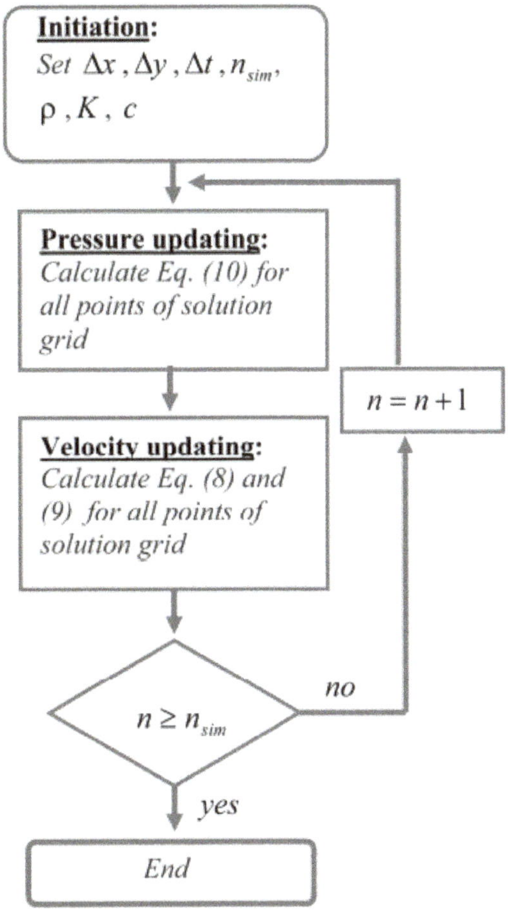

Figure 5: Basic steps of algorithm.

where, n_{sim} specifies simulation duration and V_{max} is the maximum displacement in a second. Simulation results for seismic crystal types with various scatterer geometries and lattice configurations are given in Figure 6. A decreasing in magnitude of vibration is apparent behind the seismic crystals when compared with the vibration magnitude in the unshielded region. Elliptic scatterers [22] are seen to exhibit better performance of isolating seismic waves. A serious disadvantage of the circular scatterers for the application of earthquake shielding is that the crystals with circular scatters can exhibit wave focusing effect within a certain frequency band [23,24].

Obtaining Maximum Acceleration Map from the FDTD Simulations

Maps of maximum granule acceleration are useful in estimating the destructive forces applied on buildings by seismic waves.

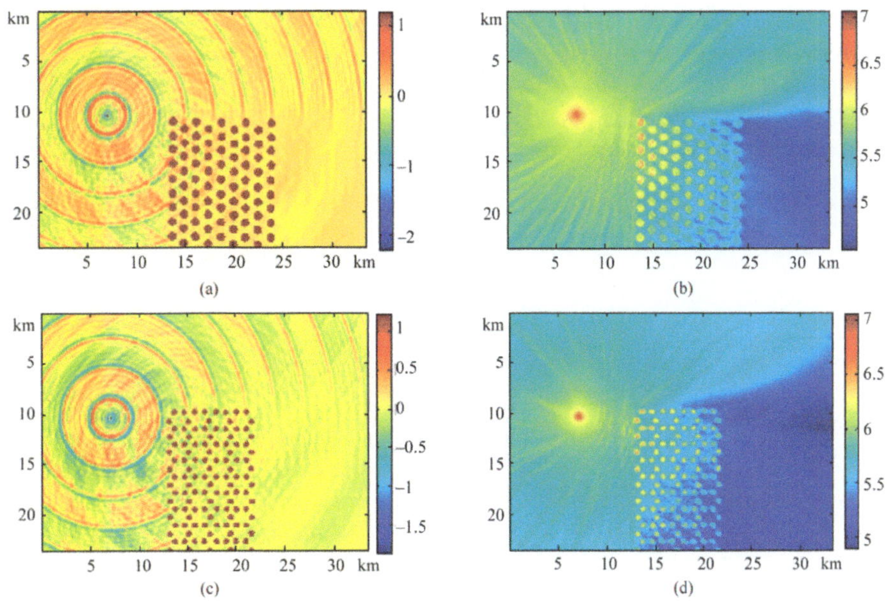

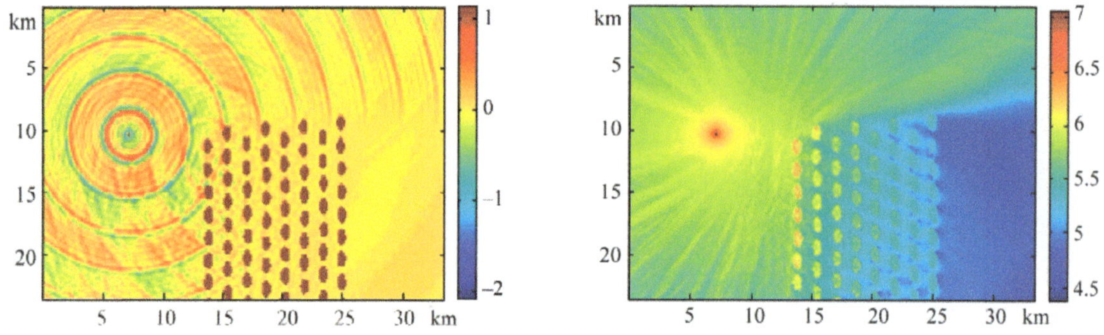

(e) (f)

Figure 6: Pressure wave (p) pattern images and vibration maps in Richter scale for triangular lattice of circular scatterer (a, b), honeycomb lattice of circular scatterer (c, d), triangular lattice of elliptic scatterer (e, f).

The granule acceleration can be simply expressed as a =∂v ∂t. By applying the backwards difference to derivatives, the amplitudes of the maximum granule acceleration are calculated in the discrete form by the following formula,

$$A_{max}(i,j) = \max\left\{ \frac{1}{\Delta u}\left(\left(v_x^{n+1/2}(i,j) - v_x^{n-1/2}(i,j) \right)^2 + \left(v_x^{n+1/2}(i,j) - v_y^{n-1/2}(i,j) \right)^2 \right)^{1/2} \right\}$$

(15)

for $n \in [0, n_{sim}]$

Figure 7 shows an estimate of the maximum acceleration maps. In these figures, the effect of the seismic crystal on the acceleration of the ground is clearly seen.

Wavelength Response Analysis for Various Seismic Crystals

In this section, the wavelength response of seismic crystal with various lattice configurations and scatterer shapes are investigated. For this purpose, we recorded pressure values ($p_{0,\lambda}(n)$, $p_{1,\lambda}(n)$) in the simulation, which is done for each sampled wavelength in the seismic spectrum. The $p_{1,\lambda}$ array stores the pressure data from the point in the shielded side and $p_{0,\lambda}$ stores the pressure data at the same point in the absence of the crystal. The reduction rate of shielding (R(l)) for each wavelength l was defined according to the recorded maximum pressure values as the following.

$$R(\lambda) = \frac{\max(p_{1,\lambda})}{\max(p_{0,\lambda})} \quad (16)$$

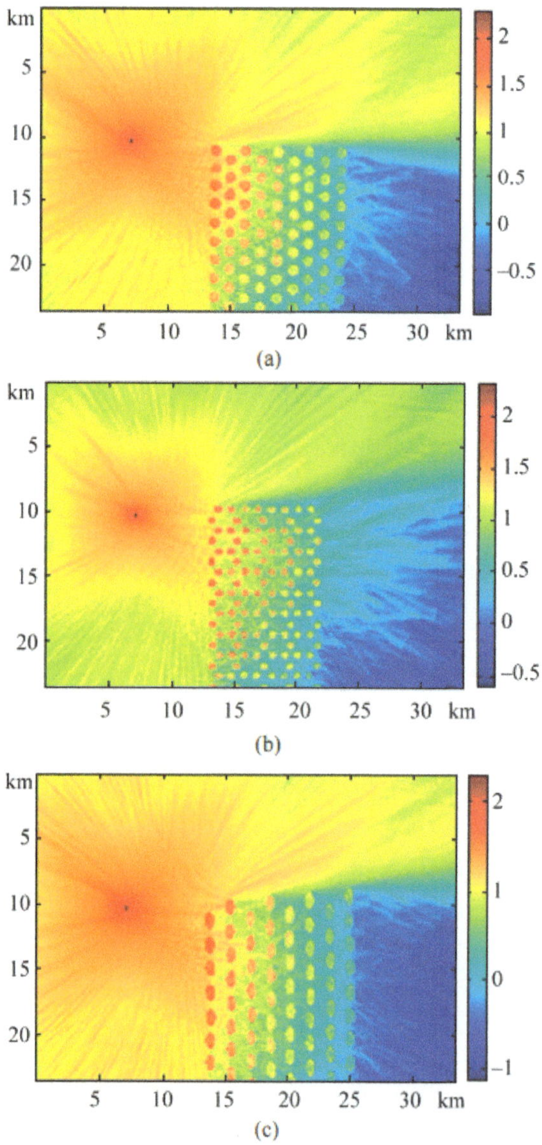

Figure 7: Maximum granule acceleration map in logarithmic scale.

In Figures 8(a)-(c), maximum pressure reduction rates versus wavelength (l) were drawn for the various seismic crystals. The seismic crystals are seen to be capable of providing a seismic attenuation at the wavelength band between nearly 1500 m and 3000 m.

CONCLUSIONS

In this paper, a possible seismic crystal application with the sizes of kilometers was investigated in an analogy with the acoustic band gaps of phononic and sonic crystals. Their possible applications to seismic wave attenuation were numerically demonstrated by means of FDTD based simulations and their possible application to earthquake shielding was theoretically discussed for various lattice configurations. We observed from the simulation results that the proposed seismic crystal can suppress higher frequency components in vibration spectrum of the earthquake and consequently lead to decreasing ground acceleration.

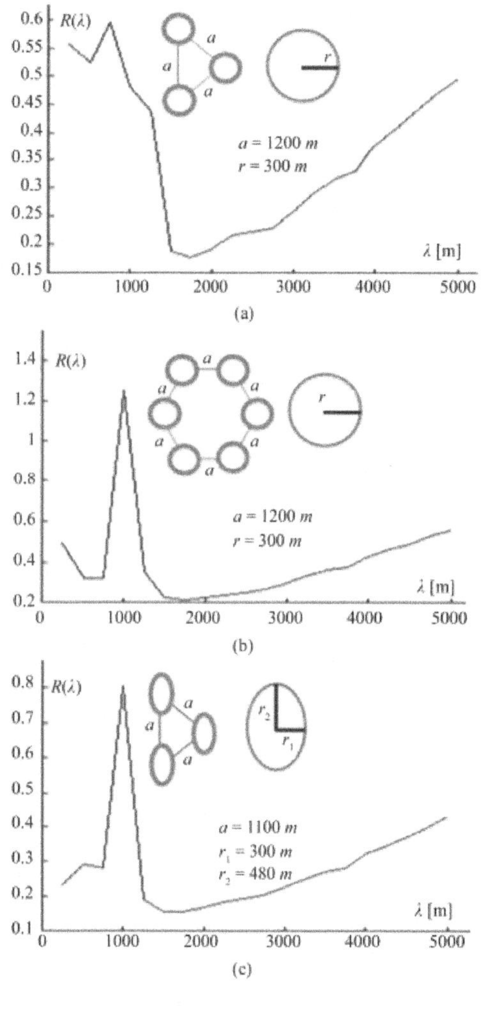

Figure 8: (a) Reduction rate for triangular lattice configuration with circular scatterer; (b) Reduction rate for honeycomb lattice configuration with circular scatterer; (c) Reduction rate for triangular lattice configuration with elliptic scatterer.

For future study, three dimensional simulation of seismic crystals should be done with more realistic parameters and the real records from earthquakes to better asses performance of earthquake shielding. Although surface wave suppression by full band gaps was theoretically and experimentally demonstrated for the hole scatterers-solid host type phononic crystal, the seismic wave attenuation via enormous size seismic crystals structures needs to be also experimentally demonstrated in future works.

REFERENCES

1. T. Miyashita, "Sonic Crystals and Sonic Wave-Guides," Measurement Science and Technology, Vol. 16, No. 5, 2005, pp. 47-63. doi:10.1088/0957-0233/16/5/R01
2. X. D. Zhanga and Z. Y. Liu, "Negative Refraction of Acoustic Waves in Two-Dimensional Phononic Crystals," Applied Physics Letters, Vol. 85, No. 2, 2004, pp. 341-343. doi:10.1063/1.1772854
3. M. M. Sigalas, "Theoretical Study of Three Dimensional Elastic Band Gaps with the Finite-Difference Time-Domain Method," Journal of Applied Physics, Vol. 87, No. 6, 2000, pp. 3122-3125. doi:10.1063/1.372308
4. T. Miyashita and C. Inoue, "Numerical Investigations of Transmission and Waveguide Properties of Sonic Crystals by Finite-Difference Time-Domain Method," Japanese Journal of Applied Physics, Vol. 40, 2001, pp. 3488- 3492. doi:10.1143/JJAP.40.3488
5. M. S. Kushwaha and B. Djafari-Rouhani, "Sonic-Stop bands for periodic Arrays of Metallics Rods: Honeycomb Structure," Journal of Sound and Vibration, Vol. 218, No. 10, 1998, pp. 697-709. doi:10.1006/jsvi.1998.1839
6. M. Hirsekorn, "Small-Size Sonic Crystals with Strong Attenuation Bands in the Audible Frequency Range," Applied Physics Letters, Vol. 84, No. 17, 2004, p.3364. doi:10.1063/1.1723688
7. E. N. Economou and M. M. Sigalas, "Classical Wave Propagation in Periodic Structures: Cermet versus Network Topology," Physical Reviews B, Vol. 48, No. 18, 1993, pp. 13434-13438. doi:10.1103/PhysRevB.48.13434
8. N. K. Batra, P. Matic and R. K. Everett, "Sonic Crystal Composites for Selective Noise Reduction," Proceedings of IEEE Ultrasonic Symposium, Vol. 1, 8-11 October 2002, pp. 547-550.
9. E. Ozbay, K. Guven, K. Aydin, "Metamaterials with Negative Permeability and Negative Refractive Index: Experiments and Simulations," Journal of Optics A: Pure Applied Optics, Vol. 9, No. 9, 2007, pp. S301-307. doi:10.1088/1464-4258/9/9/S04
10. J. Sun, C. C. Chan, X. Y. Dong and P. Shum, "Tunable Photonic Band Gaps in a Photonic Crystal Fiber Filled with Low Index Material," Journal of Optoelectronics and Advanced Materials, Vol. 8, 2006, pp. 1593-1596.
11. E. Yablonovitch, "Photonic Band-Gap Crystals," Journal of Physical Condensed Matter, Vol. 5, No. 16, 1993, p. 2443. doi:10.1088/0953-8984/5/16/004
12. M. Ciobanu, L. Preda, A. Popescu, M. Mihailescu and M. I. Rusu, "Designing Tunable Photonic Crystals with Band Gaps in Microwave Range," Journal of Computational and Theoretical Nanoscience, Vol. 7, No. 6, 2010, pp. 1032- 1034. doi:10.1166/jctn.2010.1449
13. F. Meseguer, M. Holgado, D. Caballero, N. Benaches, C. Lopez, J. Sanchez-Dehesa and J. Llinares, "Two-Dimensional Elastic Bandgap Crystal to Attenuate Surface Waves," Journal of Lightwave Technology, Vol. 17, No. 11, 1999, pp. 2196-2201. doi:10.1109/50.803011
14. F. Meseguer, M. Holgado, D. Caballero, N. Benaches, J. S'anchez-Dehesa, C. L'opez, and J. Llinares, "Attenuation

of Surface Elastic Waves (Earthquakes) by Phononic Crystals," Physical Reviews B, Vol. 59, No. 19, 1999, pp. 12169-12172. doi:10.1103/PhysRevB.59.12169

15. S. Benchabane, A. Khelif, J. Y. Rauch, L. Robert and V. Laude, "Evidence for Complete Surface Wave Band Gap in a Piezoelectric Phononic Crystal," Physical Reviews E, Vol. 73, No. 6, 2006, p. 065601(4). doi:10.1103/PhysRevE.73.065601

16. V. Laude, M. Wilm, S. Benchabane and A. Khelif, "Full Band Gap for Surface Acoustic Waves in a Piezoelectric Phononic Crystal," Physical Reviews E, Vol. 71, No. 3, 2005, p. 036607(7). doi:10.1103/PhysRevE.71.036607

17. Y. Tanaka and S. Tamura, "Two-Dimensional Phononic Crystals: Surface Acoustic Waves," Physica B: Condensed Matter, Vol. 263-264, 1999, pp. 77-80. doi:10.1016/S0921-4526(98)01197-1

18. R. Martınez-Salaa, C. Rubioa, L. M. Garcia-Raffib, J. V. Sanchez-Pereza, E. A. Sanchez-Pereza and J. Linaresa, "Control of Noise by Trees Arranged Like Sonic Crystals," Journal of Sound and Vibration, Vol. 291, No. 1-2, 2006, pp. 100-106. doi:10.1016/j.jsv.2005.05.030

19. J. V. Sanchez, D. Caballero, R. Martinez-Sala, C. Rubio, J. Sanchez-Dehesa, F. Meseguer, J. Llinares and F. Galvez, "Sound Attenuation by Two-Dimensional Array of Rigid Cylinders," Applied Physics Letters, Vol. 80, No. 24, 1998, pp. 5325-5328. doi:10.1103/PhysRevLett.80.5325

20. M. Brun, S. Guenneau and A. B. Movchan, "Achieving Control of In-Plane Elastic Waves," Applied Physics Letters, Vol. 94, No. 6, 2009, p. 061903(3). doi:10.1063/1.3068491

21. M. Farhat, S. Guenneau and S. Enoch, "Ultrabroadband Elastic Cloaking in Thin Plates," Applied Physics Letters, Vol. 103, No. 2, 2009, p. 024301(4). doi:10.1103/PhysRevLett.103.024301

22. L. Y. Wu and L. W. Chen, "The Dispersion Characteristics of Sonic Crystals Consisting of Elliptic Cylinders," Journal of Physics D: Applied Physics, Vol. 40, No. 23, 2007, pp. 7579-7583. doi:10.1088/0022-3727/40/23/051

23. C. Qiu, X. Zhang and Z. Liu, "Far-Field Imaging of Acoustic Waves by a Two-Dimensional Sonic Crystal," Phys Reviev B, Vol. 71, No. 5, 2005, p. 054302(6). doi:10.1103/PhysRevB.71.054302

24. S. Alagoz and B. B. Alagoz, "Frequency-Controlled Wave Focusing by a Sonic Crystal Lens," Applied Acoustics, Vol. 70, No. 11-12, 2009, pp. 1400-1405. doi:10.1103/PhysRevB.71.054302

25. L. E. Kinsler, A. R. Frey, B. Coppens and J. V. Sanders, "Fundamentals of Acoustics," John Wiley & Sons Inc., New York, 1982.

CHAPTER 12

EVALUATION OF STRAIN ACCUMULATION IN GLOBAL SUBDUCTION ZONES FROM SEISMICITY DATA

Ryoya Ikuta, Yuta Mitsui, Yuri Kurokawa and Masataka Ando

Keywords: Huge earthquake, Slip deficit, Subduction, Mega-thrusts, Stick–slip

INTRODUCTION

Although interplate earthquakes with magnitudes of ≥9 (M9-class earthquakes) occur infrequently, they cause severe damage and fatalities over huge areas, as shown by the 2004 Sumatra (Mw 9.2) and 2011 Tohoku (Mw 9.0) earthquakes. Such earthquakes can rupture subduction zones along hundreds of kilometers and generate destructive tsunami waves. Evaluating the probability of M9-class interplate earthquakes in global subduction zones has been an object of seismology for decades.

Geodetic data are useful for monitoring the source regions of huge interplate earthquakes during interseismic periods. For example, several geodetic inversion studies using an on-land Global Navigation Satellite System (GNSS) network reported that the source region of the 2011 Tohoku earthquake had been locked from at least 1996 to 2003 before the occurrence of an M7-class earthquake (e.g., Ito et al. 2000; Nishimura et al. 2004; Suwa et al.2006; Hashimoto et al. 2009; Loveless and Meade 2010). However, studies published prior to 2011 did not associate interplate coupling with an upcoming huge interplate earthquake because the GNSS observation period (~15 years) was thought to be too short to judge whether the interplate coupling was persistent or temporary. In fact, geodetic observations in the studies mentioned above were disrupted by the M7-class earthquake in 2003.

Citation: Ryoya Ikuta, Yuta Mitsui, Yuri Kurokawa and Masataka Ando, Evaluation of strain accumulation in global subduction zones from seismicity data, DOI: 10.1186/s40623-015-0361-5.

The degree of coupling between plates is also indicated by the spatio-temporal distribution of interplate earthquakes. For example, Yamanaka and Kikuchi (2004) analyzed source rupture processes and slip distributions for eight large earthquakes (~M7.5) occurring in the past 80 years around the source region of the Tohoku earthquake and concluded that these earthquakes did not fully release the slip expected from the relative plate motion. In the aftermath of the 2011 Tohoku earthquake, the apparent low seismic coupling can be explained in terms of slip-deficit accumulation, which should be released by a huge interplate event (Fig. 1). Ikuta et al. (2012) compiled the slip history of M7–8-class earthquakes around the source region of the Tohoku earthquake, including those analyzed by Yamanaka and Kikuchi (2004), to show that their cumulative seismic slip was smaller than expected for the relative plate motion in the region (Fig. 1). The authors showed that the center of the slip-deficit distribution coincides with a region that experienced significant slip during the M9 earthquake and that the area of the slip-deficient region was larger than that of a persistently locked region that had been identified through inversions of GNSS data for the 15 years prior to the Tohoku earthquake. This result indicates that part of the rupture area of the huge interplate earthquake had been locked for several hundred years and had elastically deformed the surrounding region of the upper plate. Consequently, the effective relative motion between the overriding plate and the subducting plate around the locked area was broadly suppressed compared with other regions, thereby reducing the cumulative seismic slip of M7–8-class earthquakes. Even if the size and persistence of the locked area are debated, it is agreed that an M9-class earthquake requires a cumulative slip-deficit equivalent to its slip over its rupture area.

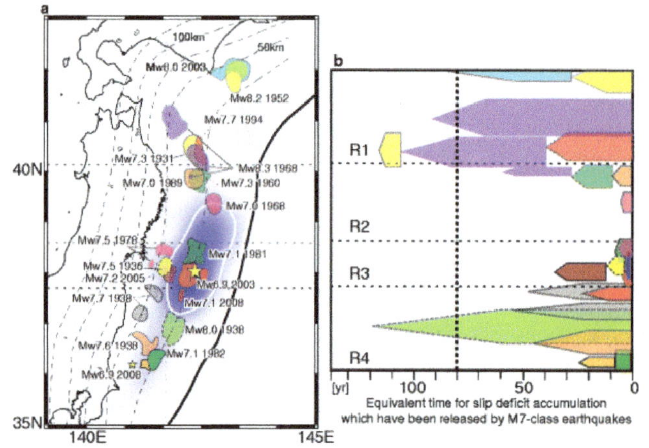

Figure. 1 Co-seismic slip of the 2011 Tohoku earthquake and previous M7-class earthquakes around the source region (after Ikuta et al. 2012). **a** Co-seismic slip distribution of the 2011 Tohoku earthquake (blue color scale); the area with slip greater than 10 m is enclosed by a white line. The two stars show the locations of the main shock and the largest after-shock (March 11, 2011). The asperity distribution for M7-class earthquakes occurring in the past 80 years is shown bycolored contours (after Murotani et al. 2004; Yamanaka and Kikuchi 2004; Yamanaka (NGY Seismology Notebook,http://www.seis.nagoya-u.ac.jp/sanchu/Seismo_Note, last updated November 25, 2015)). The contour for each asperity encloses the area in which the slip is greater than half of the maximum slip. **b** Cumulative seismic slip distribution along the trench for the earthquakes shown in **a**. The total length of each arrow represents the maximum slip of the event, and the length of the body of each arrow represents the average slip. Modified after Figure 12b in Yamanaka and Kikuchi (2004), with the addition of the following: data for the 1952 and 2003 earthquakes off Tokachi in the northern part of the trench are from Yamanaka and Kikuchi (2003); data for earthquakes in 1938 and 1982 in the R4 region are from Murotani et al. (2004) and Mochizuki et al. (2008), respectively; and data for recent earthquakes are from Yamanaka, NGY Seismology Notebook, http://www.seis.nagoya-u.ac.jp/sanchu/Seismo_Note, last updated November 25, 2015. Slip on spatially overlapping asperities is accumulated. Vertical dotted line shows the total expected tectonic slip over 80 years

Although a deficit in cumulative seismic slip relative to plate motion is necessary for an M9-class earthquake to occur, not all slip-deficient areas are locked, as the plate interface has varying frictional properties. While some areas demonstrate stick–slip instability, others slip stably. The former is defined as a velocity-weakening frictional property and the latter as velocity-strengthening in the context of the rate- and state-dependent friction law (Dieterich 1979; Ruina 1983). Some areas even behave as conditionally stable, which accompanies the rupture of an adjacent stick–slip zone (Scholz 1998). Further parts show occasional slow slip, one of which, along a subducting plate, was first reported by Hirose et al. (1999). The frictional properties of a slow-slip area have been modeled in various ways. A deficit in cumulative seismic slip can also be explained by aseismic slip on a stable or conditionally stable interface. However, if the frictional properties on the plate interface do not change with time (i.e., the stick–slipping part essentially slips seismically), we can estimate the slip deficit from the seismic slip history of the stick–slip patches.

Previous studies have shown the persistence of the stick–slip property of the plate interface for earthquakes of various sizes. For example, studies of repeating M2–4 earthquakes have revealed cumulative seismic slips that are consistent with the slips of the plate interfaces around the earthquakes (e.g., Matsuzawa et al. 2002; Igarashi et al.2003). Uchida et al. (2005) showed that the same behavior occurs for larger earthquakes and that the recurrence interval of a characteristic M4.8 earthquake off Kamaishi, Japan, is controlled by the interplate creep rate in the surrounding area. Their result is explained in terms of a time-predictable model in which the isolated stick–slip patch slips when the loaded stress exceeds its strength. They suggested that the M4.8 source area should be stuck during the interseismic period and should slip unstably when the loaded stress exceeds its strength. Based on an analysis of small repeating earthquakes, Uchida and Matsuzawa (2011) also evaluated the distribution of interplate coupling around the source area of the Tohoku earthquake. They found a good spatial correlation between the area of the coupled region prior to the Tohoku earthquake and the area of large co-seismic slip. Yamanaka and Kikuchi (2003) analyzed the source rupture processes of two much larger earthquakes: the 1952 M8.1 earthquake and the 2003 M8.0 earthquake, both off Tokachi. The rupture areas of the two earthquakes are nearly identical, and the 5-m seismic slip of the 2003 event is consistent with the slip deficit accumulated over the 51 years from 1952 to 2003, assuming a subduction rate of ~10 cm/year.

In the above studies, the slip of earthquakes of various sizes is regarded as an indicator of slip on the surrounding plate interfaces. In this paper, we apply this concept to global subduction zones to identify regions with the potential to generate M9-class earthquakes. For each interplate earthquake up to M8, we assume that a stick–slip patch exists and we evaluate the interplate slip deficit by comparing the cumulative slip of each patch with the relative plate motion. From the areas of the patches that are accumulating slip deficit, we calculate the areas in which strain is accumulating due to the locked regions that are potential source areas of huge interplate earthquakes. Note that we do not evaluate the cumulated slip deficit itself but instead evaluate the spatial extent of the area dominated by slip-deficient stick–slip patches.

METHODS: EVALUATION OF A 110-YEAR-LONG SLIP DEFICIT

To study slip deficit on a plate boundary, we first define the plate boundary configuration and then identify stick–slip patches of interplate earthquakes using the centroid moment tensor solution catalog provided by the Global CMT (GCMT) project (Dziewonski et al. 1981; Ekström et al. 2012). We then trace the slip history of these stick–slip patches using the ISC-GEM Global Instrumental Earthquake Catalogue (ISC-GEM catalog; Storchak et al. 2013) to evaluate the slip deficit. Finally, we evaluate the extent of the area in which slip deficit has accumulated by comparing the seismic slip with the relative plate motion. The spatial extent of the slip-deficient area is regarded as the size of a hidden, potential earthquake rupture area. The specific steps employed in this analysis are described below and shown in the flowchart in Fig. 2.

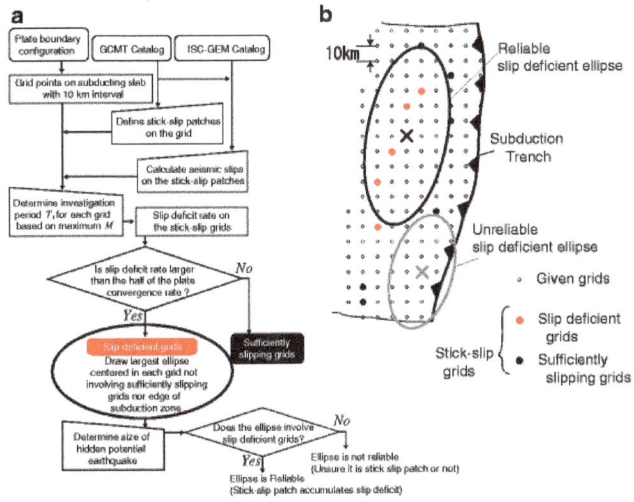

Figure. 2 a Flowchart of our analysis. We use plate boundary configurations, the GCMT catalog, and the ISC-GEM catalog as input data. Grid points are set on the plate boundary at 10-km intervals. We analyze interplate earthquakes from the GCMT catalog. On the plate boundary, we define stick–slip patches according to the magnitude of the interplate earthquakes using Eqs. 1, 3, and 4. The seismic slip of the patches is calculated at the grid points based on the magnitudes of the GCMT and the ISC-GEM catalog earthquakes. The cumulative seismic slip on the stick–slip grid points is divided by the period of investigation, yielding the slip rate. The plate convergence rate minus the slip rate is defined as the slip-deficit rate on each stick–slip grid point. If the slip-deficit rate is larger than half of the relative plate motion, the grid point is regarded as a "slip-deficient grid"; otherwise, it is a "sufficiently slipping grid". The largest elliptical area that contains no sufficiently slipping grid points is regarded as the area of hidden potential earthquakes. **b**Schematic illustration of the estimation of hidden potential earthquakes. The red and black solid circles show "slip-deficient" and "sufficiently slipping" grid points, respectively. The black open ellipse centered at the grid point shown by a cross is an example of source area estimation for hidden potential earthquakes. This ellipse contains four "slip-deficient" grid points. The gray open ellipse is an example of an unreliable ellipse as it does not contain any "slip-deficient" grid points and is terminated by the edge of the plate boundary. Each grid point has an ellipse drawn in the same way

Plate Boundary Configurations

First, we defined the plate boundary configuration using the slab surface geometry constructed by Gudmundsson and Sambridge (1998). In mapping stick–slip patches on the plate boundary, we considered patches shallower than the Moho depth of the overriding plate to be available in the slip-deficit evaluation. Even if the sub-Moho mantle was seismogenic, the viscous flow of mantle wedge beneath the Moho would be likely to release at least some tectonic stress by viscous relaxation for rheologies constrained by experimental mineral physics data (e.g., Hirth and Kohlstedt 2003). We adopted 35 km as the Moho depth. In the actual procedure, we implied the 50 km depth by Gudmundsson and Sambridge (1998) to be 35 km according to Syracuse and Abers (2006), which pointed out that the slab surface contours of Gudmundsson and Sambridge (1998) tend to be shifted downwards compared with slab seismicity by ~15 km at a depth of 50 km. We placed grid points at 10-km spacing at depths shallower than 50 km on the plate boundaries of Gudmundsson and Sambridge (1998) and calculated the interplate slip history at each of the grid points.

Stick–slip patch map constructed from the GCMT catalog

We extracted interplate earthquakes that occurred during the assigned period from the GCMT catalog, which lists earthquakes of $M \geq 5.5$ that have occurred since 1976. We define interplate earthquakes as those located on the slab surface with the following properties: dip angle of <30°; plunge of the N (neutral)-axis, which is the intersection of the two nodal planes, of <15°; and direction of the T (tension)-axis within ±45° of the direction of relative plate motion.

To construct the map of stick–slip patches from the extracted interplate earthquakes, we convert seismic moments to slip and rupture size. The seismic moment M_0 (Nm) is calculated from the moment magnitude M_w under the relation derived by Hanks and Kanamori (1979):

$$\log M_0 = 1.5M_W + 9.1 \tag{1}$$

From the definition of seismic moment (Aki 1966), the average slip D_{ave} over the rupture area is

$$M_0 = \mu S D_{ave} \tag{2}$$

where μ is rigidity expressed in Pa, and S is rupture area in square meter. The rigidity is assumed to be 30 GPa in this study. The relationship between the area and slip of interplate earthquakes was studied by Murotani et al. (2008) using strong motion seismic records, yielding the following scaling relationship between S and M_0:

$$S = 1.48 \times 10^{-4} {M_0}^{2/3} \quad (\mathrm{m}^2)\,. \tag{3}$$

Since Murotani et al. (2008) did not consider the aspect ratio of the rupture area, we derived the relation between length L (m) and width W (m) of the fault as

$$L = 0.1650W^{1.239} \tag{4}$$

from Blaser et al. (2010), in which the L and W were independently related to M_W. The width, length, and average slip of each earthquake can therefore be derived from the earthquake's magnitude using Eqs. 1, 2, 3, and 4. The slip distribution during each rupture must be assumed to evaluate the slip history of the slab surface. Though many other studies have reported a scaling relationship between M_0 and L, W, and average slip (e.g., Geller 1976; Wells and Coppersmith 1994), a characteristic shape of slip distribution has not been proposed. Murotani et al. (2008) defined the asperity as the area in which the slip of an earthquake is >1.5 times the average slip over the whole rupture area. In this study, we interpret this asperity as a stick–slip patch. This showed that the asperity occupies about 1/5 of the rupture area and its average slip is 2.2 times that of the whole rupture area. As a simple function, we define slip distribution by the following Gaussian function (Fig. 3a):

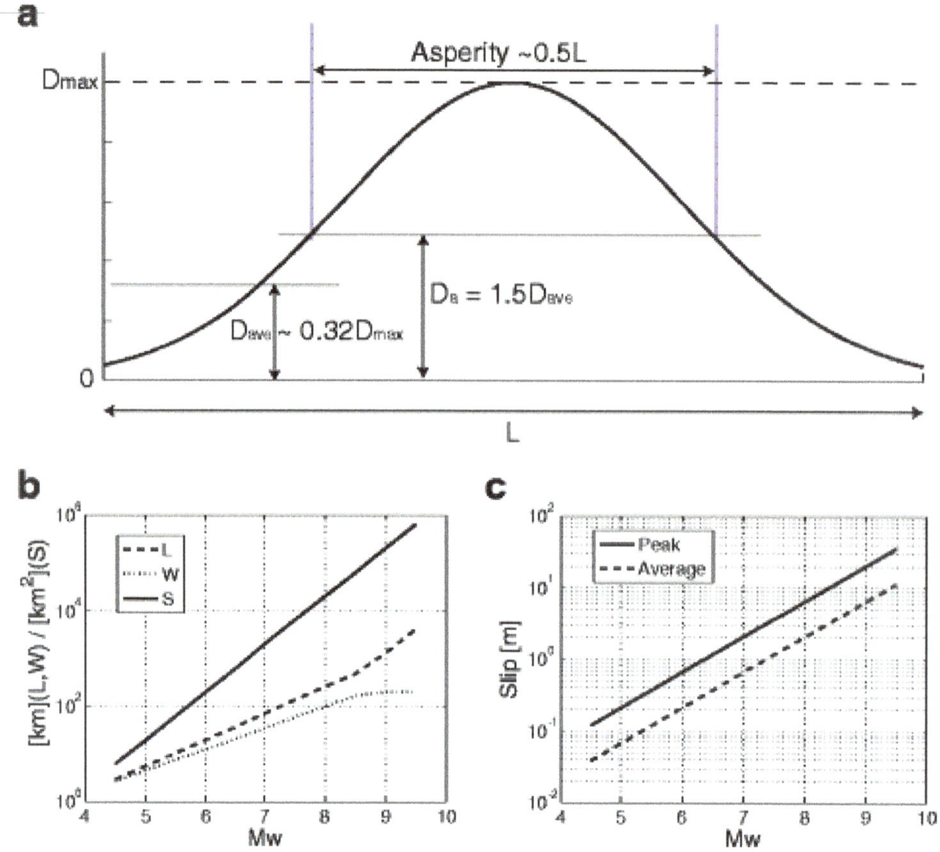

Figure. 3

Slip distribution and adopted scaling law of interplate earthquakes. **a** Gaussian slip distribution defined along fault strike. **b** Scaling relationships between fault area (solid line), fault length (dashed line), fault width (dotted line), and earthquake magnitude. Values are expressed in kilometer scale for readability. **c** Scaling relationships between the peak (solid line) and average slip (dashed line) of earthquake faults and earthquake magnitude

$$D = D_{\max} \exp\left(-\frac{x^2}{0.29^2 L^2} - \frac{y^2}{0.29^2 W^2}\right) \ (\mathrm{m}), \tag{5}$$

where x and y are the distances along strike and dip from the centroid, respectively. The coefficient 0.29^2 is adjusted to satisfy Murotani's relation. For an elliptical fault, the area S is π4LWπ4LW. From Eq. 5, D_{max} is 3.1 times D_{ave}. By substituting D_{ave} (as derived from Eqs. 2 and 3) into Eq. 5, the slip distribution can be calculated for a givenM_W (Fig. 3b, c). We define the inner area of an ellipse with long and short axes L/2 and W/2, respectively, as a stick–slip patch. In our analysis, we assume that the along-dip length of the slipping area W does not exceed 250 km, which is the largest-ever seismologically recorded width and was recorded for the 1964 Alaskan earthquake (Christensen and Beck 1994). Note that this cutoff does not affect the estimation of the cumulative seismic slip of cataloged earthquakes because all such earthquakes have W<250 km. This cutoff affects the calculation of the magnitude of the hidden potential earthquakes from the spatial extent of the slip-deficient area, as described in the "Estimation of the size of potential rupture area hidden behind slip-deficient stick–slip patches" section. However, this impact is minor, and a change in the cutoff value of 100 km only results in a change in the estimated magnitude of a ~M9 potential earthquake of 0.1. This cutoff is employed solely to avoid unrealistically large magnitude estimates.

The fault slip distribution and stick–slip patch are calculated using the above procedure using the GCMT catalog. Hereafter, grid points included in the stick–slip patches are called stick–slip grid points.

Slip accumulation and slip-deficit rate on stick–slip patches deduced from the GCMT and ISC-GEM catalogs

Cumulative seismic slip was estimated for the stick–slip grid points. The cumulative seismic slips of each stick–slip grid point for the earthquakes listed in the GCMT catalog were calculated by applying Eqs. 2, 3, 4, and 5 to each earthquake.

We used the ISC-GEM catalog to estimate the slip for the period before the beginning of the GCMT project, which started as the Harvard CMT in 1976. The ISC-GEM catalog is a global catalog of hypocenter locations and magnitudes of instrumentally recorded earthquakes from 1900. We calculated the cumulative seismic slip on each grid point for the cataloged earthquakes in the same way as for those in the GCMT catalog, although for the ISC-GEM catalog the center of slip (the origin of x and y defined in Eq. 5) is the earthquake hypocenter rather than the centroid. From the cumulative seismic slip, the slip-deficit rate U_j is defined for the jth grid point as follows:

$$U_j = C_j - \frac{\sum_i^N D_{ji}}{T_j}, \tag{6}$$

where C_j is relative plate motion speed (meters/year) on the jth grid point, derived from Bird (2003). D_{ji} is seismic slip (meters) on the grid point by the ith earthquake, and T_j is the total period of interest, which varies among different grid points. To evaluate the rate of slip deficit using the catalogs, it is necessary to consider two ways in which the catalogs are deficient. The first deficiency is in the record of smaller earthquakes during the early years of the ISC-GEM catalog, when cutoff magnitudes were 7.5, 6.25, and 5.5 for the periods prior to 1918, from 1918 to 1959, and after 1959, respectively. To compensate for this incompleteness, we defined different total periods of the earthquake history, T, for different stick–slip grid points depending on the maximum earthquake size for each grid point. The stick–slip grid points with maximum earthquake sizes smaller than 6.25, from 6.25 to 7.5, and larger than 7.5 are evaluated for the periods starting in 1960, 1918, and 1900, respectively. Since larger earthquakes have longer recurrence intervals, it is reasonable that the stick–slip grid points in stick–slip patches that are capable of producing larger earthquakes should be evaluated using longer time periods. The second deficiency is a lack of mechanism information since fault types are only known for the period when the GCMT catalog is available. Outside of this period, we selected earthquakes with estimated source areas that include stick–slip patches identified from the GCMT catalog and treated them as interplate earthquakes. This assumption results in an overestimation of cumulative slip due to miss-classification of intraplate earthquakes as interplate earthquakes and therefore results in an underestimation of the resulting slip deficit of the grid point. The implications of this underestimation are discussed in the "Validity of assumptions" section.

Errors in estimating the locations of epicenters in the cumulative slip estimation can be important. The center of the slip is simply defined by the epicenter of each earthquake; however, since errors in epicenter locations are ~5–15 km, we do not adopt the epicenter location as the center of the ellipse for small earthquakes. We define the centers of the slip distribution ellipses for earthquakes of <M6.25 with short diameters of <15 km as the closest stick–slip grid point within 15 km of the earthquake's epicenter. This procedure minimizes the apparent accumulation of slip deficit on small stick–slip patches caused by uncertainties in the epicenter location.

Based on the above procedure, we estimated the slip and then the slip deficit of the stick–slip grid points using the GCMT catalog and the ISC-GEM catalog.

Estimation of the size of potential rupture area hidden behind slip-deficient stick–slip patches

We estimated the slip deficit around the source area of the 2011 Tohoku-Oki earthquake from the GCMT and ISC-GEM earthquake catalogs. Figure 4 shows the estimated cumulative slip of the stick–slip grid points from 1900 to 2010 around the source area of the 2011 Tohoku earthquake, superposed on the slip distribution for M7-class earthquakes occurring in the past 80 years deduced from strong motion records (after Murotani et al. (2004), Yamanaka and Kikuchi (2004), and Y. Yamanaka (NGY Seismology Notebook, http://www.seis.nagoya-u.ac.jp/sanchu/Seismo_Note, last updated November 25, 2015)). The estimated slip distribution correlates well with that deduced from seismic records, although with a few exceptions such as the lack of slip estimated from strong motion around 36° N and 42° N. We attribute the difference around 36° N to the incompleteness of slip distribution analyses for old earthquakes. In this region, six large earthquakes with magnitude ≥M6.8 that occurred before 1968 are listed in the ISC-GEM catalog. We attribute the difference around 42° N to underestimation of the 1952 earthquake off Tokachi by strong motion data due to scaling out of the seismograms (Yamanaka and Kikuchi 2003). Since our estimation of cumulative slip using earthquake catalogs seems to be as accurate as previous estimations using strong motion records, at least along the Japan trench, we applied this cumulative slip estimation technique to the area for which historical strong motion records are unavailable. Using the estimated cumulative slip, we estimated the rupture area of a hidden potential earthquake from the spatial extent of slip-deficient stick–slip patches. First, we defined the stick–slip grid points with slip-deficit rates (Eq. 6) larger than half of the relative plate motion speed as "slip-deficient" and those with rates smaller than half of the relative plate motion speed as "sufficiently slipping". Next, centering each grid point, we drew the largest possible ellipse, with long and short radii corresponding to L/4 and W/4 (Eq. 5), respectively, which does not include any sufficiently slipping grid points. Each ellipse is terminated by either a sufficiently slipping grid point or by the end of the subduction zone under the condition that the diameter of the long axis of the ellipse is ≤1400 km (corresponding to the rupture size of an M9.5 earthquake by the scaling relationship of Eqs. 3 and 4). We calculated the earthquake magnitude from the size of the ellipse through the scaling relationship defined by Eqs. 1, 3, and 4 (e.g., the diameter is 440 km for M9 events and 110 km for M8 events). The calculated magnitude corresponds to that of an earthquake that ruptures the whole area enclosed by the ellipse as its asperity. If the ellipse is in contact with or contains stick–slip grid points of past earthquakes with magnitudes greater than that calculated from the size of the ellipse, the calculated magnitude is replaced by that of the past earthquake and the resulting magnitude is regarded as "reliable". Otherwise, the calculated magnitude is retained. The calculated magnitude is regarded as "unreliable" when no stick–slip grid points are included within the ellipse, since in this case it is not possible to judge whether the lack of stick–slip grid points reflects the absence of a stick–slip patch (i.e., the fault slips aseismically) or an accumulation of slip deficit on a hidden stick–slip patch, which would cause future interplate earthquakes.

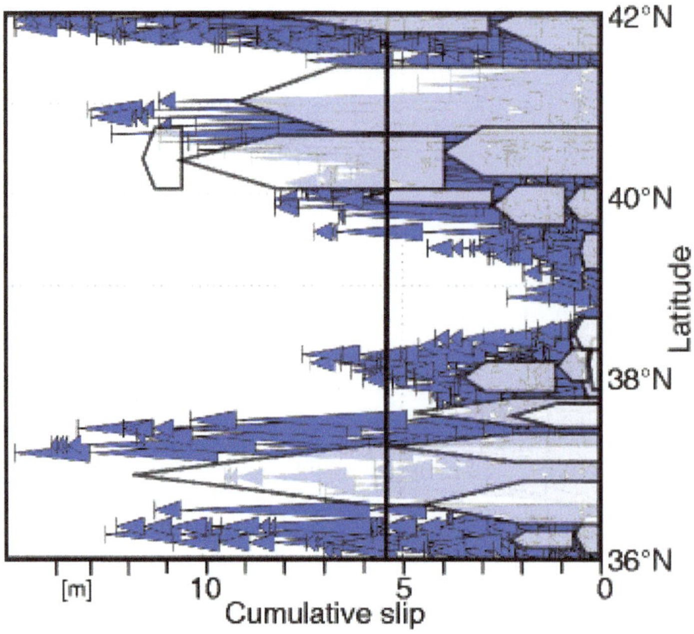

Figure. 4

Cumulative seismic slip for the 110 years preceding the 2011 Tohoku earthquake. Blue triangles are cumulative slip calculated from the GCMT and ISC-GEM catalogs. The superposed translucent pentagon-shaped patches are cumulative slip deduced from strong motion records, after Ikuta et al. (2012). The solid vertical line denotes half of the cumulative relative plate motion over the past 110 years

By applying the above procedure to all grid points on the slab surface shallower than 35 km, we estimated the locations of source areas of potential earthquakes.

RESULTS

Spatial extent of "slip-deficient" grid points prior to three M9-class earthquakes in the past decade

We tested the proposed method by applying it to the source areas of the M9-class 2011 Tohoku, 2004 Sumatra, and 2010 Chile earthquakes using the earthquake catalogs prior to the three events. The available catalog periods for the 2011 Tohoku earthquake and the 2004 Sumatra earthquake extend from 1900 to one month prior to each earthquake. For the source area of the 2010 Chile earthquake, we used the catalog for the period from January 1961 to January 2010 because the 1960 Chile earthquake (Mw9.5) occurred in the area adjacent to the 2010 event. The calculated slip for the 1960 Chile earthquake is 46 m, which corresponds to the cumulative relative plate motion for ~460 years, assuming a plate convergence rate of 10 cm/year. Our method is designed to evaluate the spatial extent of slip deficit using earthquakes with slip less than the cumulative convergence within the catalog period, as further discussed in the "Validity of assumptions" section. Therefore, we estimated the slip deficit before the 2010 earthquake using the catalog after the 1960 huge interplate event.

Figure 5a–c shows the distribution of the calculated magnitude around the rupture areas of the three M9-class events. Around the rupture area of the 2011 Tohoku earthquake (Fig. 5a), the calculated magnitude is >9.0 along the up-dip end of the plate boundary between 36° N and 39° N, which coincides with the area of large slip in the 2011 Tohoku earthquake. Other areas with a calculated magnitude of >9.0 along the Japan trench and its extension are located along the Kuril trench centered on 41° N and the Izu–Bonin trench at ~30° N. Around the source area of the 2004 Sumatra earthquake (Fig. 5b), the area with a calculated magnitude of >9.0 extends from 5° S to 15° N. However, most of these grid points do not have slip-deficient grid points inside their ellipses and the reliability of the slip deficit is low. For the 2010 Chile earthquake (Fig. 5c), the calculated magnitude is >9.0 in the area between 35° S and 45° S, which includes the source areas of the 1960 and 2010 earthquakes.

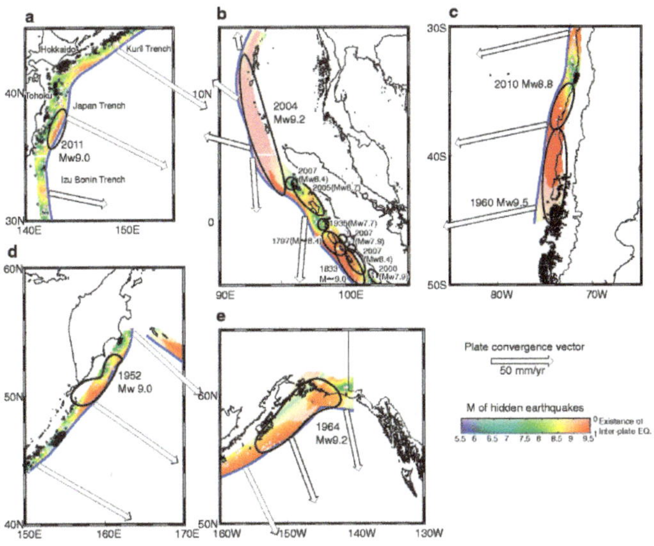

Figure. 5 The expected earthquake size around the source areas of actual M9-class events, as estimated using the seismic catalog before (**a–c**) and after (**c–e**) the seismic events. Color shows the magnitudes of the potential earthquakes that are estimated from the size of the slip-deficient ellipse for each grid point. Unreliable grid points that have no slip-deficient grid points within their ellipses are shown by translucent colors. **a** The Kuril, Japan, and Izu–Bonin trenches. The source area of the 2011 Tohoku earthquake (Mw9.0) is enclosed by a solid line, and the trench is shown by a solid blue line. Arrows show the relative motion of the overriding plate with respect to the subducting slab. **b** The Java trench. The source areas of past earthquakes including the 2004 Sumatra earthquake (Mw9.2) are enclosed by solid lines. The rupture areas of the 2005 (Mw8.7), 2007 (Mw7.9), 2007 (Mw8.4), 2007 (Mw8.4), 1935 (Mw7.7), 1833 (M~9.0), and 1797 (M~8.4) events are modified from Briggs et al. (2006). **c** The Chile trench. The source areas of the two M9-class events are shown by solid lines (e.g., Barrientos and Ward 1990; Pollitz et al. 2011). **d** Kamchatka. The source area of the 1952 Kamchatka earthquake (Mw9.0) is enclosed by a solid line. **e** Alaska. The source area of the 1964 Alaska earthquake (Mw9.2) is enclosed by a solid line. The rupture area is modified from Christensen and Beck (1994). The color scales, legend, and arrow length are the same for **a–e**

Spatial extent of "slip-deficient" grid points after M9-class earthquakes of the 20th century

We applied the same procedure as that detailed in the previous section to the rupture areas of the two other M9-class earthquakes that occurred in the 20th century; i.e., the 1952 Mw9.0 Kamchatka earthquake and the 1964 M9.2 Alaska earthquake (in addition to the 1960 Mw9.5 Chile earthquake). However, in this analysis we used the catalog following the M9-class events instead of preceding them.

Figure 5c–e shows the spatial distributions of calculated magnitudes around the rupture areas. Around the rupture area of the 1952 Kamchatka earthquake (Fig. 5d), the calculated magnitude is >9.0 between 48° N and 52° N, which coincides with the rupture area of the 1952 Kamchatka earthquake. Around the rupture area of the 1964 Alaska earthquake (Fig. 5e), the magnitude is >9.0 from 160° W to 140° W and significantly protrudes westward from the rupture area of the earthquake defined by Christensen and Beck (1994). This result may indicate the existence of another source area for a huge earthquake adjacent to the area that ruptured in 1964. The result around the source area of the 1960 Chile earthquake is described in the previous subsection (Fig. 5c). The magnitudes of the six M9-class events are consistent with those expected from the spatial extents of the slip-deficient grid points.

Spatial extents of "slip-deficient" grid points in present global subduction zones

We applied the same procedure as that described above to circum-Pacific subduction zones using earthquake catalogs prior to 2011. The rate of relative plate motion at each grid point was derived from Galgana et al. (2007), Ohkura (2013), and Bird (2003). The catalog period is from 1900 to 2010, except for the Ecuador–Peru–Chile trench, for which we chose a start time of 1907 in order to evaluate the rupture area of the 1906 Ecuador earthquake (M8.8). In Fig. 6, the areas in which calculated magnitudes exceed 9.0 are enclosed by ellipses. In addition to the six known M9-class source areas, the following 25 fault segments are enclosed by ellipses in Fig. 6: the border of the Kuril and Japan trenches; northern, central, and southern Ryukyu; southern Izu–Bonin; northern, central, and southern Mariana; western and central Aleutian; South Sandwich; northern, central, and southern Central America; western, central, and eastern South Java; Manila; Ecuador-Peru; Northern Chile; New Britain; New Hebrides; Tonga; and the northern and southern Kermadec trenches (Table 1).

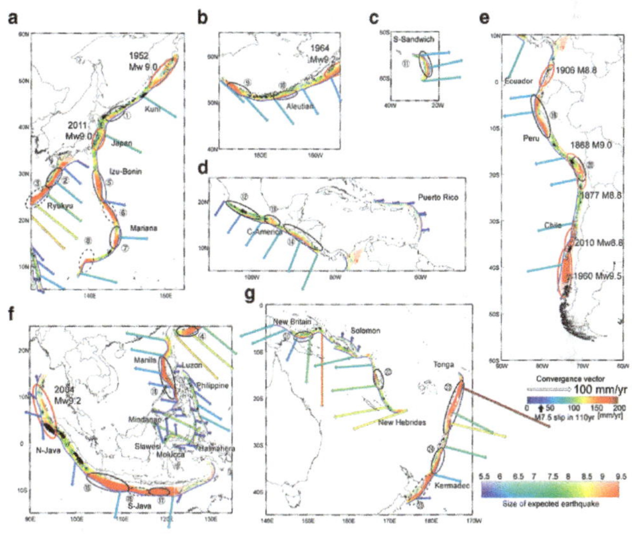

Figure. 6 Expected size distribution of hidden potential earthquakes in global subduction zones. Red colors correspond to larger magnitudes for the expected earthquakes. Areas with low reliability have translucent colors. The lengths and colors of the vectors show the convergence rate of the plate at the subduction zone (showing the displacement vector of the overriding plate). Red circles with white outlines and open circles denote slip-deficient and sufficiently slipping grid points, respectively. Solid and broken ellipses show categories A and B areas, respectively. **a** The Kuril, Japan, Izu–Bonin, Mariana, and Ryukyu trenches. The areas with expected magnitudes of >9.0 are indicated by numbered ellipses. The source areas of the 1952 Kamchatka (Mw9.0) and 2011 Tohoku (Mw9.0) earthquakes are enclosed byred ellipses. **b** The Aleutian trench. The source area of the 1964 Alaska earthquake (Mw9.2) is enclosed by a red ellipse. **c** The South Sandwich trench. **d** The Central America (C-America) and Puerto Rico trenches. **e** The Peru–Chile trench. The source areas of the huge 1960 and 2010 events are enclosed by red ellipses. **f** The North Java, South Java, Manila, East Philippine, Molucca, Halmahera, and Sulawesi trenches. The Manila trench and a huge area in the southern part of the Java trench show expected magnitudes of >9.0. Focal areas of the huge historical earthquakes that occurred in 1868 (M9.0), 1877

(M8.8), and 1906 (M8.8) are also shown. The areas for the 1868 and the 1877 events were derived from intensity distributions presented by Comte and Pardo (1991), and the area of the 1906 event is from Kanamori and McNally (1982). **g** The New Britain, Solomon, New Hebrides, Tonga, and Kermadec trenches

Table 1 Subduction zone, plate identifiers and Euler poles/convergence rate

Subduction zone	Plate pair	λ, °N	ϕ, °E	ω°/Myr	Convergence rate [mm/year] (/at latitude/direction [deg])	Author	# in Fig. 7categories: A–C	Uncertain because (S) convergence is short/(N) no interplate earthquakes
Kuril–Japan	PA–OK	−55.42	97.14	0.845	–	Bird2003	① A	–[b]
Ryukyu	PS–ON	−36.69	−40.42	3.603	–	Bird2003	②A, ③B, ④A	–[c]
Izu–Bonin	PA–PS	1.20	134.20	1.000	–	Bird2003	⑤A	–
Mariana	PA–MA	−43.78	−30.80	1.278	–	Bird2003	⑥B, ⑦A, ⑧B	–
Aleutians	PA–NA	−48.71	101.83	−0.749	–	Bird2003	⑨A, ⑩A	–[d]
South Sandwich	SA–SW	36.60	150.40	1.880	–	Bird2003	⑪A	–
C-America (92° W<)	CO–NA	27.88	−120.68	1.357	–	Bird2003	⑫A, ⑬A	–
C-America (<92° W)	CO–CA	21.63	−121.36	1.256	–	Bird2003	⑭A	–
Puerto Rico (17° N<)	NA–CA	−78.41	82.54	0.175	–	Bird2003	–	S/N
Puerto Rico (<17° N)	SA–CA	−51.50	114.30	0.272	–	Bird2003	–	S/N
Cascadia	JF–NA	−20.71	67.78	0.769	–	Bird2003	–	N
Halmahera	SU–BH	42.79	−79.05	1.24	–	Bird2003	–	–
Philippine (<5° N)[a]	PS–		–		9/2° N/N 90° W	Ohkura2013	–	S
Philippine (5° N<)[a]	PS–		–		40/13° N/N 45° W 62/8° N / N 60° W	Ohkura2013	C	–
Luzon[a]	PS–		–		9/17° N/N 90° W 15/15° N/N 90° W	Ohkura2013	–	S
Molucca (<2° N)[a]	–SU		–		26/1° N/N 80° W	Ohkura2013	–	S
Molucca (2° N<)[a]	–SU		–		17/5° N/N 70° W	Ohkura2013	–	S
Mindanao[a]	SU–		–		25/6° N/N 90° E	Ohkura2013	–	S
Sulawesi	SU–MS	2.10,	126.20	3.40	–	Bird2003	–	–
South Java (<5° N, <122° E)	AU–SU	1.85	60.18	0.709	–	Bird2003	⑮A, ⑯B, ⑰A	–
South Java (122° E<)	AU–TI	14.12	−50.54	1.743	–	Bird2003	–	S/N
North Java (5° N<)	IN–BU	13.49	94.79	2.349	–	Bird2003	–	–[e]
Manila[a]	SU–				62/5° N /N 100° E 37/ 8° N/N 100° E	Ohkura2013	⑱A	–
Ecuador–Peru(<4° S)	NZ–ND	52.30	−91.08	0.661	–	Bird2003	–	–
Ecuador–Peru (4° S–15° S)	NZ–SA	55.96	−94.01	0.725	–	Bird2003	⑲A	–
Peru-Chile (15° S-21° S)	NZ–AP	79.02	−172.70	0.626	–	Bird2003	⑳A	–
Chile (21° S<)	NZ–SA	55.96	−94.01	0.725	–	Bird2003	–	–[f]
Solomon	NB–PA	4.00	−41.00	0.330	–	Bird2003	–	S
New Britain (<147° E)	WL–NB	28.71	130.35	1.261	–	Bird2003	–	–
New Britain (147° E-154)	WL–SB	−5.73	144.86	9.736	–	Bird2003	㉑A	–

New Britain (154° E<)	PS–SU	−59.99	−1.24	1.047	–	Bird2003	–	–
New Hebrides (<14° S)	AU–PA	60.08	1.74	1.074	–	Bird2003	–	–
New Hebrides (14° S<)	AU–NH	8.71	164.55	2.139	–	Bird2003	㉒A	–
Tonga	PA–TO	−28.81	−177.74	9.30	–	Bird2003	㉓A	–
Kermadec	PA–KE	−47.52	176.89	2.831	–	Bird2003	㉔A, ㉕B, C (~30° S)	–

λ and ϕ are the latitude and longitude of the Euler poles. Plate abbreviations follow Bird (2003): PA = Pacific, OK = Okhotsk, PS = Philippine Sea, MA = Mariana, NA = North America, SA = South America, SW = Sandwich, CO = Cocos, CA = Caribbean, JF = Juan de Fuca, SU = Sunda, BH = Birds Head, MS = Molucca Sea, AU = Australia, TI = Timor, BU = Burma, NZ = Nazca, ND = North Andes, AP = Altiplano, NB = North Bismarck, WL = Woodlark, NH = New Hebrides, TO = Tonga, KE = Kermadec

[a]Convergence rate is applied rather than the plate rotation parameters

[b]Experienced two great earthquakes (1952, M9.0; 2011, M9.0)

[c]N for the Nankai Trough—northern extension of the Ryukyu trench

[d]Experienced one great earthquake (1964, M9.2)

[e]Experienced one great earthquake (2004, M9.2)

[f]Experienced two great earthquakes (1960, M9.5; 2010, M8.8)

DISCUSSION

Validity of assumptions

We first discuss the validity of the assumptions made in this study in order to categorize the calculated slip-deficient areas. We rate the reliability of these areas as follows: category A—strong candidate for strain-accumulated area; category B—a candidate with some uncertainty; and category C—a minor candidate that does not show a large area of slip deficit and has a large uncertainty.

This study was based on the following assumptions:

1. Each interplate earthquake has a stick–slip patch around its centroid. Once a stick–slip patch is defined by an interplate earthquake, it will only slip by earthquake motion, not by creep
2. Stick–slip patches within the rupture area of potential huge earthquakes should accumulate a slip deficit during the interseismic period. Grid points that have slip deficits larger than half of the relative plate displacement during this period are regarded as slip-deficient grid points. In contrast, grid points with cumulative slips larger than half of the slip expected from the relative plate motion are regarded as sufficiently slipping grid points. Areas that include slip-deficient grid points and that do not include any sufficiently slipping points are regarded as accumulating strain
3. The magnitude of each hidden potential earthquake is calculated from the size of the area that contains slip-deficient grid points and that contains no sufficiently slipping grid points, rather than from the slip-deficit value
4. The stick–slip patches are defined only for interplate earthquakes listed in the GCMT catalog. All earthquakes that slipped on the stick–slip patches prior to the GCMT catalog are regarded as interplate earthquakes

The accumulated relative plate displacement during the period of investigation is sufficient to evaluate slip deficit

The first and second assumptions are the basic principles of this study and are supported by previous studies, as described in the "Introduction" section. Since the results of the tests conducted for the source areas of the six huge earthquakes that have occurred during the past 110 years support these basic principles, we consider that these assumptions are valid. It could be argued that stick–slip patches for M7-class earthquakes should be detected by the geodetic observations if the patches are locked during the interseismic period, but this is not always the case. For example, geodetic studies of the interseismic period around the Tohoku-Oki rupture area (e.g., Loveless and Meade 2011) show only one large coupling area around the centroid of the Tohoku-Oki earthquake. Loveless and Meade (2011) carried out a checkerboard test and showed that it is difficult to resolve off-shore locked patches with diameters of <100 km using geodetic inversions, as all stations are located on-shore. Conversely, results from small repeating earthquakes do show coupling around some past M7-class earthquakes (Uchida and Matsuzawa 2013).

The third assumption could lead to large errors because one slip-deficient grid point combined with a vast surrounding aseismic region can generate large areas of apparent accumulated strain. In terms of the fourth assumption, the misclassification of intraplate earthquakes as interplate earthquakes would result in an overestimate of the total seismic slip. The fifth assumption is also critical, as the short duration of the investigated periods results in large uncertainties in determining whether the plate motion is sufficient for the analysis. The impact of these three assumptions on the uncertainty of the results is evaluated below.

First, we discuss the case where a few slip-deficient grid points are surrounded by a vast aseismic region, which generates a large calculated apparent magnitude. We classified the areas showing a large calculated magnitude (as delineated by the ellipses in Fig. 6) into categories A and B based on the localization of slip-deficient grid points. Category A areas contain a good coverage of slip-deficient grid points and do not contain any vacant spaces larger than the rupture ellipse of a magnitude 9.0 earthquake on the plate boundary. The following 20 areas were classified as category A: the southern part of the Kuril trench; the northern part of the Ryukyu trench; the southern part of the Izu–Bonin trench; western and central Aleutian trench; the South Sandwich trench; northern, central, and southern Central America trench; western and eastern South Java trench; the Manila trench; the Ecuador-

Peru trench; northern Chile trench; the New Britain trench; the New Hebrides trench; the Tonga trench; and northern Kermadec trench. It is probably that the interplate seismic slip in these areas is suppressed due to the locked regions that will likely be responsible for future huge earthquakes. The remaining five areas contain large vacant spaces equivalent to the rupture ellipses of M9.0 earthquakes and are recognized as category B: the central part of the Ryukyu trench; the northern and southern parts of the Mariana trench; central South Java trench; and southern Kermadec trench. For these areas, we cannot deny the possibility that the few "slip-deficient" grid points and the surrounding vast aseismic regions have generated the large calculated magnitudes. Although category A areas are relatively reliable, there is still some uncertainty in the calculated expected earthquake magnitudes, as older interplate earthquakes that occurred outside the defined stick–slip patches are ignored. In the Nankai Trough, though which is "unreliable" area, two M8-class interplate earthquakes that occurred in 1944 and 1946 are ignored in our analysis because no stick–slip patches were defined through the GCMT catalog calculations. We cannot discount the possible existence of such missed stick–slip (and sufficiently slipping) patches in the category A areas. In such a case, the magnitude of the hidden potential earthquake would be overestimated. However, significant overestimation is unlikely in subduction zones with high convergence rates because if M8 (not M9) is the largest capable event, such an event with a slip of ~3 m should have occurred within the past 110 years.

Second, we evaluate the impact of misclassification of the earthquake mechanism. All earthquakes that occurred before the beginning of the GCMT catalog with slip areas on the plate interface that include stick–slip grid points are considered to be interplate earthquakes, which should overestimate interplate cumulative slip to some extent. We evaluate the degree of this overestimation using the GCMT catalog for earthquakes that occurred after 1976. Figure 7 shows the spatial distribution of the ratio of the total slip of interplate earthquakes to the total slip of all earthquakes. The ratio λ_j for the jth grid point is calculated using all earthquakes within 220 km of the grid point:

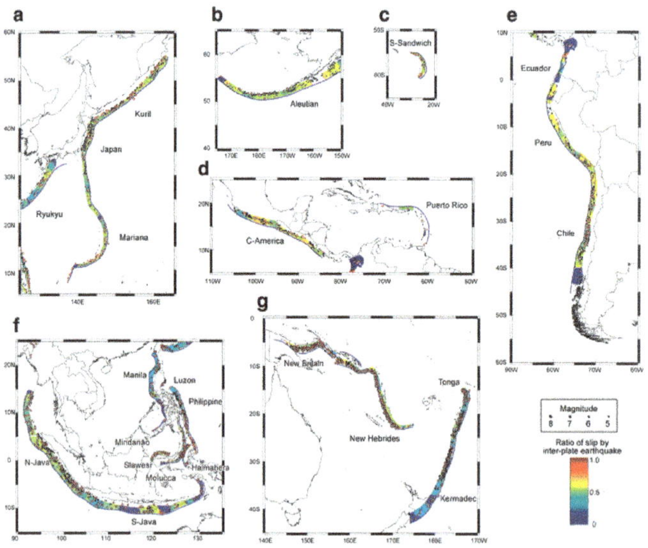

Figure. 7 Spatial distribution of the ratio of the averaged interplate slip to the sum of averaged interplate and intraplate slip. Theblack and red focal mechanism solutions are for interplate and intraplate earthquakes, respectively

$$\lambda_j = \frac{\sum_I D_I}{\sum_I D_I + \sum_K D_K}, \quad (7)$$

where D_I and D_K represent D_{max} (Eq. 5) for each interplate and intraplate earthquake, respectively. The distance of 220 km corresponds to the long axis of the M9.0 stick–slip patch. In this calculation, we omit the earthquakes with slips (from Eqs. 2, 3, 4, and 5) larger than their respective relative plate displacements over the 34 years of GCMT data used in this study. The inclusion of such large and rare earthquakes would cause the results to be unstable and unreliable. The resulting ratios are generally low in the Southwest Pacific (Fig. 7). Figure 7 shows that interplate earthquakes are not prevalent in the Ryukyu trench, most trenches around the Philippine Islands, and the southern Kermadec trench. Consequently, slip deficits based on these results would be underestimated. This is especially true around the Philippine Islands, where calculated magnitudes based on the spatial extent of slip-deficient grid points do not exceed 9.0 except for the Manila trench (Fig. 6e). Therefore, the calculated magnitude may be underestimated around the Philippines. In the Kermadec trench, the calculated magnitude exceeds 9.0 in the south but is smaller in the central part of the trench at ~30° S and is therefore underestimated in this region. We categorize the areas in which the calculated magnitudes do not exceed 9.0 and interplate slip is less than half of the total slip as category C. Such areas include the northern Kermadec trench (~30° S) and the northern Philippine trench (Fig. 6e, g). This evaluation contains additional uncertainties due to the short duration of the test period using GCMT solutions.

Finally, we evaluate the uncertainty of the results due to the short duration of measurements of relative plate displacements. Regions with relatively low plate velocities have larger uncertainties in slip deficit. Figure 8 shows a schematic illustration of the slip history of a huge stick–slip patch (M9) and surrounding smaller patches. The section of the hanging wall around the patch is dragged by the subducting slab (via the stick–slip patch) and its motion relative to the subducting slab is consequently reduced. Therefore, the cumulative seismic slip of the surrounding small stick–slip patches becomes smaller due to this reduced relative motion. Our analysis is designed to detect this slip-deficient area by summing the seismic slip (white areas in Fig. 8); however, we cannot extract the hidden source area using earthquakes with slips larger than the relative plate displacement (Fig. 8b). If we focus on a stick–slip patch with a unit slip smaller than the cumulative plate motion within a given period, the absence of seismic slip might indicate that some larger stick–slip patches also exist behind. However, if we focus on a stick–slip patch with a slip larger than cumulative plate motion, neither the absence nor the presence of seismic slip would provide information on hidden potential earthquakes because the occurrence of an earthquake within the given period is simply a matter of probability. It would be ideal to make observations until the cumulative relative plate motion exceeds the slip of the stick–slip patch used as the slip-deficit indicator. However, it is not easy to know how much displacement of the relative plate motion is sufficient for the estimation because we do not know the size of the potential earthquakes hidden behind the apparent seismicity. As a practical measure of the sufficiency of the period, we compare the total relative plate motion over 110 years with the seismic slip of a single M7.5 earthquake for various subduction zones. Peak slip for an M7.5 earthquake is 4.28 m. Relative plate motions smaller than this are shown in dark blue in Fig. 6 and include those at the Puerto Rico, Solomon, and Northern Java trenches, as well as the trenches around the Philippine Islands, except for the Manila trench. In these regions, the estimated slip-deficit distribution should be affected by relatively small earthquakes and the uncertainties in the results are large compared with other regions. In Table 1, these areas, which display large uncertainties in slip deficit due to small relative plate motions, are listed as uncertain areas, in addition to the Nankai Trough in Japan and Cascadia in North America where no stick–slip patches have been defined.

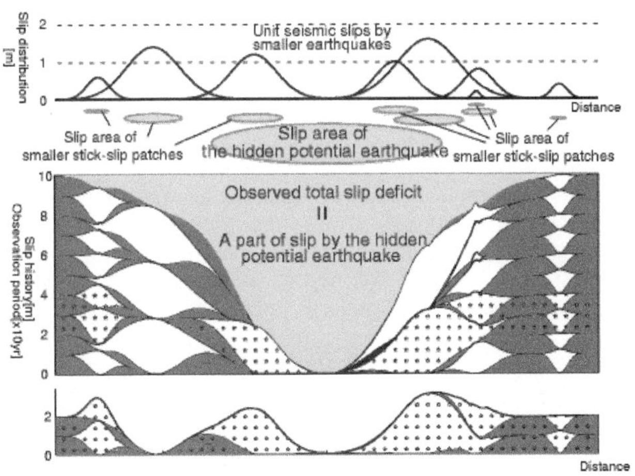

Figure. 8 Schematic illustration of the interplate slip model. Top: locations of stick–slip patches on the plate boundary (ellipses) and their amount of seismic slip (Gaussian curves) by respective earthquakes. Eight stick–slip patches have their own earthquake sizes. A stick–slip patch on the plate boundary has the potential to produce an earthquake with an area and slip much larger than those of the other eight patches. Middle: slip history along the plate boundary at a constant time interval. Seismic and aseismic slip are shown by white and dark gray colors, respectively. Each stick–slip patch releases its slip deficit mainly by seismic slip. The accumulated slip increases with increasing distance from the center of the hidden potential stick–slip patch. In this case, each time interval corresponds to a relative plate motion of 1 m. The area covered by small stars corresponds to the slip during periods from time steps #2 to #4. Bottom: Slip history for the period from time steps #2 to #4 extracted from the middle figure. The slip deficit due to the hidden potential earthquake is not apparent

Focal areas of M9-class earthquakes in the past 110 years

As described above, we developed a method of evaluating strain accumulation on subduction plate boundaries and we applied this method to the six known M9-class earthquakes that have occurred in the past 110 years. We also applied the method to circum-Pacific subduction zones and calculated earthquake magnitudes from the spatial extents of slip-deficient areas. In this section we discuss the correlation between the slip-deficient area and the rupture area of the known M9-class earthquakes. The areas that were calculated from the catalog to have magnitudes of ≥9 prior to the 2011 Tohoku, 2004 Sumatra, and 2010 Chile M9-class interplate earthquakes correlate closely with the rupture areas of the actual M9-class events (Fig. 5). The rupture area of the 2004 Sumatra earthquake is unreliable because the absence of interplate thrust earthquakes in this region during the time when GCMT solutions are available means that we cannot judge whether the area contains a stick–slip patch. In contrast, the results around the source areas of the 2011 Tohoku and the 2010 Chile earthquakes are regarded as being highly reliable because these areas contain widespread stick–slip grid points with slip deficits compared with the relative plate motions. Large magnitudes are estimated in the southern extension of the rupture area of the 2004 Sumatra earthquake from 5° S to 5° N (Fig. 5b). In this area, huge earthquakes occurred in 1833 (~M9.0; Briggs et al. 2006) and 1797 (~M8.4), which correspond to the area with the large calculated magnitude. Six M8-class events have also occurred since 2004 with source areas that correspond to those of the expected large-magnitude earthquakes.

The slip deficits after the 1960 Chile, 1952 Kamchatka, and 1964 Alaska M9-class events evaluated from the catalog also show good correlations between the calculated large-magnitude areas and the rupture areas (Fig. 5c, e). This result reflects a lull in seismicity after an almost full release of seismic moment around the rupture area and may also suggest that the strong and/or large stick–slip patches capable of producing M9-class earthquakes are persistent and recover their locking behavior immediately after huge interplate earthquakes.

Correspondence between historical huge events and the areas where huge earthquakes are expected from the slip-deficit analysis

Our result over the circum-Pacific region using the catalog prior to 2011 shows 25 category A and B segments (Fig. 6). Historical studies point to several more M9-class interplate earthquakes than the six huge events shown in Fig. 5. For example, the 1700 Cascadia (M9.0), 1868 Peru (M9.0), 1877 North Chile (M8.8), and 1906 Ecuador (M8.8) earthquakes are listed as ≥M8.8 by the USGS (http://earthquake.usgs.gov/earthquakes/world/historical_mag.php last updated November

25, 2015). For Cascadia, our method cannot evaluate the slip deficit because no interplate earthquakes occurred during the time covered by the GCMT catalog. Instead, our results indicate that the region along the Ecuador–Peru–Chile trench is classified as category A. Figure 6f shows our result along the trench, as estimated from the catalog for the period from 1907 to 2010 together with the rupture areas of the historical M9-class earthquakes. Around the rupture areas of the 1906 M8.8, 1868 M9.0, and 1877 M8.8 earthquakes, our analysis anticipates M9 or larger earthquakes, although the spatial extents and the locations of these events do not agree exactly with those of the actual earthquakes. The reliability of the expected magnitude around the 1906 source area is low because only a few stick–slip grid points exist in this region. The area with expected magnitudes of >M9 for the 1868 and 1877 events is localized along only one border of the actual rupture areas, which were widely spread along the trench.

Along the Ecuador–Peru trench, a huge area extending from 2° S to 10° S has a large calculated magnitude. Historical records give no indication that a great earthquake has occurred in this region during the last 400 years (Kelleher 1972). Although McCann et al. (1979) classified this region as the site of no great earthquakes based on its earthquake history, our results show that it contains many stick–slip patches with cumulative seismic slips smaller than the relative plate motion, indicating that this region is a likely location of future M9-class events.

Subduction with back-arc spreading

Our results show large expected magnitudes of >9.0 in several subduction zones that have back-arc spreading, including the Mariana, Ryukyu, South Sandwich, and Izu–Bonin trenches, which are classified as Mariana-type trenches. Uyeda (1982) described the differences between Mariana-type and Chilean-type trenches and pointed out that earthquakes with magnitudes significantly larger than 8.0 do not occur in Mariana-type trenches. Scholz and Campos (1995) evaluated this relationship quantitatively and proposed that seismic coupling in Mariana-type trenches is suppressed due to a reduction in the normal force at the plate interface, which is related to the absolute velocity of the upper plate and the length of the subducting slab. The seismic coupling correlates with the calculated normal force, which was explained in the context of the rate- and state-dependent friction law (e.g., Ruina 1983), which states that a reduction in normal force increases the sliding stability of the interface. We should note that their theory only evaluates the likelihood that a plate interface is a region of stable slip, and it does not exclude the possibility that great earthquakes may occur if unstable sliding patches exist. Our results show that stick–slip patches with cumulative seismic slips smaller than the relative plate motions occur in the Mariana trench and other Mariana-type trenches. These patches should compensate for the slip deficit by slipping seismically in the future, unless the frictional properties of these patches change from stick–slip to stable slip; we cannot exclude the possibility of M9-class earthquakes in these regions.

Frequency of M9-class earthquakes

Seismologists have attempted to correlate the history of earthquakes over the past century with subduction zone properties such as convergence rate, slab age, type of overriding plate, and fault temperature, in order to understand the limits on earthquake size. For example, Ruff and Kanamori (1980) showed significant correlations between characteristic earthquake size in subduction zones and the convergence rates and ages of the subducting slab. They suggested that a young and fast-moving slab would be subducted into the mantle at a lower angle and would stick more strongly to the overriding plate, resulting in larger earthquakes. However, later studies (e.g., Pacheco et al. 1993; Stein and Okal 2007) that incorporated more recent earthquake data and improved data show weaker correlations than those proposed by Ruff and Kanamori (1980). Furthermore, the underlying idea that some subduction zones may never produce an M9-class earthquake has been called into question with the occurrence of the 2004 Sumatra Mw9.2 and the 2011 Tohoku Mw9.0 earthquakes. Our study does not consider the physical properties of subduction zones and is based purely on the idea of slip deficit. As a result, the calculated magnitudes in 31 areas exceed M9.0. This number may be considered too high in terms of the number of areas that are likely to be candidates for M9-class earthquakes. However, it should not be surprising from a statistical viewpoint. Figure 9shows the magnitude–frequency relationship for earthquakes occurring between 1900 and 2010 with magnitudes of >7.5. For magnitudes between 7.5 and 9.5, the relationship seems to follow the Gutenberg–Richter law (GR law), suggesting that M9-class earthquakes are not unusual ones in global subduction zones.

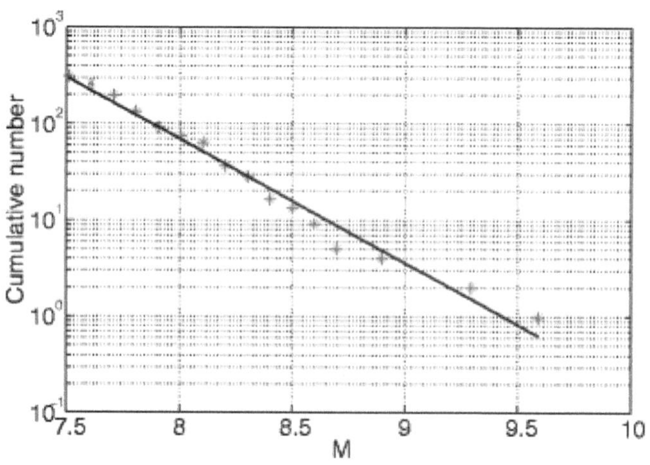

Figure. 9 Magnitude-cumulative frequency curve for earthquakes of >M7.5 in 1900–2010 derived from the ISC-GEM catalog. The asterisks show actual data and the solid line is the linear fit. The 2011 Tohoku earthquake (Mw9.0) is not listed, and the 2010 Chile earthquake (Mw8.8) is listed as M8.3 in the catalog

If we accept that the observed frequency of M9-class events, at six in the past 110 years, is reasonable in terms of the GR law, we can assume that the same number of events will occur each 110 years. Upcoming M9-class events should occur in regions other than the six known areas because the recurrence interval of each M9-class event should be longer than several hundred years. Consequently, it would not be surprising if the number of areas capable of producing an M9-class earthquake is several times greater than this value of six. Indeed, 31 such areas are defined by our study, which is within the range of this estimate.

CONCLUSIONS

We evaluated the interplate slip deficit for the 110 years prior to the 2004 Sumatra (Mw9.2), 2010 Chile (Mw8.8), and 2011 Tohoku (Mw9.0) M9-class earthquakes from a seismic catalog. We found that large areas with sizes equivalent to source areas of M9.0 events around the rupture areas of the earthquakes accumulated slip deficit. We also investigated the slip deficit that accumulated after the 1952 Kamchatka (Mw9.0), 1960 Chile (Mw9.5), and 1964 Alaska (Mw9.2) earthquakes and found that slip deficit accumulated around the rupture areas after the occurrence of the M9-class earthquakes. We applied the same approach to global subduction zones and found that 31 regions, including the 6 listed above, show slip deficits over large areas equivalent to the size of an M9-class earthquake source area.

DECLARATIONS

Acknowledgements

We are grateful to J. Loveless and T. Matsuzawa for critical reviews and encouraging comments that improved the manuscript. We thank the Global CMT project and the International Seismological Center (ISC) for providing the earthquake catalog. The manuscript was improved following helpful discussions on slip-deficit distribution with Y. Morishita, T. Watanabe, and K. Yamaoka.

REFERENCES

1. Aki K (1966) Generation and propagation of G waves from the Niigata earthquake of June 16, 1964. Part 2. Estimation of earthquake moment, released energy, and stress-strain drop from the G wave spectrum. Bull Earth Res Inst 44:73–88
2. Barrientos SE, Ward SN (1990) The 1960 Chile earthquake: inversion for slip distribution from surface deformation. Geophys J Int 103:589–598
3. Bird P (2003) An updated digital model of plate boundaries. Geochem Geophys Geosyst 4(3):1027. doi:10.1029/2001GC000252
4. Blaser L, Kruger F, Ohmberger M, Scherbaum F (2010) Scaling relations of earthquake source parameter estimates with special focus on subduction environment. Bull Seis Soc Am 100(6):2914–2926
5. Briggs RW, Sieh K, Meltzner AJ, Natawidjaja D, Galetzka J, Suwargadi B et al (2006) Deformation and slip along the Sunda megathrust in the Great 2005 Nias-Simeulue Earthquake. Science 311:1897–1901
6. Christensen DH, Beck SL (1994) The rupture process and tectonic implications of the Great 1964 Prince William Sound Earthquake. PureAppl Geophys 142:29–53
7. Comte D, Pardo M (1991) Reappraisal of great historical earthquakes in the Northern Chile and Southern Peru seismic gaps. Nat Hazards 4:23–44
8. Dieterich JH (1979) Modeling of rock friction 1. Experimental results and constitutive equations. J Geophys Res 84(B5):2161–21681
9. Dziewonski AM, Chou T-A, Woodhouse JH (1981) Determination of earthquake source parameters from waveform data for studies of global and regional seismicity. J Geophys Res 86:2825–2852. doi:10.1029/JB086iB04p02825
10. Ekström G, Nettles M, Dziewonski AM (2012) The global CMT project 2004–2010: centroid-moment tensors for 13,017 earthquakes. Phys Earth Planet Inter 200–201:1–9. doi:10.1016/j.pepi.2012.04.002
11. Galgana G, Hamburger M, McCaffrey R, Corpuz E, Chen Q (2007) Analysis of crustal deformation in Luzon, Philippines using geodetic observations and earthquake focal mechanisms. Tectonophysics 432:63–87
12. Geller RJ (1976) Scaling relations for earthquake source parameters and magnitudes. Bull Seis Soc Am 66(5):1501–1523
13. Gudmundsson O, Sambridge M (1998) A regionalized upper mantle (RUM) seismic model. J Geophys Res 103(B4):7121–7136
14. Hanks T, Kanamori H (1979) A moment magnitude scale. J Geophys Res 84(B5):2348–2350
15. Hashimoto C, Noda A, Sagiya T, Matsu'ura M (2009) Interplate seismogenic zones along the Kuril-Japan trench inferred from GPS data inversion. Nat Geosci 2:141–144
16. Hirose H, Hirahara K, Kimata F, Fujii N, Miyazaki S (1999) A slow thrust slip event following the two 1996 Hyuganada earthquakes beneath the Bungo Channel, southwest Japan. Geophys Res Lett 26(21):3237–3240
17. Hirth G, Kohlstedt D (2003) Rheology of the upper mantle and the mantle wedge: a view from the experimentalists, inside the subduction factory. Geophys Monogr Ser 138:83–105. doi:10.1029/138GM06
18. Igarashi T, Matsuzawa T, Hasegawa A (2003) Repeating earthquakes and interplate aseismic slip in the northeastern Japan subduction zone. J Geophys Res 108(B5):2249. doi:10.1029/2002JB001920
19. Ikuta R, Satomura M, Fujita A, Shimada S, Ando M (2012) A small persistent locked area associated with the 2011 Mw9.0 Tohoku-Oki earthquake, deduced from GPS data. J Geophys Res 117, B11408. doi:10.1029/2012JB009335
20. Ito T, Yoshioka S, Miyazaki S (2000) Interplate coupling in northeast Japan deduced from inversion analysis of GPS data. EarthPlanet Sci Lett 176:117–130
21. Kanamori H, McNally KC (1982) Variable rupture mode of the subduction zone along the Ecuador-Colombia coast. Bull Seis Soc Am 72(4):1241–1253
22. Kelleher JA (1972) Rupture zones of large South American earthquakes and some predictions. J Geophys Res 77(11):2087–2103
23. Loveless JP, Meade BJ (2010) Geodetic imaging of plate motions, slip rates, and partitioning of deformation in Japan. J Geophys Res 115, B02410. doi:10.1029/2008JB006248
24. Loveless JP, Meade BJ (2011) Spatial correlation of interseismic coupling and coseismic rupture extent of the 2011 MW=9.0 Tohoku-oki earthquake.Geophys Res Lett 38, L17306. doi:10.1029/2011GL048561

25. Matsuzawa T, Igarashi T, Hasegawa A (2002) Characteristic small earthquake sequence off Sanriku, northeastern Honshu, Japan. Geophys Res Lett 29(11):1543. doi:10.1029/2001GL014632
26. McCann WR, Nishenko SP, Sykes LR, Krause J (1979) Seismic gaps and plate tectonics: seismic potential for major boundaries. Pure Appl Geophys 117:1082–1147
27. Mochizuki K, Yamada T, Shinohara M, Yamanaka Y, Kanazawa T (2008) Weak interplate coupling by seamounts and repeating M ~ 7 earthquakes. Science 321:1194-1197
28. Murotani S, Kikuchi M, Yamanaka Y (2004) Rupture processes oflarge Fukushima-Oki earthquakes in 1938(II) (in Japanese), paper presented at Fall Meeting 2004. Seismological Society of Japan, Fukuoka
29. Murotani S, Miyake H, Koketsu K (2008) Scaling of characterized slip models for plate-boundary earthquakes. Earth Planets Space 60:987–991
30. Nishimura T, Hirasawa T, Miyazaki S, Sagiya T, Tada T, Miura S et al (2004) Temporal change of interplate coupling in northeastern Japan during 1995–2002 estimated from continuous GPS observations. Geophys J Int 157:901–916
31. Ohkura T (2013) GPS observation in Mindanao, Philippine—for evaluation of earthquake potential, Abstract for Japan Geoscience Union 2013 Meeting., S-SS34-P11 (in Japanese)
32. Pacheco JF, Sykes LR, Sholz CH (1993) Nature of seismic coupling along simple plate boundaries of the subduction type. J Geophys Res 98(B8):14,133–14,159
33. Pollitz FF, Brooks B, Tong X, Bevis MG, Foster JH, Bürgmann R et al (2011) Coseismic slip distribution of the February 27, 2010 Mw 8.8 Maule, Chile earthquake. Geophys Res Lett 38, L09309. doi:10.1029/2011GL047065
34. Ruff L, Kanamori H (1980) Seismicity and the subduction process. Phys Earth Planet Int 23:240–252
35. Ruina A (1983) Slip instability and state variable friction laws. J Geophys Res 88(B12):10,359–10,370
36. Scholz CH (1998) Earthquakes and friction laws. Nature 391:37–42
37. Scholz CH, Campos J (1995) On the mechanism of seismic decoupling and back arc spreading at subduction zones. J Geophys Res 100(B11):22,103–22,115
38. Stein S, Okal EA (2007) Ultralong period seismic study of the December 2004 Indian Ocean earthquake and implications for regional tectonics and the subduction process. Bull Seis Soc Am 97(1A):S279–S295
39. Storchak DA, Di Giacomo D, Bondár I, Engdahl ER, Harris J, Lee WHK et al (2013) Public release of the ISC-GEM global instrumental earthquake catalogue (1900–2009). Seismol Res Lett 84:810–815
40. Suwa Y, Miura S, Hasegawa A, Sato T, Tachibana K (2006) Interplate coupling beneath NE Japan inferred from three-dimensional displacement field. J Geophys Res. doi:10.1029/2004JB003203
41. Syracuse EM, Abers GA (2006) Global compilation of variations in slab depth beneath arc volcanoes and implications. Geochem Geophys Geosyst 7, Q05017. doi:10.1029/2005GC001045
42. Uchida N, Matsuzawa T (2013) Pre- and postseismic slow slip surrounding the 2011 Tohoku-oki earthquake rupture. Earth Planet Sci Lett 374:81-91
43. Uchida N, Matsuzawa T (2011) Coupling coefficient, hierarchical structure, and earthquake cycle for the source area of the 2011 off the Pacific coast of Tohoku earthquake inferred from small repeating earthquake data. Earth Planets Space 63:675–679
44. Uchida N, Matsuzawa T, Hasegawa A, Igarashi T (2005) Recurrence intervals of characteristic M4.8 ± 0.1 earthquakes off-Kamaishi, NE Japan—comparison with creep rate estimated from small repeating earthquake data, Earth Planet. Sci Lett 233:155–165
45. Uyeda S (1982) Subduction zones: an introduction to comparative subductology. Tectonophysics 81:133–159
46. Wells DL, Coppersmith KJ (1994) New empirical relationships among magnitude, rupture length, rupture width, rupture area, and surface displacement. Bull Seis Soc Am 84(4):974–1002
47. Yamanaka Y, Kikuchi M (2003) Source process of the recurrent Tokachi-oki earthquake on September 26, 2003, inferred from teleseismic body waves. EarthPlanetsSpace 55:e21–e24
48. Yamanaka Y, Kikuchi M (2004) Asperity map along the subduction zone in northeastern Japan inferred from regional seismic data. J Geophys Res 109, B07307. doi:10.1029/2003JB002683

CHAPTER 13 MODIFICATION IN ATMOSPHERIC REFRACTIVITY AND GPS BASED TEC AS EARTHQUAKE PRECURSORS

Sheetal P. Karia, Kamlesh N. Pathak, Kunvar S. Yadav, Nikita P. Chaudhary, Nilesh C. Patel. Ranjan Jana

Keywords: Total Electron Content, Refractivity, Earthquake

INTRODUCTION

A major earthquake (M = 7.8) struck Iran near the border with Pakistan on 16 April 2013 at 10:44 GMT. Tremors from this earthquake were also felt in India. This made us do some intriguing research about whether large earthquakes are associated with ionospheric/atmospheric changes caused due to earthquake preparation above the epicenter or throughout the fault line. If so, then this might be useful in identifying precursory signature. Variations in ground, atmosphere and space parameters have been studied by many. Some of them have shown [1] [2] that studying atmospheric variability could also reveal active tectonic processes in the Earth's crust. The atmospheric refractivity is proportional to the pressure and temperature gradient of the troposphere. With respect to the earthquake preparation the vertical profiles of humidity, pressure and temperature are changed. These variations in the local parameters cause the atmospheric refractivity to change and can potentially affect the ray path of any electromagnetic wave. It is evident from [3] -[8] that seismic events affect the local atmosphere and the ionospheric parameters. The variation in ionospheric TEC (total electron content) prior to an earthquake is conducted by many researchers from above references. The GPS signal propagating through the ionosphere is enhanced in phase and delayed in time. This time delay being a function of electron density when measured by using a dual frequency receiver can compute the TEC (the total of no electron encountered along a particular line-of sight by the propagating wave). From the difference between the two-frequency measurements, the TEC along the signal path between GPS and the receiver can be calculated. Unit of TEC is 10^{16} per square meter. The TEC is variable in space and time and is a function of local time season and geomagnetic latitude.

Research of seismogenic-origin anomalous effects in different ionospheric parameters has been carried out in the past decade by [2] [7] [9] , stress-activated positive hole charge carriers in rocks and the generation of preearthquake signals. Freund 2000 [10] has provided the reason for enlargement in the scope of research in this field. Examination of the validity of the preseismic anomalous transmission of very high frequency (VHF) electromagnetic waves beyond the line of sight was originally proposed by Kushida and Kushida [11] in Japan. Fujiwara et al. [12] statistically demonstrated the existence of atmospheric anomalies lasting for a few minutes to several hours before earthquakes. Fujiwara and colleagues monitored the VHF waves (FM radio; around 80 megahertz) in eastern Tokyo that were transmitted from an over-the horizon radio station in northern Honshu, and thoroughly compared the waves with the seismicity in the surrounding regions. They found that the transmission anomalies were significantly enhanced within five days prior to $M \geq 4.8$ earthquakes. Increase in propagation horizon of VHF has been reported and the role of seismic effect on the refractive index has been examined to explain the observed TEC features for some earthquakes recently [8] [13] .

In the present report, the TEC profile is obtained for the period prior to and after the earthquake day (one month). Further the modification in GPS signal path is observed by counting the total number of satellite passes registered by the receiver (in other way, it is to estimate the Field Of View, FOV

Citation: P. Karia S. , N. Pathak, K. , S. Yadav K. , P. Chaudhary N. and C. Patel. Ranjan Jana N. (2014) Modification in Atmospheric Refractivity and GPS Based TEC as Earthquake Precursors. Positioning, 5, 46-52. doi: 10.4236/pos.2014.52006.

of the receiver) for the same period. In order to investigate that the modification in the GPS signal is caused due to sudden changes in atmospheric refractivity, its profile is obtained from the radiosonde observations.

DATA AND ANALYSIS

The TEC data are collected by a GPS receiver at Surat (21.16°N, 72.78°E) and IGS station Lucknow (26.91°N, 80.95°E) available in RINEX Format. We select Lucknow station as it is one of the IGS station around the fault line of the Iran earthquake (shown in the Map of Figure 1). In order to attempt to identify earthquake-induced imprints on TEC profiles, it is necessary to have knowledge on the relation of TEC with solar and geomagnetic parameters; we show in Figure 2 the variation in Dst-index for the month of April 2013. It is observed that the Dst-index did not vary throughout the observation period and so the geomagnetic condition is quite. Also the F 10.7 did not show much variation and its value represents a medium solar activity. Under quite geomagnetic condition the variation in TEC could be attributed to the earthquake preparation. The vertical TEC were calculated first as explained in [14]. In the second step the percentage deviation from the monthly mean was calculated:

$$\Delta TEC = 100(TEC - TEC_{av})/TEC_{av} \quad (1)$$

as a background level TEC_{av} the monthly mean was taken for the month of April 2013, every hour of UT.

The Atmospheric profiles for Temperature, Pressure and Humidity are derived from the measured data at given station for different altitude from the radiosonde balloon. These data are available athttp://weather.uwyo.edu/upperair/sounding.html. Radiosonde observations are a key data set in operational weather forecasting and upper air climate research. Due to their global distribution and high vertical resolution radiosonde observations have also been used for calibration and validation of satellite temperature and water vapour retrievals [15] -[17] .

In the present paper the variations in refractivity profile dN/dh are investigated using radisonde balloon measurements of Temperature, Pressure and Humidity from two different stations, Shiraz in Iran and New Delhi in India, along the faultline of the Iran-Pakistan border earthquake (see yellow dots in Figure 1).

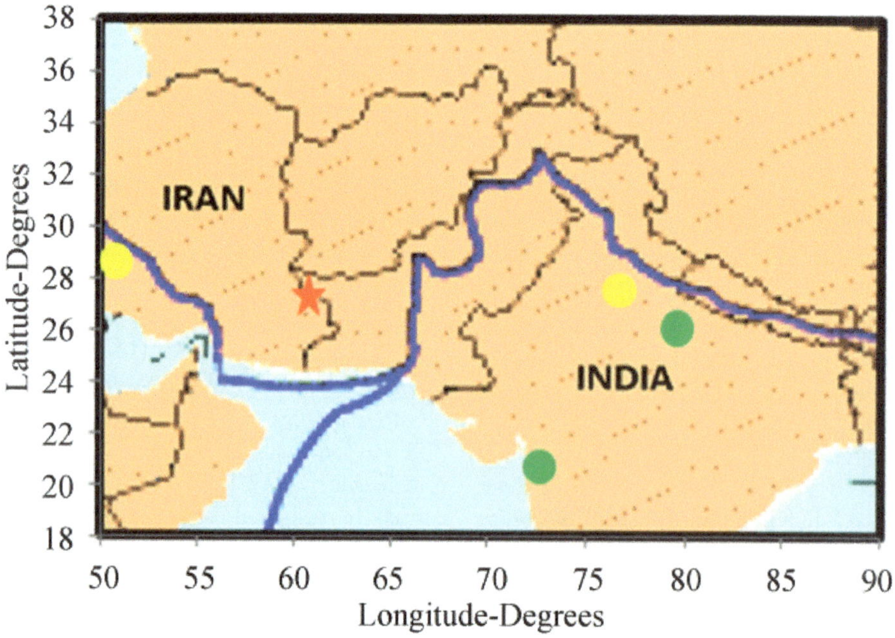

Figure 1: Show the Map of the faultline (blue line) of the Iran earthquake. The red star indicates the epicenter and the yellow and green circles show the radiosonde and GPS stations. Both the radiosonde stations are very close to the faultline and so can be the best choice to study the atmosphere around the faultline prior to an earthquake.

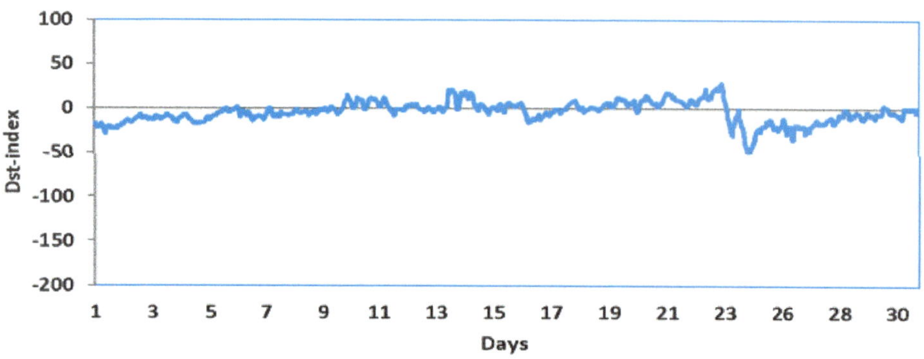

Figure 2: Dst-index for the month of April 2013.

RESULTS AND DISCUSSIONS

Figure 1 shows the Map of the faultline (blue line) of the Iran earthquake. The red star indicates the epicenter and the yellow and green circles show the radiosonde and GPS stations. Both the radiosonde stations are very close to the faultline and so can be the best choice to study the atmosphere around the faultline prior to an earthquake. Both Lucknow and Surat TEC have shown some variations prior to the earthquake (see Figure 3(a) and Figure 4(a)). The percentage deviation ΔTEC from the monthly mean are presented in Figure 3(b) and Figure 4(b) for both the stations Lucknow and Surat respectively. It is seen that percentage deviation for Lucknow reaches up to 100% on 9 April and 15 April that is 7 and 1 day prior to the main shock on 16 April. However, the percentage deviation for Surat TEC is up to 60% which may be due to its distance from the faultline.

The troposphere modifies the propagation of VHF ray paths [18] through variations in the refractive index n. In a normal troposphere condition where n decreases monotonically with height (h) radio rays are slightly refracted towards the surface of the Earth. The ray path thus takes on a curvature C = −dn/dh. The bending of a ray relative to the Earth's surface is therefore C − (1/a), where "a" is the radius of the Earth. As ray curvature C changes with refractive index gradient, its relation with earth's curvature does also change and therefore the concept of "effective earth's radius" ae is introduced and ratio of effective to true earth radius is given by:

$$K = \frac{ae}{a} = \frac{1}{\left(1 + a\frac{dn}{dh}\right)} \quad (2)$$

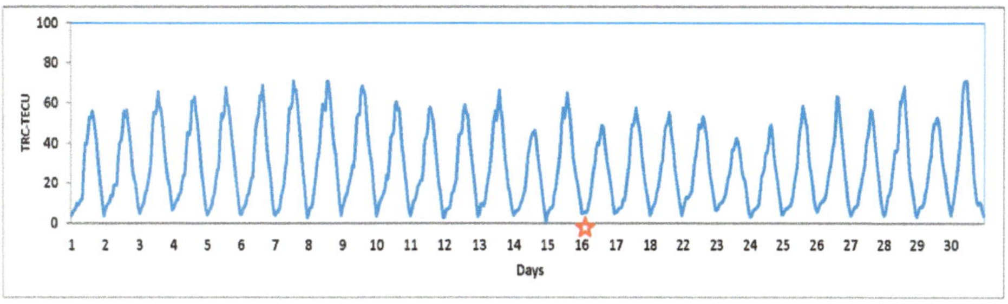

(a)

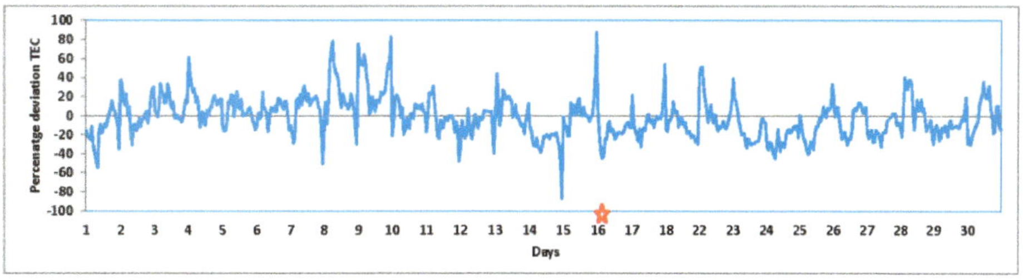

(b)

Figure 3: (a) The daily TEC variation for the month of April 2013 obtained at Lucknow station; (b) Percentage deviation in TEC from monthly mean for Lucknow station. Star indicates the Earthquake day.

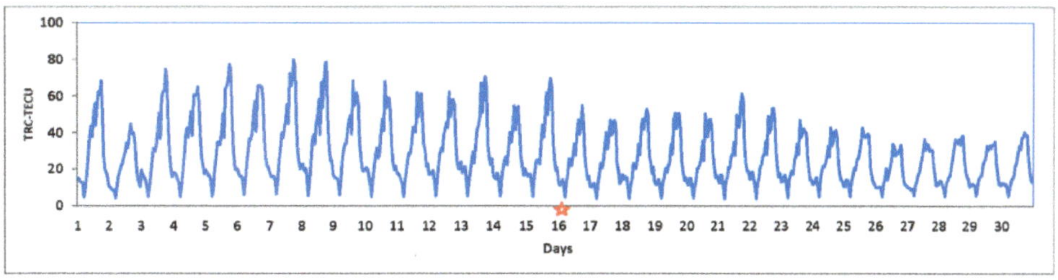

(a)

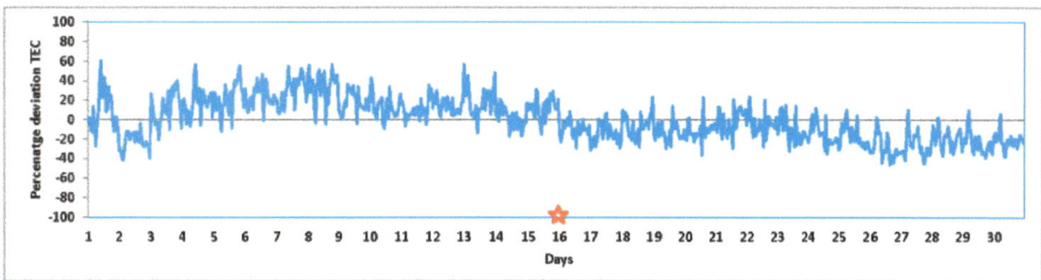

(b)

Figure 4: (a) The daily TEC variation for the month of April 2013 obtained at Surat station; (b) Percentage deviation in TEC from monthly mean for Surat Station. Star indicates the Earthquake day.

Generally, in the communication field dn/dh is expressed by a parameter known as the Radio refractive Index, RRI = N given by:

$$\left[N = (n-1)\times 10^{6}\right] \tag{3}$$

The relation of N with temperature, pressure and water vapor is expressed as [19] :

$$\frac{dN}{dh} = 77.6\left[\left(\frac{1}{T}\right)\frac{dP}{dh} - \left\{\left(\frac{P}{T^{2}}\right) + \left(9620\frac{e}{T^{3}}\right)\right\}\frac{dT}{dh} + \frac{4810}{T^{2}}\left(\frac{de}{dh}\right)\right]\mathrm{m}^{-1} \tag{4}$$

For a normal lapse rate dN/dh = −40 Km^{-1} and K = 4/3, this factor defines a standard field of view of line of sight propagation links. For any change in dN/dh from the above value results in increase or decrease of such path link hops [18] . Modification in dN/dh values is probable at the time of impending earthquake as release of heat, if any, in the earthquake preparatory processes could reduce normal lapse rate on dT/dh and so the second term of equation (4) becomes less negative. This makes dN/dh more negative compared to normal lapse rate. A critical condition occurs when dN/dh = −157 Km^{-1} making C = 1/a, i.e., the earth appears as a flat surface to the ray path thereby increasing the line of sight radio horizon. Further when dN/dh becomes <−157 Km^{-1}, the ray takes a guided path. This explains why a large number of satellites (beyond the standard LOS) are seen by the receiving antenna for elevation of zones covering 10° - 30°. Now, to understand why an earthquake precursory signature on radio refractive index should exceed the critical limit of −157 km^{-1} one needs to understand the role of Earth's surface in the heat exchange processes. As mentioned in Devi et al. [20] the Earth's surface act as a catalyst of energy exchange processes with the near-Earth environment, the troposphere. In order to understand the heat exchange processes, the changes in specific latent heat flux (SLHF) were evaluated prior to the Earthquake [21] . They found an area of enhanced SLHF very close to the epicenter. A significant amount of work has been done associating modifications observed in SLHF, SST and water vapour to an impending EQ Singh et al. [22] [23] . These changes suggest the existence of interactions between the lithosphere and the troposphere. Ozounou et al. [24] mentioned that the processes nearest to the ground surface observed in atmosphere are the changes of air humidity and temperature. Further they explain that because gas/radon is emitted from underground, the increase of the temperature is marked immediately on the ground surface. Temperature difference between faults area and far from the faults leads to horizontal air movements, air mixture and air temperature rise over all earthquake preparation area. The relative humidity drop often accompanies it: water vapor attachment to the ions decreases the content of free water vapor in air. This may lead to a modification in atmospheric refractivity above the epicenter region and subsequently decreases the dN/dh some days prior to the earthquake.

The sounding data have been used to determine the dN/dh for the stations around the faultline. The graph of dN/dh vs days of the April month is obtained (see Figure 5(a) and Figure 5(b)). It can be seen that the dN/dh decreased beyond the critical condition (<−157 Km^{-1}) for both the radiosonde stations Shiraz in Iran and New Delhi in India prior to the earthquake. This attributes to the modification in FOV (field of view) of GPS receiver at Lucknow. FOV is estimated by counting the number of satellite passes registered high TEC for different elevation (see Figure 6). It is interesting to see that the day of decrease in dN/dh beyond the critical condition and the day of modification in FOV of GPS receiver is same (8 April). As Lucknow and New Delhi are closer to the faultline, it is simple to depict that the modification in atmosphere around the faultline resulted in increase in FOV of the receiver.

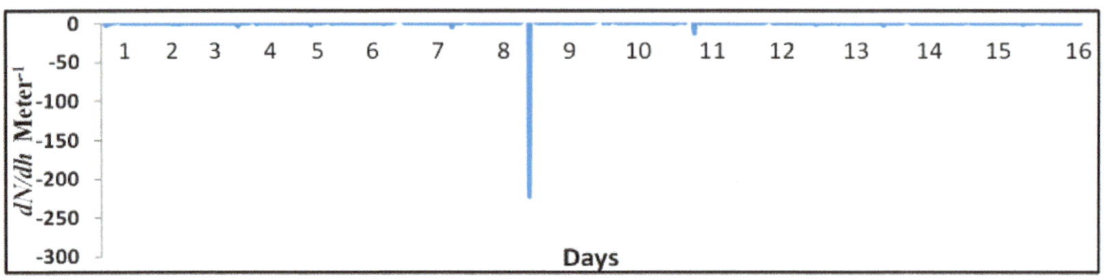

(a)

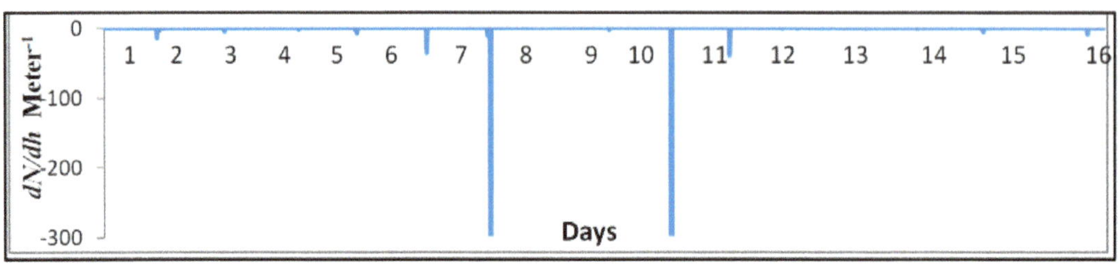

(b)

Figure 5: (a) The radio refractive index gradient dN/dh for Shiraz, Iran. It is seen that the dN/dh reduced beyond the critical limit <−157 Km^{-1} on 8 April 2013; (b) The radio refractive index gradient dN/dh for New Delhi. It is seen that the dN/dh reduced beyond the critical limit <−157 Km^{-1} on 8 and 11 April 2013.

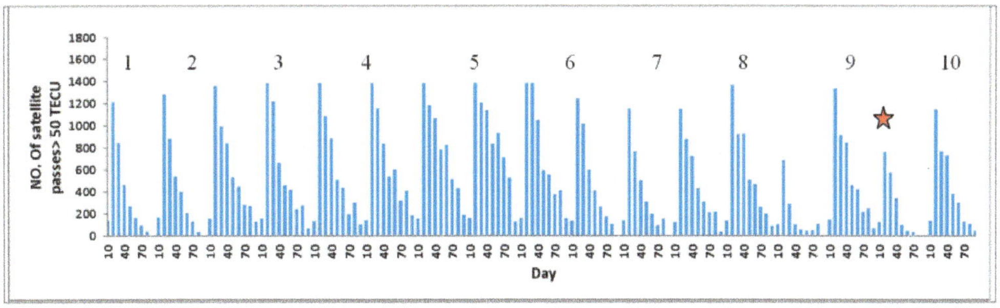

Figure 6: The change in number of satellite passes that registered higher TEC. It is observed that the GPS FOV modified on 8 April 2013. Star indicates the Earthquake day (i.e., 16 April 2013).

With respect to the above results from the GPS TEC data, it is seen that it may be possible to realize atmospheric changes prior to a strong earthquake within its preparatory zone through the variation in the parameters like TEC.

CONCLUSION

In conclusion, for the Iran earthquake, there is a visible change observed in refractivity gradient and satellite paths. However, anomaly-identifying standards should be established based on long-term regional analysis to eliminate subjectivity. Furthermore, other factors that may result in refractivity variations should also be carefully considered in order to apply these data for earthquake prediction. From the preliminary results, an association between ionospheric/atmospheric effects and its epicenter position cannot be ruled out. We can conclude that, in search of an earthquake precursory signature, atmospheric refractivity and ionospheric TEC are two important parameters.

ACKNOWLEDGEMENTS

We are thankful to the Department of Science and Technology Government of India to provide necessary funds to conduct the research.

REFERENCES

1. Liu, J.Y., Chuo, Y.J., Shan, S.J., Tsai, Y.B., Pulinets, S.A. and Yu, S.B. (2004) Pre-Earthquake Ionospheric Anomalies Monitored by GPS TEC. Annals of Geophysics, 22, 1585-1593. http://dx.doi.org/10.5194/angeo-22-1585-2004
2. Pulinets, S. and Boyarchuk, K. (2004) Ionospheric Precursors of Earthquake. Springer, New York. [Paper reference:2]
3. Liu, J.Y, Tsai, C.H.F. and Chen, Y.I. (2001) Variations of Ionospheric Total Electron Content during the Chi-Chi Earthquake. Geophysical Research Letters, 7, 1383-1386.http://dx.doi.org/10.1029/2000GL012511
4. Plotkin, V.V. (2001) GPS Detection of Ionospheric Perturbation before the 13 February 2001, El Salvador Earthquake. Natural Hazards and Earth System Sciences, 3, 249-253.
5. Pulinets, S.A., Leyva, C.A., Bisiacchi-Giraldi, G. and Ciraolo, L. (2005) Total Electron Content Variations in the Ionosphere before the Colima, Mexico, Earthquake of 21 January 2003. Geofisica International, 44, 369-377.
6. Krankowski, A., Zakharenkova, I.E. and Shagimuratov, I.I. (2006) Response of the Ionosphere to the Baltic Sea Earthquake of 21 September 2004. Acta Geophysica, 54, 90-101.
7. Zakharenkova, E., Shagimuratov, I.I., Krankowski, A. and Lagovsky, A.F. (2007) Precursory Phenomena Observed in the Total Electron Content Measurements before Great Hokkaido Earthquake of September 25 2003. Studia Geophysica et Geodaetica, 51, 267-278.
8. Karia, S.P. and Pathak, K.N. (2011) A Change in Refractivity of the Atmosphere and Large Variation in TEC Associated with Some Earthquakes, Observed from GPS Receiver. Advances in Space Research, 47, 867-876. http://dx.doi.org/10.1016/j.asr.2010.09.019 [Paper reference:2]
9. Hayakawa, M. and Molchanov, O.A. (2002) Seismo-Electromagnetics: Lithosphere-Atmosphere-Ionosphere Coupling. Terrapub, Tokyo.
10. Freund, F. (2000) Time-Resolved Study of Charge Generation and Propagation in Igneous Rocks. Journal of Goephysical Research, 105, 11001-1019.
11. Kushida, Y. and Kushida, R. (2002) Possibility of Earthquake Forecast by Radio Observations in the VHF Band. Journal of Atmopheric Electricity, 22, 239-255.
12. Fujiwara, H., Kamogawa, M., Ikeda, M., Liu, J.Y., Sakata, H., Chen, Y.I., Ofuruton, H., Muramatsu, S., Chuo, Y.J. and Ohtsuki, Y.H. (2004) Atmospheric Anomalies Observed during Earthquake Occurrences. Geophysical Research Letters, 31, L17110.http://dx.doi.org/10.1029/2004GL019865
13. Devia, M., Barbaraa, A.K., Depuevab, A.H., Ruzhinb, Y.Y. and Depuevb, V. (2010) Anomalous Total Electron Content (TEC) and Atmospheric Refractivity Prior to the Very Strong China Earthquake of May 2008. International Journal of Remote Sensing, 31, 3589-3599.
14. Karia, S.P. and Pathak, K.N. (2011) GPS Based TEC Measurements for a Period August 2008-December 2009 near the Northern Crest of Indian Equatorial Ionospheric Anomaly Region. Journal of Earth System Science, 120, 851-858.http://dx.doi.org/10.1007/s12040-011-0114-1
15. Divakarla, M.G., Barnet, C.D., Goldberg, M., McMillan, L.M., Maddy, E., Wolf, W., Zhou, L. and Liu, X. (2006) Validation of Atmospheric Infrared Sounder Temperature and Water Vapor Retrievals with Matched Radiosonde

Measurements and Forecasts. Journal of Geophysical Research, 111, D09S15.

16. Tobin, D.C., Revercomb, H.E., Knutson, R.O., Lesht, B.M., Strow, L.L., Hannon, S.E., Feltz, W.F., Moy, L.A., Fetzer, E.J. and Cress, T.S. (2006) Atmospheric Radiation Measurement Site Atmospheric State Best Estimates for Atmospheric Infrared Sounder Temperature and Water Vapor Retrieval Validation. Journal of Geophysical Research, 111, Article ID: D09S14. http://dx.doi.org/10.1029/2005JD006103
17. Reale, A., Tilley, F., Ferguson, M. and Allegrino, A. (2008) NOAA Operational Sounding Products for Advanced TOVS. International Journal of Remote Sensing, 29, 4615-4651.http://dx.doi.org/10.1080/01431160802020502
18. Dolukahnov, M. (1971) Propagation of Radio Waves. MIR Publisher, Moscow, 140-167. [Paper reference:2]
19. Hall, M.P. and Bradley, P.A. (1998) Troposphere and Ionosphere, Section 13. In: Mazda, F., Ed., Telecommunications Engineer's Reference Book, 2nd Edition, Reed Educational and Professional Publishing Ltd., Oxford.
20. Devi, M., Barbara, A.K., Ruzhin, Y.Y. and Hayakawa, M. (2012) Over-the-Horizon Anomalous VHF Propagation and Earthquake Precursors. Surveys in Geophysics, 33, 1081-1106.
21. Xu, T., Chen, Z., Li, C.B., Wu, J., Hu, Y.L. and Wu, Z.S. (2011) GPS Total Electron Content and Surface Latent Heat Flux Variations before the 11 March 2011 M9.0 Sendai Earthquake. Advances in Space Research, 48, 1311-1317.http://dx.doi.org/10.1016/j.asr.2011.06.024
22. Singh, R.P., Bhoi, S. and Sahoo, A.K. (2001) Significant Changes in the Ocean Parameters after the Gujarat Earthquake. Current Science, 80, 1376-1377.
23. Singh, R.P., Cervone, G., Kafatos, M., Prasad, A.K., Sahoo, A.K., Sun, D., Tang, D.L. and Yang, R. (2007) Multi-Sensor Studies of the Sumatra Earthquake and Tsunami of 26 December 2004. International Journal of Remote Sensing, 28, 2885-2896.http://dx.doi.org/10.1080/01431160701237405

CHAPTER 14 INTELLIGENT SEISMIC-ACOUSTIC SYSTEM FOR IDENTIFYING THE LOCATION OF THE AREAS OF AN EXPECTED EARTHQUAKE

Telman Aliev, Qambar Quluyev, Farhad Pashayev, Ulkar Sattarova and Narmin Rzayeva

Keywords: Seismic-Acoustic Signal, Robust Noise Technologies, RNM ASP Station, Seismic-Acoustic Monitoring Centre, Earthquake Forecasting

INTRODUCTION

It is known that much research has been carried out on the causes and nature of earthquakes [1] . The possibility of receiving various types of seismic information from the deep strata of the earth also remains a popular research subject [2] [3] . Methods, such as noise analysis [4] - [9] , continuous monitoring system designing [10] [11] , earthquake damage assessment and damage minimization [12] - [14] , wavelet transform and finite elements [15] - [17] , are employed in analysing seismic signals received during earthquakes. In all of these research papers, the problems associated with earthquake prediction are the basic research trend [18] - [21] . Different models and technologies have been and are being developed [22] - [24] ; numerous population-oriented early warning systems, models and technologies for the rapid response of rescue groups of relevant agencies have been developed and commissioned [25] - [29] . Regardless of the aforementioned research papers, earthquakes are not predicted in good time, which leads to massive disastrous consequences.

Papers [11] [30] [31] propose a seismic-acoustic system for monitoring the earthquake origin process. The system consists of a network of nine seismic-acoustic stations for the robust noise monitoring of anomalous seismic processes (RNM ASPs). The experiments that have been conducted on these stations since 2010.07.01 establish that a cross-correlation emerges between the useful signal and the noise of the seismic-acoustic information during the ASP origin.

The results of the operation of these stations have demonstrated that each of them separately performs reliable indication of ASP origin processes that precede earthquakes based on the variation of the cross-correlation function between the useful signal and the noise. However, it is impossible to determine the coordinates of an expected earthquake with sufficient accuracy by using these stations. The experimental research has demonstrated that it is possible, however, to create an intelligent system, which would allow locating the ASP area by means of these stations. We consider one possible way to create such a system in the following paragraphs.

PROBLEM STATEMENT

It is known that in seismically active regions, after a certain period T_0 of the normal seismic state and at the end of a certain period T_1, an earthquake usually occurs due to originating ASPs.

Despite the difference in duration of T_0 and T_1, the problem of monitoring the beginning of the ASP origin comes down to providing a reliable indicator of the beginning of period T_1.

This matter has been considered in detail in [9] [11] [31] .

Research [11] sets the problem for creating a technology and a system allowing one to register the starting instant of the period T_1. However, the experimental research [9] [11] [31] demonstrated that the beginning of T_1 does not take place only during the ASP origin. For this reason, besides the

Citation: Aliev T. , Quluyev Q. , Pashayev F. , Sattarova U. and Rzayeva N. (2016) Intelligent Seismic-Acoustic System for Identifying the Location of the Areas of an Expected Earthquake. Journal of Geoscience and Environment Protection, 4, 147-162. doi: 10.4236/gep.2016.44018.

registration of the beginning of the period T_1, the monitoring of the beginning of the ASP origin also requires an indication of the change in the estimate of the cross-correlation function $R_{X\varepsilon}(\mu)$ between the useful signal $X(i\Delta t)$ and the noise $\varepsilon(i\Delta t)$.

Thereby, the present paper sets up the problem of using the estimate $R_{X\varepsilon}(\mu)$ of the seismic-acoustic signal $g(i\Delta t)$ as an informative attribute for indicating the beginning of the ASP origin, which requires calculation of the estimate $R_{X\varepsilon}(\mu)$ during the monitoring process.

Furthermore, the practical application of a network of RNM ASP stations also requires developing a technology for determining the location of the earthquake area.

Therefore, the present paper poses the problem of developing an intelligent system for monitoring the ASP origin, identifying the location of the area of an expected earthquake.

TECHNOLOGY FOR DETERMINING THE INFORMATIVE ATTRIBUTES OF THE LATENT PERIOD OF ASP ORIGIN

Our research demonstrated that when ASP arises at the start of period T_1, estimates of the cross-correlation function $R_{X\varepsilon}(\mu=0)$ between the useful signal $X(i\Delta t)$ and noise $\varepsilon(i\Delta t)$, the noise variance D_ε and the noise correlation $R_{X\varepsilon\varepsilon}(\mu=0)$ change in the first place [9] [11] [31] . The reason is that in the beginning of period T_0, noise $\varepsilon(i\Delta t)$ forms due to the influence of the ASP. Therefore, in period T_1, correlation arises between the useful signal $X(i\Delta t)$ and noise $\varepsilon(i\Delta t)$ and the estimate $R_{X\varepsilon}(\mu)$ grows sharply. For this reason, the estimate $R_{X\varepsilon}(\mu)$ can be regarded as the main informative attribute, whose use is reasonable during the monitoring of the latent period of ASP origin.

Since 2010.07.01, both conventional technologies and robust noise technologies have been used to register the beginning of the latent period of ASP origin on RNM ASP stations. We were unable to register period T_1 with sufficient reliability and authenticity by using estimates of conventional correlation and spectral technologies. When robust noise technology was put into practice, the estimate of the cross-correlation function $R_{X\varepsilon}(\mu)$ would change abruptly in the beginning of period T_1. This turned out to be the crucial factor, which allows the monitoring of the beginning of ASP origin with sufficient reliability. With this in mind, the estimate $R_{X\varepsilon}(\mu)$ was taken as an informative attribute in solving the problem of monitoring the ASP origin during the building of the network of RNM ASP stations.

The expression for determining the estimates of the relay correlation function $R^*_{g\varepsilon}(\mu=0)$ between the useful signal $X(i\Delta t)$ and the noise $\varepsilon(i\Delta t)$ is given in [11] in the following form:

$$R^*_{X\varepsilon}(\mu=0)\approx\frac{1}{N}\sum_{i=1}^{N}\left[\operatorname{sgn} g(i\Delta t)g(i\Delta t)-2\operatorname{sgn} g(i\Delta t)g((i+1)\Delta t)+\operatorname{sgn} g(i\Delta t)g((i+2)\Delta t)\right]. \tag{1}$$

It has also been shown that, with the estimates $R^*_{X\varepsilon}(\mu=0), R^*_{gg}(\mu=1), R_{gg}(\mu=1)$ available and using the equality of the relationship between $R^*_{gg}(\mu=1)$ and $R_{gg}(\mu=1)$ and $R^*_{X\varepsilon}(\mu=0)$ and $R_{X\varepsilon}(\mu=0)$

$$\frac{R^*_{gg}(\mu=1)}{R_{gg}(\mu=1)}=\frac{R^*_{X\varepsilon}(\mu=0)}{R_{X\varepsilon}(\mu=0)}, \tag{2}$$

and the estimate $R_{X\varepsilon}(\mu=0)$ can be determined by the formula:

$$R_{X\varepsilon}(\mu=0)=\frac{R_{gg}(\mu=1)R^*_{X\varepsilon}(\mu=0)}{R^*_{gg}(\mu=1)} \tag{3}$$

Our experiments demonstrated that to raise the reliability and authenticity of the monitoring results, it is reasonable to also use the estimates of the noise correlation value $R_{X\varepsilon\varepsilon}(\mu=0)$ and the noise variance D_ε, which are determined by the following expressions [9] [11] [31] as additional informative attributes:

$$R_{X\varepsilon\varepsilon}(\mu)=R_{X\varepsilon}(\mu)+D_\varepsilon=\frac{1}{N}\sum_{i=1}^{N}\left[g^2(i\Delta t)+g(i\Delta t)g((i+2)\Delta t)-2g(i\Delta t)g((i+1)\Delta t)\right] \tag{4}$$

$$D_\varepsilon=R_{X\varepsilon\varepsilon}(\mu=0)-R_{X\varepsilon}(\mu=0) \tag{5}$$

Thus, it is possible to use formulas (1), (3), (4) and (5) to determine the estimates $R^*_{X\varepsilon}(\mu=0)$, $R_{X\varepsilon}(\mu=0)$, $R_{X\varepsilon\varepsilon}(\mu=0)$, and D_ε, by means of which the monitoring of the ASP can be undertaken with sufficient reliability.

INTELLIGENT TECHNOLOGY AND SYSTEMS FOR IDENTIFYING THE LOCATION OF THE AREA OF ASP ORIGIN

It is known that when an ASP enters its critical state, an earthquake occurs. Boundaries of the earthquake's focus and magnitude depend on the structure and nature of the strain-stress distribution in the rocks in the particular place. Rock deformation is uneven and transmits elastic waves. The volume of deformed rocks is an important factor determining the strength of the seismic impact and the formation of seismic-acoustic noise $g(i\Delta t)$. Each main burst is preceded by quite a long period of time T_1 of earthquake preparation. This period can last up to several dozens of hours.

The analysis of seismic information, received by means of acoustic sensors installed at the heads of suspended oil wells, has demonstrated that when ASPs arise, seismic-acoustic noise spreads in the deep strata of the Earth for several dozens of hours T_1 before the expected earthquake [9] [11] [31] . It has been experimentally established that monitoring the beginning of time T_1 by means of the technology described above is carried out quite reliably by RNM ASP stations (Figure 1). In the following paragraphs, we discuss the possibility of developing intelligent technology to identify the location of the ASP area by means of the information received from the stations built in nine seismically active Caspian regions (Figure 1). Those stations have the following geographical coordinates and their wells have the following depths:

Qum Island:	40.310425˚	50.008392˚	3500 m	July	2010
Siazan:	41.046217˚	49.172058˚	3145 m	November	2011
Naftalan:	40.609521˚	46.791458˚	4000 m	May	2012
Shirvan:	39.933170˚	48.920745˚	4900 m	November	2011
Neftchala:	39.358333˚	49.246667˚	1430 m	June	2012
Nakhchivan:	39.718000˚	44.876000˚	1800 m	March	2013
Qazakh	41.311889˚	45.108611˚	200 m	August	2013
Turkmenistan	38.530089˚	56.654472˚	300 m	August	2013
Cybernetic	40.223252˚	49.800833˚	10 m	February	2014

The experiments carried out on the RNM ASP stations have demonstrated that the seismic-acoustic noises received by hydrophones from the deep strata of the Earth are direct precursors of the earthquake preparation process.

The results of the measuring and analysis of those noises were sent from each station to the server of the seismic-acoustic monitoring centre (MC) via the satellite communication. The system is also capable of transferring received data to servers of other MCs in the neighbouring countries of seismic-acoustic regions.

To carry out large-scale experiments on ASP monitoring, as Figure 1 shows, since 2010.07.01 RNM ASP stations have been made operational, one by one, at Qum Island, in Shirvan, Siazan, Naftalan, Neftchala, Nakhchivan (on the borders with Turkey and Iran), Turkmen01 (Turkmenistan), Qazakh (on the border with Georgia), and Cybernetic (Baku). The last three stations have been built on 300-m, 200-m and 10-m deep water wells, respectively. The stations were built on the wells, in which pipes are naturally filled with water. Hydrophones were placed inside at the depth of 10 - 20 m from the water level. The analysis of seismic-acoustic signals received by these stations has demonstrated that seismic-acoustic noises emerge during ASP origin. Those noises spread within a radius of 300 - 500 km dozens of hours before the seismic waves are registered by ground seismic stations.

Synchronous robust analysis of seismic-acoustic signals received from all stations via satellite communication is undertaken in the process of the network operation. Estimates of the noise characteristics $R_{X\varepsilon}(\mu)$, $R_{X\varepsilon\varepsilon}(\mu), D_\varepsilon$ are transmitted to the server of the MC from the stations. The changes in them are used to indicate, for instance, the beginning of the ASP origin T_{1i} and T_{1j} for the i-th and j-th stations.

The operating experience of the stations demonstrates that each of them separately allows for reliable

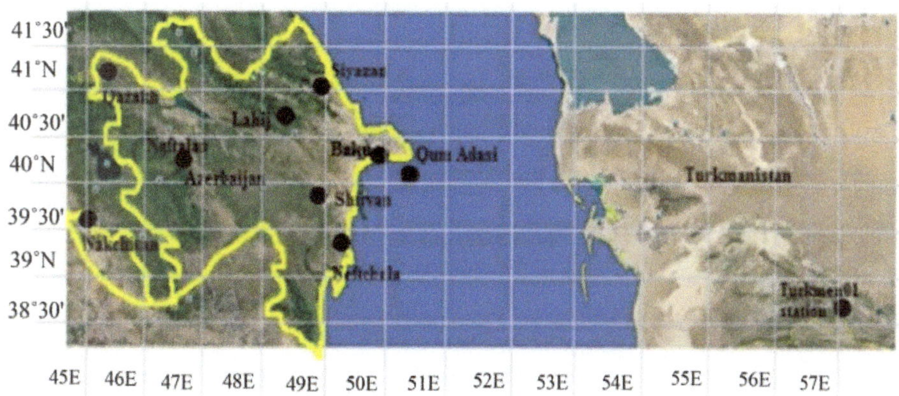

Figure 1: Map of the locations of RNM ASP stations in the Caspian seismically active region.

indication of the process of origin of ASP, which precedes an earthquake. It also demonstrates that the results obtained by means of the network of those stations opens the way to developing an intelligent technology for identifying the location of the area of an expected earthquake. To do that, we first determine the combinations of moments of indication T_{1i} and T_{1j} by means of the network of these stations. Those combinations and the coordinates of their location are source data to solve the problem of identifying the location of ASP origin. To raise the level of authenticity and reliability of the obtained results, it is reasonable to use time differences

$T_{1i} - T_{1j}$ for each selected pair of stations, alongside the combination of time of indication. In other words, solving the problem at hand requires determining not only the combination T_{1i}, T_{1j} but also the time of indication of ASP between the stations, i.e. the difference $\Delta\tau_{ij} = (T_{1i} - T_{1j})$, as source data.

Our experiments have demonstrated that it is difficult to determine the beginning of the time of indication T_{1i} with sufficient accuracy using the values of the estimates of noise characteristics. Thus, considering the importance and necessity of enhancing the accuracy, the proposed system provides for duplication of the process of determining $\Delta\tau_{ij}$. It was also found to be reasonable to use the extreme value of the estimate of the cross-correlation function $R_{g_ig_j}(\mu_{max})$ between the signals $g_i(i\Delta t)$ and $g_j(i\Delta t)$ obtained through different combinations of the RNM ASP stations to determine the time difference $\Delta\tau_{ij} = (T_{1i} - T_{1j})$ by the following expressions:

$$R_{g_ig_j}(\mu_{max}) = \frac{1}{N}\sum_{i=1}^{N} g_i(i\Delta t)g_j(i+\mu)\Delta t \quad (6)$$

$$R^*_{g_ig_j}(\mu_{max}) = \frac{1}{N}\sum_{i=1}^{N} g_i^2(i\Delta t)g_j^2(i+\mu)\Delta t \quad (7)$$

$$R^*_{g_ig_j}(\mu_{max}) = \frac{1}{N}\sum_{i=1}^{N} g_i(i\Delta t)g_j^2(i\Delta t) \quad (8)$$

In that case, the procedure for determining the difference in the time of monitoring between different stations on the MC server comes down to the following:

Determining when the beginning of period T_{1i} of the ASP origin is registered by the first station Qum Island;

Determining the time of monitoring for the second (Shirvan), third (Siazan), fourth (Naftalan), etc. stations;

Determining the sets of estimates of cross-correlation functions $R_{g_vg_j}(i\Delta t)$, $R^*_{g_vg_j}(i\Delta t)$ by the Expressions (6)-(8) and further, selecting from the obtained estimates those time shifts $\mu \cdot \Delta t$, at which the curve of the cross-correlation function has the peak value, i.e. the extreme value; these time shifts are used to determine $\Delta\tau_{ij} = (T_{1i} - T_{1j})$, i.e. the time difference in registration of ASP by the i-th and j-th stations, respectively; using the found time differences $\Delta\tau_{ij} = (T_{1i} - T_{1j})$ as source data to identify the location of the ASP area.

Thus, in the proposed system (Figure 2), the estimates of the noise characteristics $R_{X\varepsilon}(\mu)$, $R_{X\varepsilon\varepsilon}(\mu)$ and D_ε obtained as a result of ASP monitoring performed by the Qum Island, Shirvan, Siazan, Naftalan, Neftchala, Nakhchivan, Turkmen01 (Turkmenistan), Qazakh and Cybernetic (Baku) RNM ASP stations are synchronously transmitted via satellite communication to the MC server. The obtained results are used to form the combination of sequences of indication times T_{1i} T_{1j} and the combinations of time differences $\Delta\tau_{ij}$, which are used as source data to identify the location of an expected earthquake.

The experiments on the said stations carried out in the period from 2010.07.01 to 2014.06.01 have demon- strated the following active earthquake areas in Azerbaijan and neighbouring regions within a radius of 500 - 600 km around the network of the RNM ASP stations.

Turkmen coast of the Caspian Sea;

In the Caspian Sea south of the Absheron peninsula;

In the Caspian Sea north of the Absheron peninsula;

In the Shirvan region of Azerbaijan;

In the north-western regions of Azerbaijan;

In the southern regions of Azerbaijan;

South of the Caucasus region of the Russian Federation;

In the north-eastern regions of Iran;

In the north-western regions of Iran (near Tabriz);

On the Iranian-Iraqi-Turkish border;

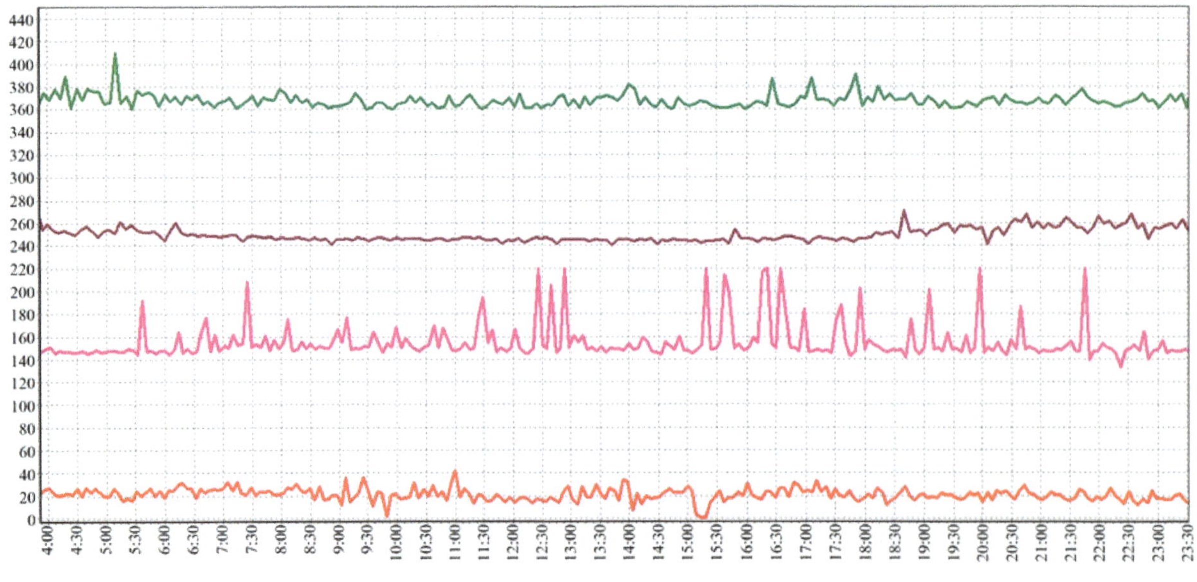

Figure 2: VII—Siazan, Qum Island, Shirvan, Neftchala 2013-03-26 Georgia-Russia.

In the northern regions of Iran;

In the eastern regions of Turkey;

In the western regions of Georgia (Black Sea).

Some of the results of the registration of ASP in those areas by the RNM ASP stations are given in [11] .

Numerous earthquakes with magnitudes of 3 - 4 have occurred in those regions in the last 1.5 - 2 years. For each area, the combinations of the sequence of the times of the ASP registration by Qum Island, Shirvan, Siazan, Neftchala, Naftalan and Nakhchivan by the RNM ASP stations practically repeated themselves. Our analysis of the recorded charts has demonstrated that each sequence combination of the time of the indication of the current ASP corresponds to one of the concrete earthquake areas. After studying the problem of interpretation of the experimental materials for over 2 years, we have learned to identify the location of the area of an expected earthquake, error-free intuitively, using these combinations of time. It then became obvious that the problem of identifying the location of an expected earthquake can be solved by using expert systems (ESs). This, in its turn, demonstrated the possibility of creating an ES which will allow seismologists to use the network of the RNM ASP stations as a tool to determine the area of an expected earthquake.

The experimental version of the ES for identifying the location of the ASP area (ESILA) proposed in this paper is based on the knowledge base (KB) comprising the totality of the sets $W_1, W_2, W_3, \cdots, W_{13}$ of the locations of the respective areas. The elements of each of these sets are formed from the data of the charts recording the parameters of all earthquakes registered by the RNM ASP stations in all 13 areas from 2010.07.01 to the present day. Each element of the KB consists of the combination of the sequence of times of the ASP indication by the stations T_{1i}, T_{1j}, the combination of the differences in times of the indication $\Delta\tau_{ij}$, and from the combination of the estimates of the cross-correlation function $R_{X\varepsilon}(\mu=0)$. Each element of the KB also contains the value of magnitude M_i determined during the corresponding earthquakes by ground seismic stations. The date of the earthquake is also specified in each element. In the case when there is only one element, the KB appears as follows:

$$W_1\begin{cases} T_{11}^{1(1)} & T_{11}^{2(1)} & \cdots & T_{11}^{6(1)} & M_1 \\ \Delta\tau_{11}^{1} & \Delta\tau_{21}^{1} & \ddots & \Delta\tau_{61}^{1} & M_1 \\ R_{X\varepsilon}^{1(1)}(\mu=0) & R_{X\varepsilon}^{2(1)}(\mu=0) & \cdots & R_{X\varepsilon}^{6(1)}(\mu=0) & M_1 \end{cases}$$

$$W_2\begin{cases} T_{11}^{1(2)} & T_{11}^{2(2)} & \cdots & T_{11}^{6(2)} & M_2 \\ \Delta\tau_{11}^{2} & \Delta\tau_{21}^{2} & \ddots & \Delta\tau_{61}^{2} & M_2 \\ R_{X\varepsilon}^{1(2)}(\mu=0) & R_{X\varepsilon}^{2(2)}(\mu=0) & \cdots & R_{X\varepsilon}^{6(2)}(\mu=0) & M_2 \end{cases}$$

$$W_{13}\begin{cases} T_{11}^{1(13)} & T_{11}^{2(13)} & \cdots & T_{11}^{6(13)} & M_{13} \\ \Delta\tau_{11}^{13} & \Delta\tau_{21}^{13} & \ddots & \Delta\tau_{61}^{13} & M_{13} \\ R_{X\varepsilon}^{1(13)}(\mu=0) & R_{X\varepsilon}^{2(13)}(\mu=0) & \cdots & R_{X\varepsilon}^{6(13)}(\mu=0) & M_{13} \end{cases}$$

Each set $W_1 - W_{13}$ of the experimental version of the KB consists of several dozens of such elements and new elements are added during each new earthquake. During the operation of the ES, after the monitoring and indication of the time of the beginning of the current ASP has been completed, the stations form the current combinations of the sequence of times of indication T_{1i}, T_{1j}, the combinations of differences in times of indication $\Delta\tau_{ij}$, and the combinations of the values of estimates $R_{X\varepsilon}(\mu)$.

On 2014.01.05, the experimental phase of the identification of the location of earthquake areas by ESILA was launched. This phase is carried out as follows. The current element is formed based on the results of monitoring of the network of RNM ASP stations. After that, the current element is compared with all elements of the sets $W_1, W_2, W_3, \cdots, W_{13}$ in the identification unit of the expert system (IUES). If it matches any element of any set, the location of the area of an expected earthquake is identified based on the number of the current element. The number of the ASP area is saved in the decision-making unit (DMU) of the ES. At the same time, the current element is entered into the set of the KB. Thus, new elements are continuously written into the KB during the ESILA operation and the network of RNM ASP stations and ESILA operate as a single system.

To check the authenticity and reliability of the identification of the location of the ASP area, the described ESILA was tested during all subsequent earthquakes. The obtained results have demonstrated the real possibility of practical application of the experimental version of ESILA to identify the location of the ASP area, which creates prerequisites for using the system in question as a tool in determining the location of the area of an expected earthquake. Taking this prospect into account, a function of forming the following information and providing it to seismologists was included in the list of basic functions of the DMU of ESILA: date of the current ASP and the number of the area of the expected earthquake; results of the current monitoring performed by the RNM ASP stations; estimated lead time at the beginning of ASP monitoring compared with the time of registration of the expected earthquake by ground seismic stations; list of all elements previously registered in the corresponding set during the origin of the previous ASP in the estimated location of the area of the expected earthquake (with dates); the amount of elements matching the current elements; magnitudes of previous earthquakes; minimum magnitude of the expected earthquake; if the KB contains no elements matching at least some of the elements in the sets $W_1 \div W_{13}$, information on the impossibility of identifying the earthquake area is formed in the DMU.

Analysing the experience of the use of the ES in identifying the location of the area of the expected earthquake has demonstrated that improved reliability and authenticity of the obtained results requires increasing the number of RNM ASP stations. Taking this into account, since the beginning of 2013, a station in the Nakhchivan Autonomous Republic near the border with Turkey and Iran and Turkmen01station in Turkmenistan have been commissioned. On 2013.07.01, another station was built in the Qazakh region on the border with Georgia and at the Institute of Cybernetics (Baku) (Figure 1).

It should be noted that the results of the experimental monitoring of the station built in the basement of the Institute of Cybernetics in a 10-m deep well practically match the readings of the Qum Island station built in a 3500-m deep well.

RESULTS OBTAINED BY THE INTELLIGENT SYSTEM IDENTIFYING AREAS OF EXPECTED EARTHQUAKES FROM 2013.01.01 TO 2014.07.02

As was mentioned earlier, the test operation of the system under discussion started on 2014.01.06. In this period, some identification errors were detected during weaker earthquakes. Besides, we also detected erroneous identification results at 2 or 3 RNM ASP stations simultaneously during a malfunction of the power supply system, communication systems and hydrophone, controller and other units. In the normal state of operation of all RNM ASP stations, no errors were detected in the results of the identification of the location of areas of expected earthquakes with strength exceeding 5 points.

The list of the location of areas of expected earthquakes identified from the archived monitoring results in 2013-2014 is very long, which is why Table 1 contains only the results of 10 identified areas of expected earthquakes with magnitude over 5 points from 2013.01.01 to 2014.07.06. Figures 2-11 are the recorded ASP charts that preceded those earthquakes. The data in the first column in Table 1 are taken from the website of the Euro-Mediterranean Seismological Centre (EMSC) (http://www.emsc-csem.org/#2).

Table 1: Identified areas of expected earthquakes

(a)

№	Date, time, coordinates, magnitudes and depth of earthquake epicentre	$\Delta\tau_{12}$	$\Delta\tau_{13}$	$\Delta\tau_{14}$	$\Delta\tau_{15}$	$\Delta\tau_{16}$	$\Delta\tau_{17}$	$\Delta\tau_{18}$	$\Delta\tau_{19}$	$\Delta\tau_{1,10}$
1	2	4	5	6	7	8	9	10	11	12
1	2013-03-26 23:35:25.0 UTC 43.19N; 41.67E mag 4.8; 10 km	35	−120	-	135	•	-	•	-	300
2	2013-05-28 00:09:52.0 UTC 43.22N; 41.58E mag 5.2; 2 km	−115	−150	−250	-	•	-	•	-	150
3	2013-09-17 04:09:13.0 UTC 42.13N; 45.80E mag 5.1; 2 km	-	−150	-	390	•	120	•	-	100
4	2013-11-24 18:05:41.0 UTC 34.06N; 45.52E mag 5.6; 10 km	-	•	-	−10	−60	•	•	-	160
5	2014-01-10 00:45:31.0 UTC 41.86N; 49.41E mag 4.8; 80 km	-	20	-	110	•	-	−10	-	110
6	2014-01-14 13:55:02.0 UTC 40.33N; 52.95E mag 5.2; 48 km	-	−45	•	−120	•	•	-	-	160
7	2014-01-28 23:47:35.0 UTC 32.45N; 50.02E mag 4.9; 10 km	−135	-	-	100	−300	-	-	-	120
8	2014-02-10 12:06:48.0 UTC 40.23N; 48.63E mag 5.4; 55 km	−300	-	-	-	45	75	-	-	75
9	2014-06-07 06:05:32.4 UTC 40.32N; 51.58E mag 5.4; 44 km	145	20	-	−70	-	-	-	120	80
10	2014-06-29 17:26:10.4 UTC 41.62N; 46.68E mag 5.1; 20 km	305	−85	-	-	-	315	•	-	100

(b)

№	R_{2X_k}	R_{3X_k}	R_{4X_k}	R_{5X_k}	R_{6X_k}	R_{7X_k}	R_{8X_k}	R_{9X_k}	Number and location of the area of expected earthquake
13	14	15	16	17	18	19	20	21	22
1	50	100	-	140	-	-	-	-	Georgia (Sak'art'velo)
2	150	160	250	-	-	-	-	-	Georgia (Sak'art'velo)
3	-	40	-	80	-	80	-	-	Caucasus Region, Russia
4	-	-	-	150	250	-	-	-	Iran-Iraq Border
5	-	110	-	110	-	-	40	-	Caspian Sea, Offshore Azerbaijan
6	-	100	-	120	-	-	-	-	Turkmenistan
7	110	-	-	-	180	-	-	-	Western Iran
8	130	-	-	-	260	230	-	-	Azerbaijan
9	25	40	-	100	-	-	-	80	Offshore Turkmenistan
10	20	120	-	-	-	25	-	-	Azerbaijan

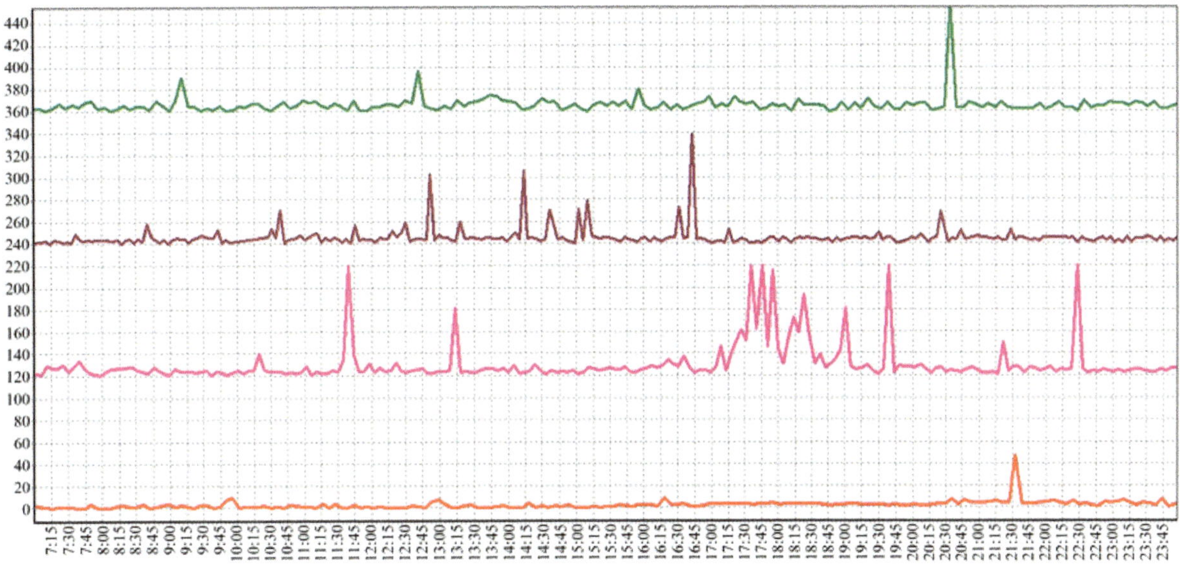

Figure 3: VII—Siazan, Naftalan, Shirvan, Qum Island 2013-05-27 Georgia (Sak'art'velo).

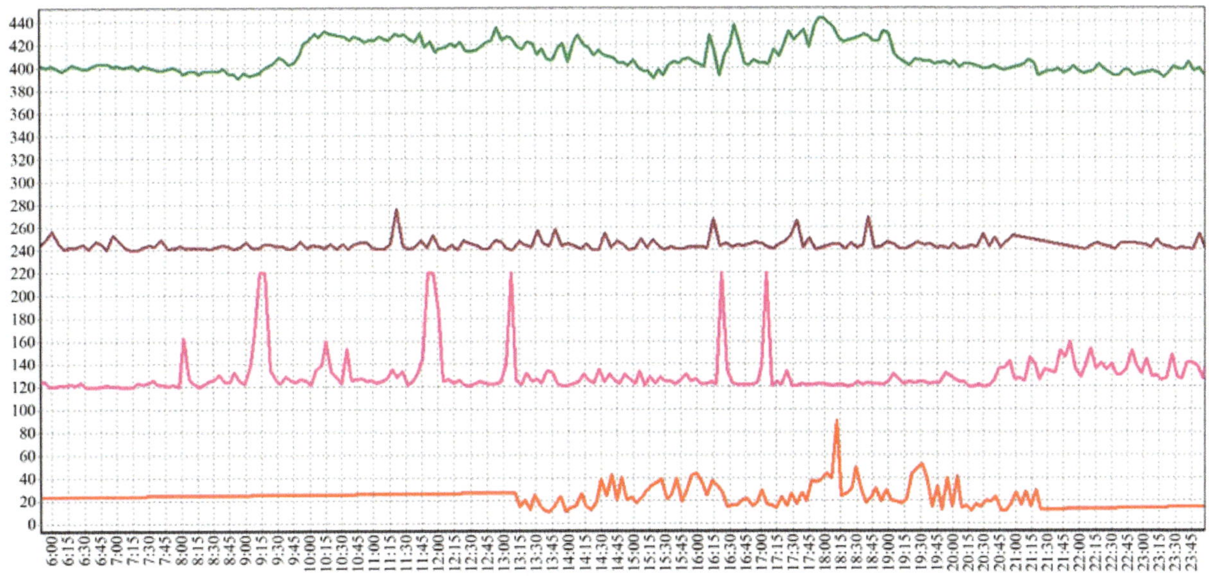

Figure 4: VII—Qazakh, Siazan, Qum Island, Neftchala 2013-09-16 Russia.

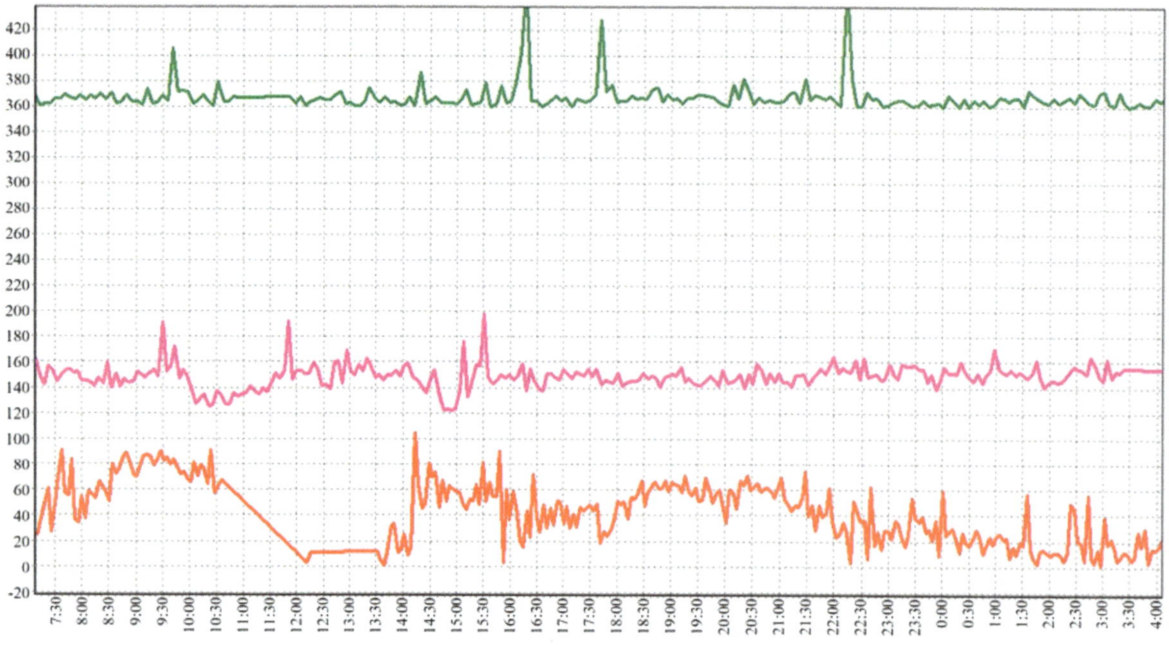

Figure 5: X—Qum Island, Neftchala, Nakhchivan 2013-11-21 Iran-Iraq border.

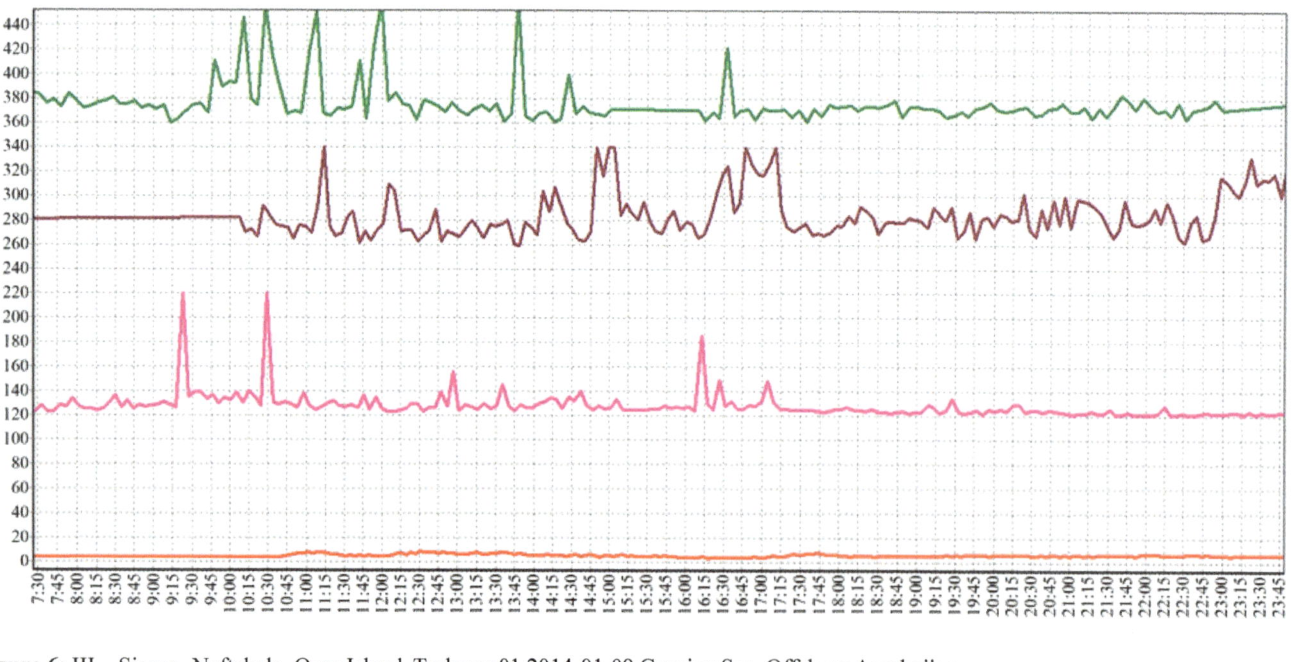

Figure 6: III—Siazan, Neftchala, Qum Island, Turkmen01 2014-01-09 Caspian Sea, Offshore Azerbaijan.

The time of the earthquakes in Table 1 is given in UTC as provided by the EMSC website. The charts show the local time (Baku time), which is UTC+4 in the winter and UTC+5 in the summertime.

Column 22 of Table 1 provides information on the locations of the identified areas of expected earthquakes. To demonstrate the validity of those results, each row of the table is accompanied by relevant charts (Figures 2-11) recorded by the RNM ASP stations during the origin of the respective ASPs.

Sign "*" in Table 1 means that the station gave a weak reaction to the origin of the ASP of the expected earthquake. Sign "-" means that the value of the registered value of $R_{X\varepsilon}(\mu)$ is lower than the threshold level.

The first row of the table shows the results of the identification of the area of the earthquake that occurred in Georgia on 2013.03.26. It follows from Figure 2 that its beginning was indicated by the RNM ASP stations in the following order: Siazan-04:15; Qum Island-04:30; Shirvan-06:50; Neftchala-08:30. Even though the Naftalan station was not operating at the time, the system identified that such an indication corresponded to area VII. The indication was 8 - 10 hours in advance of the earthquake.

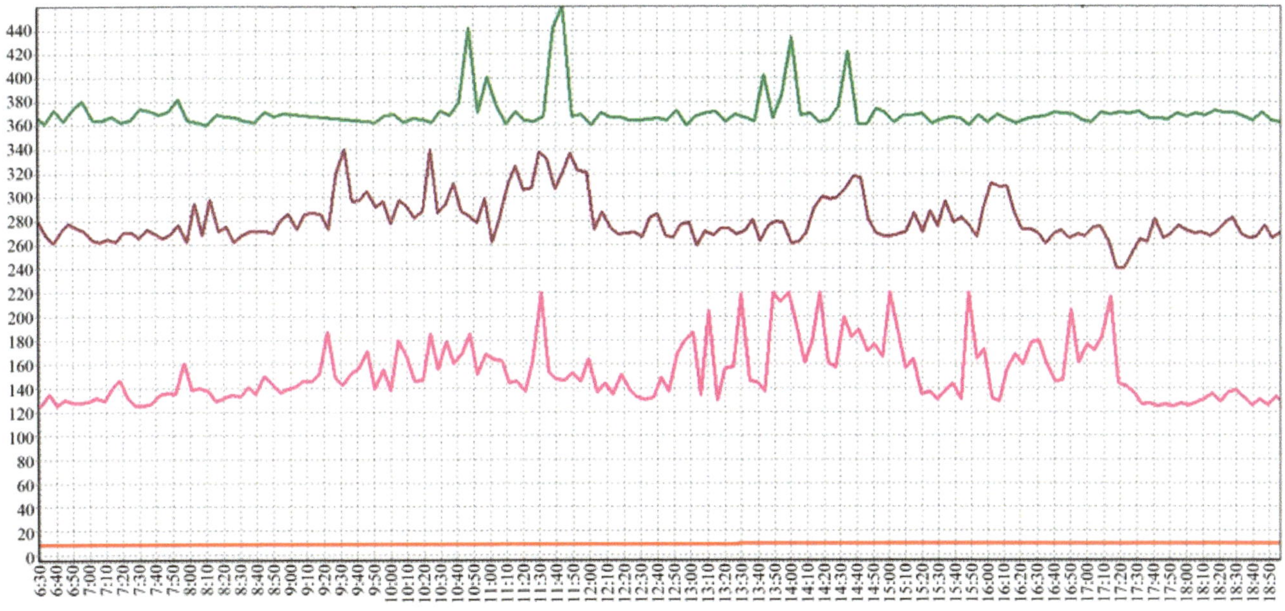

Figure 7: I—Siazan, Neftchala, Qum Island, Turkmen01 2014-01-13 Turkmenistan.

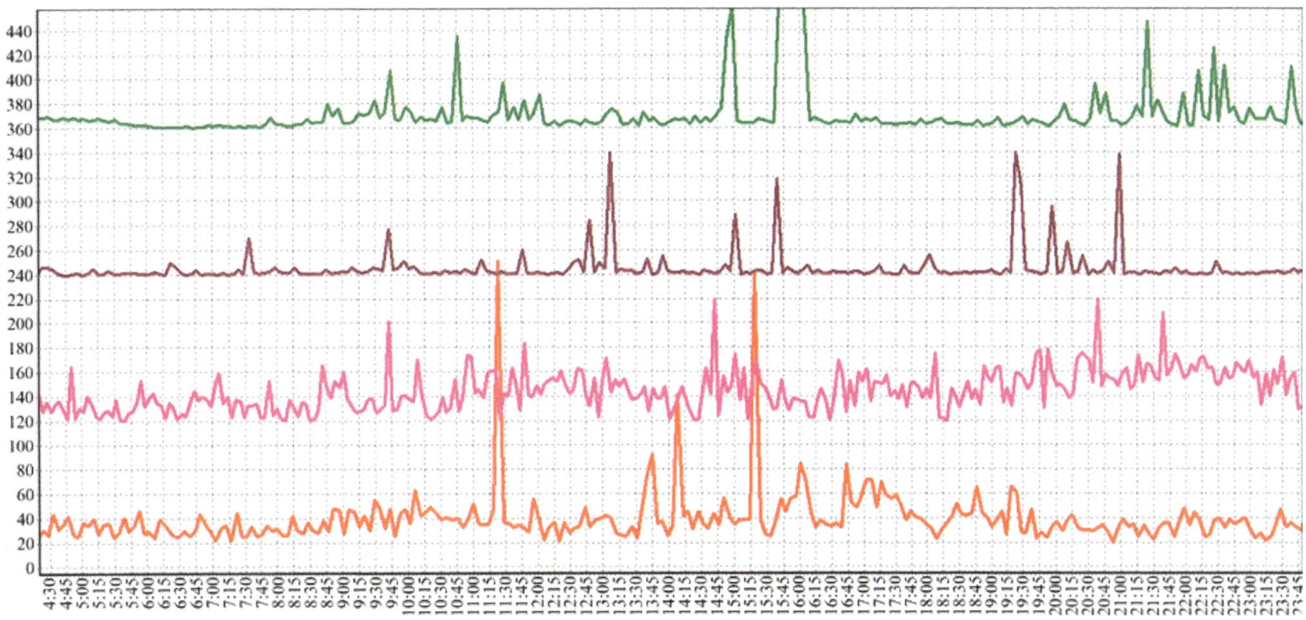

Figure 8: IX—Qum Island, Shirvan, Nakhchivan, Neftchala 2014-01-28 western Iran.

Row 2 of the table gives the results of the identification of the area of the earthquake that occurred in Georgia on 2013.05.27, 2013.05.28. According to the chart, the RNM ASP stations Siazan, Naftalan, Shirvan and Qum Island registered the ASP origin more than 20 hours before the earthquake, despite the malfunction of the Naftalan station. It is clear from the charts that the northern (Siazan) and north-western (Qum Island) stations detected an anomaly earlier than the rest of the stations. The beginning of the anomalous processes was indicated by the RNM ASP stations in the following order: Naftalan-07:30; Siazan-09:10; Shirvan-09:45; Qum Island-11:40 (Figure 3). Thus, the earthquake area was identified by the system by approximately 18:00 Baku time, which is almost 10 - 11 hours earlier than the time of registration by the ground stations.

Rows 3 of the table show the results of the identification of the area of the earthquakes that occurred in South Russia on2013.09.16.

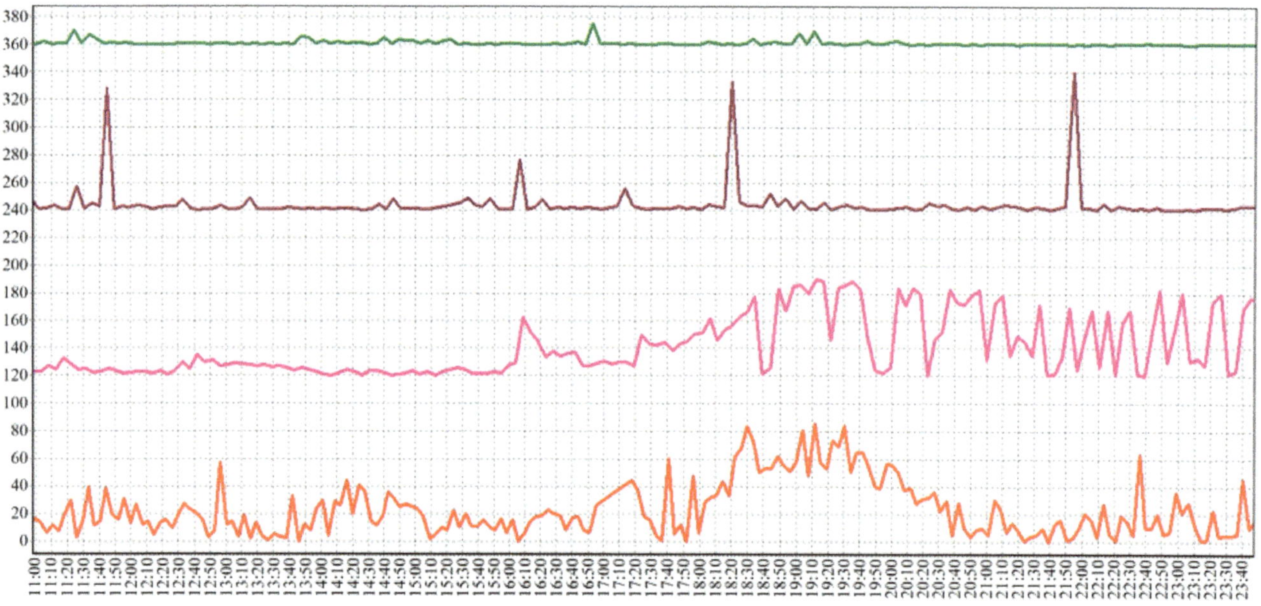

Figure 9: IV—Qum Island, Shirvan, Qazakh, Nakhchivan 2014-02-09 Azerbaijan.

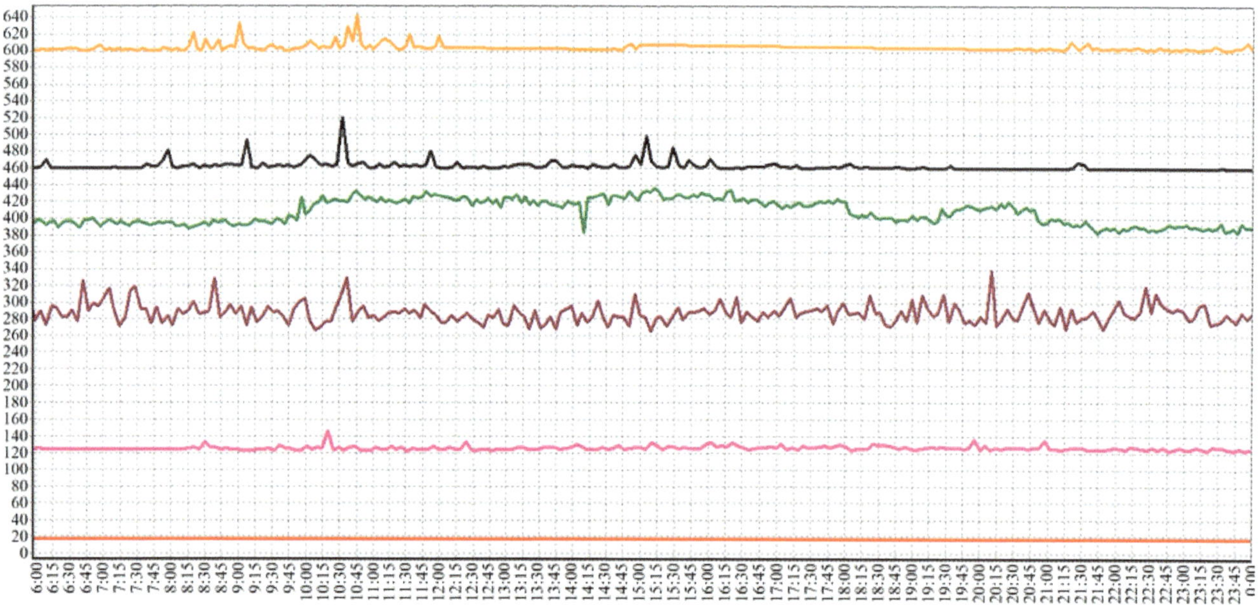

Figure 10: I—Siazan, Qum Island, Cybernetic, Neftchala, Shirvan, Turkmen01 2014-06-06 Caspian Sea, Offshore Turkmenistan.

According to the charts of the third and fourth earthquakes (Figure 4), the ASP originated in the south-east of the Caucasus region and registered in the following order: Siazan—05:30, Qum Island—08:00, Qazakh—10:00, Neftchala—14:30. Based on this sequence of registration times, the system identified earthquake area VII, which corresponds to the north-east of Azerbaijan, where an earthquake actually occurred at 16:00/17:00 Baku time. The time of the area identification was approximately 15 hours in advance of the earthquake.

According to the chart in Figure 5, based on the combination of times of registration by the stations Nakhchivan—08:00; Qum Island—09:00; Neftchala—08:50, the system identified the area of the expected earthquake on the Iran-Iraq border 12 hours beforehand.

Figure 6 shows the results of the identification of the area of the earthquake that occurred on 2014-01-09, at about 12:00, in the Caspian Sea, Offshore Azerbaijan, and was registered by the stations Turkmen01—09:15, Qum Island—09:25, Siazan—09:45, Neftchala—11:15, 16 hours in advance of the earthquake.

Row 6 of the table shows that the system identified the area of the expected earthquake in Turkmenistan. According to the chart in Figure 7, based on its sequence of registrations by the stations Neftchala—09:30,

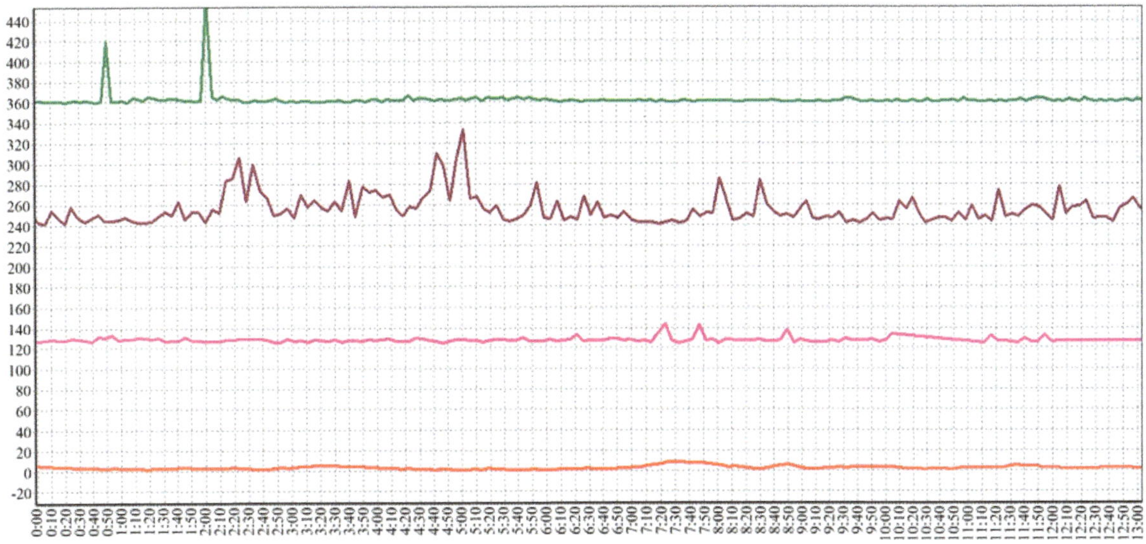

Figure 11: Siazan, Qum Island, Shirvan, Qazakh 2014-06-29 Azerbaijan.

Siazan—10:45, Qum Island—11:30, the system demonstrated that the location of the expected earthquake was in Turkmenistan, i.e. in area I east of Azerbaijan. The area of the expected earthquake was identified more than 24 hours before the earthquake was registered.

Row 7 of Table 1 shows the results of the identification of the area of the earthquake that occurred in Western Iran. According to Figure 8, the sequence of registrations by the stations Qum Island—09:45, Shirvan—07:30, Nakhchivan—04:50 and Neftchala—11:20, allowed the system to identify the location of the earthquake in area IX-Western Iran.

Row 8 of Table 1 shows the results of the identification of the area of the earthquake that occurred in Azerbaijan. According to the charts in Figure 9, the stations registered the corresponding ASP in the following order: Qum Island-17:45, Shirvan-12:45, Qazakh-19:00 and Nakhchivan-18:30, which allowed the system to identify the number (IV) of the area of the expected earthquake with a lead time of 19 hours.

Row 9 of Table 1 shows the results of identification of the area of the earthquake that occurred in Offshore Turkmenistan. According to the charts in Figure 10, this event was registered by the stations Neftchala—06:45, Qum Island—07:55, Siazan—08:15, Cybernetic—09:55, Shirvan—10:20.

Row 10 of Table 1 contains the information on the identification of the location of the expected earthquake that occurred in area I in Azerbaijan. It follows from Figure 11 that the anomaly was indicated by the stations in the following order: Siazan—00:50, Qum Island—02:15, Shirvan—07:20, Qazakh—07:30, which allowed the system to identify the location of the area of the expected earthquake (Azerbaijan).

CONCLUSIONS

1) The intelligent system based on the network of the RNM ASP stations can be used as a tool for identifying the location of the area of an expected earthquake. The seismologist is supplied with the information containing the direction and the number of the area of the expected earthquake, current combinations of the ASP, and amount, list, date and magnitude of similar combinations registered in the given area during previous earthquakes. This information will allow the seismologist to evaluate the degree of authenticity of the obtained information on the location of the area of the expected earthquake. Having enough time before the earthquake occurs, the seismologist can involve other specialists in the decision-making process if there is any doubt, ruling out an accidental error.

2) The RNM ASP stations in the network of the proposed system are built on wells with different depths and, consequently, different characteristics. These characteristics are difficult to take into account in identifying the location of the area of an expected earthquake and in determining its magnitude.

   Moreover, as the depth of a well increases, its cost increases sharply, too. This makes the building of the RNM ASP stations in the countries that have no suspended oil wells quite challenging.

   Considering the above-mentioned arguments, we recommend forming a network of stations built in 50 - 100-m deep water wells in the future, with hydrophones placed in the water column at a depth of 10 - 20 m. To improve the authenticity and reliability of the identification of the location of the area of the expected earthquake, we found it appropriate to build a network consisting of a large number of stations (over 10 - 15) in wells of equal depth located at an equal distance from one another. Integration of the network of the RNM ASP stations of the countries in several seismically active regions via satellite communication can, in the long term, allow for increasing the authenticity and reliability of determining the coordinates of the location of an expected earthquake.

3) Our experiments have demonstrated that the reliability of the ASP monitoring results and the validity of the results of identification of the location of the area of an expected earthquake grow with the growth in earthquake strength. With the strength exceeding 5 points, the results of the identification of the earthquake location proved to be valid in almost all cases. The

value of the estimate of the cross-correlation function $R_{X\varepsilon}(\mu)$ between the useful signal $X(i\Delta t)$ and the noise $\varepsilon(i\Delta t)$ decreases as the distance from the earthquake area grows. The value of the estimate of noise variance D_ε increases as the distance from the area grows; the correlation $R_{X\varepsilon}(\mu)/R_{X\varepsilon\varepsilon}(\mu)$ decreases with distance and $D_\varepsilon/R_{X\varepsilon\varepsilon}$ increases. The propagation velocity of the seismic-acoustic noise in different types of medium, e.g. water, sand or clay, significantly varies. There is a correlation between the well depth and the radius of the ASP monitoring.

4) The experiments at the Qum Island station in the Caspian Sea have demonstrated that the monitoring range of that station is much wider than that of the stations located far from the Caspian Sea. Other stations in Siazan and Neftchala located near the Caspian Sea also have a wide monitoring range compared with other stations. Practically all seismic processes reaching the Caspian Sea are clearly registered by those stations. Therefore, in building networks of new stations, one should consider the fact that the sea is a "perfect conductor" for seismic- acoustic noises emerging during ASP origin in the region.

5) The results obtained from the experimental data allow for the assumption that the lead time of the registration of the ASP origin by a seismic-acoustic RNM ASP station over standard seismic equipment is due to two factors.

   First, seismic-acoustic waves that arise in the beginning of the ASP origin do not reach the Earth's surface due to the frequency characteristics of certain upper strata, which furthers their horizontal spread in deep strata as noise. Reaching the steel pipes of the well, seismic-acoustic waves transform into acoustic signals and go to the ground surface at the velocity of sound, where they are detected by a hydrophone. At the same time, low frequency seismic waves from seismic processes are perceived at the surface after a certain amount of time, when the earthquake is already in progress, and are registered by seismic receivers of standard ground equipment much later.

   Second, the use of noise technologies by analysing seismic-acoustic noise allows one, with the emergence of correlation between the useful signal and the noise, to register the ASPs in the beginning of their origin. These two factors make it possible for RNM ASP stations to indicate the time of the beginning of the ASP origin much earlier than is done by stations of the seismic survey service.

6) Seismic-acoustic stations of ASP monitoring can also be used for monitoring the latent period of volcano formation well before the eruption. Their use will also allow the monitoring, on a regional basis, of the testing of minor and major nuclear bombs and other experiments related to the manufacture of military equipment.

ACKNOWLEDGEMENTS

The authors express their gratitude to the SOCAR for providing suspended oil wells, as well as all the assistance and support in the construction of RNM ASP stations on those wells. This study was financed partially in the course of implementation of the following projects, for which the authors are immensely grateful. SOCAR project "Development of a network of seismic-acoustic stations for robust noise monitoring of anomalous seismic processes in the South Caucasus and Caspian basin using suspended oil wells as seismic-acoustic channels". SCIENCE & TECHNOLOGY CENTER IN UKRAINE project. "Development of the technology for determination of coordinates of anomalous seismic processes". The study has been discussed among researchers from many countries during conferences and amicable meetings, and the authors wish to thank them for their interest and support. The researchers who participated actively in those discussions are: Nikolay Filimonov, Aleksandr Sidorin, Konstantin Kislov, Valeriy Mitko, Marina Minina, Lyubov Ebert, Orxan Polat, Alamdar Bayrakdar, Mehmet Kuruoglu, Afandi Nasibov, Ali Pinar, Dogan Kalafat, Aleksandr Kandzera, Murat Chariyev, Guljamal Chariyeva, Ioannis Baskoutas, Branden Kristensen, Taimi Mulder, Lutfulla Hamidov.

REFERENCES

1. Kanamori, H. and Brodsky, E.E. (2004) The Physics of Earthquakes. Reports on Progress in Physics. Reports on Progress in Physics, 67, 1429-1496. http://dx.doi.org/10.1088/0034-4885/67/8/R03
2. Ghahari, F., Jahankhah, H. and Ghannad, M.A. (2010) Study on Elastic Response of Structures to Near-Fault Ground Motions through Record Decomposition. Soil Dynamics and Earthquake Engineering, 30, 536-546. http://dx.doi.org/10.1016/j.soildyn.2010.01.009
3. Aliev, T.A., Abbasov, A.M., Aliev, E.R. and Guluyev, G.A. (2007) Digital Technology and the System for Receiving and Analyzing the Information from Deep Earth Layers for Noise-Monitoring the Technical State of the SociallySignificant Objects. Automatic Control and Computer Sciences, 41, 59-67. http://dx.doi.org/10.3103/S0146411607020010
4. Yalcinkaya, E., Tekebas, S. and Pinar, A. (2013) Analysis of Ambient Noise in Yalova, Turkey: Discrimination between Artificial and Natural Excitations. Journal of Seismology, 17, 1021-1039. http://dx.doi.org/10.1007/s10950-013-9370-7
5. Fischer, T. and Bachura, M. (2014) Detection Capability of Seismic Network Based on Noise Analysis and Magnitude of Completeness. Journal of Seismology, 18, 137-150. http://dx.doi.org/10.1007/s10950-013-9407-y
6. Kislov, K.V. and Gravirov, V.V. (2011) Earthquake Arrival Identification in a Record with Technogenic Noise. Seismic Instruments, 47, 66-79. http://dx.doi.org/10.3103/S0747923911010129
7. Yee, E., Stewart, J.P. and Schoenberg, F.P. (2011) Characterization and Utilization of Noisy Displacement Signals from Simple Shear Device Using Linear and Kernel Regression Methods. Soil Dynamics and

Earthquake Engineering, 31, 25-32. http://dx.doi.org/10.1016/j.soildyn.2010.07.011

8. Aliev, T. (2007) Digital Noise Monitoring of Defect Origin. Springer, London, 223 p.
9. Aliev, T.A., Guluyev, G.A., Pashayev, F.H. and Sadygov, A.B. (2012) Noise Monitoring Technology for Objects in Transition to the Emergency State. Mechanical Systems and Signal Processing, 27, 755-762. http://dx.doi.org/10.1016/j.ymssp.2011.09.005
10. Moser, P. and Moaveni, B. (2013) Design and Deployment of a Continuous Monitoring System for the Dowling Hall Footbridge. Experimental Techniques Society for Experimental Mechanics, 37, 15-26.
11. Aliev, T.A., Abbasov, A.M., Guluyev, G.A., Pashayev, F.H. and Sattarova, U.E. (2013) System of Robust Noise Monitoring of Anomalous Seismic Processes. Soil Dynamics and Earthquake Engineering, 53, 11-25. http://dx.doi.org/10.1016/j.soildyn.2012.12.013
12. Aliev, T.A., Abbasov, A.M., Guluyev, G.A., Pashayev, F.H. and Sattarova, U.E. (2012) Technologies and Systems for Minimization of Damage from Destructive Earthquakes. In: Etirmishli, G., Ed., Seismoforecasting Researches Carried out in the Azerbaijan Territory, Nafta-Press, Baku, 449-464.
13. Hashemi, M. and Alesheikh, A.A. (2011) A GIS-Based Earthquake Damage Assessment and Settlement Methodology. Soil Dynamics and Earthquake Engineering, 31, 1607-1617. http://dx.doi.org/10.1016/j.soildyn.2011.07.003
14. Kanamori, H. (2005) Real-Time Seismology and Earthquake Damage Mitigation. Annual Review of Earth and Planetary Sciences, 33, 195-214. http://dx.doi.org/10.1146/annurev.earth.33.092203.122626
15. Colak, O.H., Destici, T.C., Ozen, S., Arman, H. and Cerezci, O. (2006) Frequency-Energy Characteristics of Local Earthquakes Using Discrete Wavelet Transform (DWT). World Academy of Science, Engineering and Technology, 20, 38-41.
16. Hutton, D.V. (2004) Fundamentals of Finite Element Analysis. The McGraw-Hill Companies, New York, 494 p.
17. Lockwood, O.G. and Kanamori, H. (2006) Wavelet Analysis of the Seismograms of the 2004 Sumatra-Andaman Earthquake and Its Application to Tsunami Early Warning. Geochemistry Geophysics Geosystems, 7, 1-10. http://www.agu.org/journals/abs/2006/2006GC001272.shtml
18. Descherevsky, A.V., Lukk, A.A., Sidorin, A.Y., Vstovsky, G.V. and Timashev, S.F. (2003) Flicker-Noise Spectroscopy in Earthquake Prediction Research. Natural Hazards and Earth System Sciences, 3, 159-164. http://dx.doi.org/10.5194/nhess-3-159-2003
19. Shebalin, P., Keilis-Borok, P., Gabrielov, A., Zaliapin, I. and Turcotte, D. (2006) Short-Term Earthquake Prediction by Reverse Analysis of Lithosphere Dynamics. Tectonophysics, 413, 63-75. http://dx.doi.org/10.1016/j.tecto.2005.10.033
20. Aliev, T.A., Alizadeh, T.A. and Abbasov, A.M. (2005) Method for Monitoring the Beginning of Anomalous Seismic Process. International Application No. PCT/AZ2005/000006, Pub. No. WO2006/130933, International Filling Date— 19 December 2005.
21. Aliev, T.A., Abbasov, A.M., Aliev, E.R. and Guluyev, G.A. (2006) Method for Monitoring and Forecasting Earthquakes. International Application No PCT/AZ2006/00000, Pub. No. WO2007/143799, International Filling Date—16 June 2006.
22. Papagiannopoulos, G.A. and Beskos, D.E. (2006) On a Modal Damping Identification Model for Building Structures. Archive of Applied Mechanics, 76, 443-463. http://dx.doi.org/10.1007/s00419-006-0046-4
23. Papagiannopoulos, G.A. and Beskos, D.E. (2009) On a Modal Damping Identification Model for Non-Classically Damped Structures Subjected to Earthquakes. Soil Dynamics and Earthquake Engineering, 29, 583-589. http://dx.doi.org/10.1016/j.soildyn.2008.10.005
24. Zafarani, H., Noorzad, A., Ansari, A. and Bargi, K. (2009) Stochastic Modeling of Iranian Earthquakes and Estimation of Ground Motion for Future Earthquakes in Greater Tehran. Soil Dynamics and Earthquake Engineering, 29, 722-741. http://dx.doi.org/10.1016/j.soildyn.2008.08.002
25. Rydelek, P. and Pujol, J. (2004) Real-Time Seismic Warning with a Two-Station Subarray. Bulletin of the Seismological Society of America, 94, 1546-1550. http://dx.doi.org/10.1785/012003197
26. Stankiewicz, J., Bindi, D., Oth, A. and Parolai, S. (2013) Designing Efficient Earthquake Early Warning Systems: Case Study of Almaty, Kazakhstan. Journal of Seismology, 17, 1125-1137. http://dx.doi.org/10.1007/s10950-013-9381-4
27. Alcik, H., Ozel, O., Wu, Y.M., Ozel, N.M. and Erdik, M. (2011) An Alternative Approach for the Istanbul Earthquake Early Warning System. Soil Dynamics and Earthquake Engineering, 31, 181-187. http://dx.doi.org/10.1016/j.soildyn.2010.03.007
28. Satriano, C., Wub, Y.-M., Zollo, A. and Kanamori, H. (2011) Earthquake Early Warning: Concepts, Methods and Physical Grounds. Soil Dynamics and Earthquake Engineering, 31, 106-118. http://dx.doi.org/10.1016/j.soildyn.2010.07.007
29. Tsuboi, S., Saito, M. and Kikuchi, M. (2002) Real-Time Earthquake Warning by Using Broadband P Waveform. Geophysical Research Letters, 29, 2187-2191. http://dx.doi.org/10.1029/2002GL016101
30. Aliev, T.A., Alizadeh, A.A., Etirmishli, G.D., Guluyev, G.A., Pashayev, F.G. and Rzaev, A.G. (2011) Intelligent Seismoacoustic System for Monitoring the Beginning of Anomalous Seismic Process. Seismic Instruments, 47, 15-23. http://dx.doi.org/10.3103/S0747923911010026
31. Aliev, T.A., Abbasov, A.M., Mamedova, G.G., Guluyev, G.A. and Pashayev, F.G. (2013) Technologies for Noise Monitoring of Abnormal Seismic Processes. Seismic Instruments, 49, 64-80. http://dx.doi.org/10.3103/S0747923913010015

CHAPTER 15

HYDRODYNAMICAL PHENOMENA AT PREPARATION OF EARTHQUAKE

Igor Petrovich Dobrovolsky

Keywords: Earth Tides, Short-Term Precursors of Earthquakes, Level of Liquid in a Well

INTRODUCTION

Often before an earthquake (during a period of short-term precursors) there are variations in level of fluid in wells and boreholes, dimness of water in wells, variations in fluid temperature and disturbances in the regime of natural fountains. If they are really caused by an earthquake preparation all these phenomena are defined as hydrodynamical earthquake precursors. Short-term precursors, the earthquake itself and its aftershocks are the processes of the same physical sense which enables their identical phenomenological description and calculation. We comment this note. The cycle of the tectonic earthquake separates into two different stages: accumulation of energy and its release. Phenomena accompanying the stage accumulation of energy are defined as long-term precursors of an earthquake. This stage is conditioned by the appearance and development of the heterogeneity in the bowels of the Earth [1]. After that the disintegration of the heterogeneity begins and the release of energy stored. This stage consists of three phases: relatively short phase of short-term precursors, almost instantaneous act of earthquake's focus and the prolonged aftershocks activity. Thus these phases are physically identical in the sense that they change the elastic potential energy of the medium in the same direction. If the state of the medium after the period of long-term precursors is treated as initial position then the following behavior displays three decreases of elastic potential energy. Action of the hydrodynamical phenomena due to an earthquake depends on the properties of the rocks in the vicinity of a well (borehole) and its location with respect to the hypocenter of the earthquake. However the first mechanism also characterizes the variations of level in wells due of the Earth tides. Therefore we can discuss hydrodynamical phenomena in general and describe them by identical equations. Hydrodynamical precursors are known for a long time and their description is usually included into monographs (for example [2]). A noticeable paper in this field is [3] that contains the plots of variations of water level by 15 cm before the Przhevalsk earthquake of 1970 (M = 6.8) at an epicentral distance of 30 km. But there are also other examples. The striking example of the precursory anomaly is given in [4]: groundwater level decreased by 57 m before the earthquake of magnitude 5.4 at epicentral distance of 70 km. In the second case the epicentral distance is more, earthquake's magnitude is less but effect in 380 times more. Seismologists undoubtedly accept the existence of hydrodynamical precursors; however, their quantitative characteristics are unclear. The reliable data about the distribution of hydrodynamical precursors in time and space are absent. Geophysicists are not able to calculate of hydrodynamical precursors. The aim of the present paper is to calculate hydrodynamical phenomena: Earth's tidal and precursory variations in level of liquid in wells (boreholes) using identical systems of equations and to clarify data on distribution of hydrodynamical precursors on the Earth's surface.

THE EQUATIONS OF TWO-PHASE MEDIUM

We analyze the problem of filtration of fluids in porous two-phase medium. In the linear approximation, in the case of slow processes, when inertial forces and compressibility of the solid phase can be neglected with respect to the compressibility of

Citation: Dobrovolsky, I. (2014) Hydrodynamical Phenomena at Preparation of Earthquake. Open Access Library Journal, 1, 1-11. doi: 10.4236/oalib.1100896.

fluid, the system of equations of dynamics of a porous liquid has the form [1]

$$\mathrm{grad}P + \frac{\mu m_0}{k}(\mathbf{V} - \mathbf{U}) = 0,$$

$$\frac{1}{m_0}\frac{\partial m}{\partial t} + \beta\frac{\partial P}{\partial t} + \mathrm{div}\mathbf{V} = 0. \tag{1}$$

where P(x,y,z,t) is the abundant porous pressure, V and U are velocities of fluid and solid phases, respectively. The first equation of system (1) is Darsy law, the second equation reflects deformation of the fluid phase in elementary volume of a two-phase medium. Transform the first term in the second equation of system (1). We consider representative volume v_0 of a medium large enough with respect to the sizes of pores, within which pores take the volume v_{m_0}. Then the current porosity m is expressed by the formula

$$m = \frac{v_{m_0}(1+e_m)}{v_0(1+e)} = m_0\frac{1+e_m}{1+e}. \tag{2}$$

Obviously, e and e_m are connected among themselves and we write this relationship in the form e_m = qe. The value of q depends on geometry of pores, stress state, properties of solid phase and properties of fluid. Note that the compressibility of fluid is higher than that of rocks in the Earth's crust by an order of magnitude. Calculations give the value q = 2 in the case of spherical pores with water at all-round compression and porosity smaller than 20%, when pores can be regarded as almost noninteracting. When the rocks contain an infinite water layer (fissure or disk-tape pore), and the system is compressed in the direction normal to the plane of the layer, calculations give q = 7 -12 for the porosity in the range of 10% - 3%. These results are easy to receive by solution of simple problems of the elasticity theory for a spherical pore and a fluid layer in an unlimited elastic medium. In the granulated medium of the type of sands, the total variation volume takes place due to redistribution of practically solid grains, and simple calculations give $q = 1/m_0$. Usually these media has a high porosity reaching 40% in the; for dry sands at $m_0 = 0.2$, $q = 5$. Thereby, we can accept q in the range q = 5-8.

Using expression (2), where we set $e_m = qe$, $q = \mathrm{const}$ and $e = e(t)$, we obtain

$$\frac{1}{m_0}\frac{\partial m}{\partial t} = \frac{q-1}{(1+e)^2}\frac{\partial e}{\partial t} \approx (q-1)\frac{\partial e}{\partial t}. \tag{3}$$

here, e is neglected with respect to 1, because $e \le 10^{-4}$.

Now system (1) takes form

$$\mathrm{grad}P + \frac{\mu m_0}{k}(\mathbf{V} - \mathbf{U}) = 0,$$

$$\beta\frac{\partial P}{\partial t} + \mathrm{div}\mathbf{V} = -(q-1)\frac{\partial e}{\partial t}. \tag{4}$$

Excluding V from (4), we obtain the equation for P

$$\mathrm{div}\frac{k}{\mu m_0}\mathrm{grad}P - \beta\frac{\partial P}{\partial t} = (q-1)\frac{\partial e}{\partial t} + \mathrm{div}\mathbf{U}. \tag{5}$$

As U is velocity of the solid phase, then div U $=\partial e/\partial t$. Substituting this expression into (5), we finally obtain

$$\mathrm{div}\frac{k}{\mu m_0}\mathrm{grad}P - \beta\frac{\partial P}{\partial t} = q\frac{\partial e}{\partial t}. \tag{6}$$

When $k/\mu m_0 =$ const, Equation (6) becomes the classic equation of diffusion with constant coefficients

$$\frac{k}{\mu m_0}\Delta P - \beta\frac{\partial P}{\partial t} = q\frac{\partial e}{\partial t}. \tag{7}$$

General Statement of the Problem The Earth is modeled as a half-space. We consider an infinite thin horizontal water-saturated layer located at a depth within impenetrable rocks. The water-saturated layer is thin and all parameters and processes are homogeneous within it with respect to its thickness and, therefore, the filtration problem is two-dimensional. A well (borehole) is located in this layer and the liquid is at initial level. We should calculate deviations from this initial state and these deviations can be of any sign. From Equation (7), equation for the porous pressure $P(r,\varphi,t)$ in a homogeneous medium (β, μ, q, k, m_0 are constant) is

$$\frac{\partial P}{\partial t} - \kappa\Delta P = -\frac{q}{\beta}\frac{\partial e}{\partial t}, \tag{8}$$

Where $\kappa = k/\mu\beta m_0$.

The right-hand side of this equation contains the volume strain e . By the natural definition the tensile deformation is positive and the compressible deformation is negative. It is necessary to mean it while solving the problem. The first boundary condition is the condition in the well. The pressure variations in well are determined by the flow of liquid from the layer. By the Darsy law the condition takes the form

$$\left.\left(P - \frac{\kappa\chi}{r}\int_0^t \frac{\partial P}{\partial r}\mathrm{d}t'\right)\right|_{r=r_0} = 0, \tag{9}$$

Where $\chi = 2\sigma gh\beta m_0 = 2\sigma ghk/\mu\kappa$.

The expression obtained is rather curious: it contains the dependence on angle ϕ. Borehole is as though divided into infinite number of sectors. The procedure of averaging over an angle ϕ eliminates this shortcoming and the boundary condition becomes real

$$\left.\left(\bar{P} - \frac{\kappa\chi}{r}\int_0^t \frac{\partial \bar{P}}{\partial r}\mathrm{d}t'\right)\right|_{r=r_0} = 0, \tag{10}$$

where the line denotes the averaging over an angle $\bar{P} = \frac{1}{2\pi}\int_0^{2\pi} P\mathrm{d}\varphi$.

The second boundary condition (boundary condition at infinity) is $P(\infty,\varphi,t) = F$. The form of function F depends on the certain problem but the general recipe of its definition is following. If $P'(r,\varphi,t)$ is the solution of Equation (8) for the continuous layer (without a well) then $F = P'(\infty,\varphi,t)$. The differential Equation (8), the boundary condition (10), the boundary condition at infinity and the zero initial condition create the system for determination of function P and pressure in the well P_w (t)

$$\frac{\partial P}{\partial t}-\kappa\left(\frac{\partial^2 P}{\partial r^2}+\frac{1}{r}\frac{\partial P}{\partial r}+\frac{1}{r^2}\frac{\partial^2 P}{\partial\varphi^2}\right)=-\frac{q}{\beta}\frac{\partial e}{\partial t},\quad \left.\left(\bar{P}-\frac{\kappa\chi}{r}\int_0^t\frac{\partial\bar{P}}{\partial r}dt'\right)\right|_{r=r_0}=0,$$

$$P(\infty,\varphi,t)=F,\quad P(r,\varphi,0)=0,\quad P_w(t)=\bar{P}(r_0,t). \tag{11}$$

Tidal Variations of Level In the case of the Earth tides the investigated problem is the typical problem without initial conditions. We put e = Asinωt. Then the problem becomes axisymmetric. If we will introduce dimensionless variables using expressions

$$p=-\frac{\beta}{qA}P,\quad \rho=r\sqrt{\frac{\omega}{\kappa}},\quad \tau=\omega t \tag{12}$$

then from (11) we obtain the following system

$$\frac{\partial p}{\partial\tau}-\left(\frac{\partial^2 p}{\partial\rho^2}+\frac{1}{\rho}\frac{\partial p}{\partial\rho}\right)=\frac{d\sin\tau}{d\tau},\quad \left.\left(\frac{\partial p}{\partial\tau}-\frac{\chi}{\rho}\frac{\partial p}{\partial\rho}\right)\right|_{\rho=\rho_0}=0,\quad p(\infty,\tau)=\sin\tau. \tag{13}$$

Solution of (13) is based on the fact that the homogeneous equation

$$\frac{\partial u}{\partial\tau}-\left(\frac{\partial^2 u}{\partial\rho^2}+\frac{1}{\rho}\frac{\partial u}{\partial\rho}\right)=0 \tag{14}$$

has the family of solutions via Kelvin functions with arbitrary constants

$$u(\rho,\tau)=C_1\left(\mathrm{bei}(\rho)\cos(\tau-\alpha)+\mathrm{ber}(\rho)\sin(\tau-\alpha)\right)+C_2\left(\mathrm{kei}(\rho)\cos(\tau-\alpha)+\mathrm{ker}(\rho)\sin(\tau-\alpha)\right). \tag{15}$$

To obtain this solution we search the solution in form $u=f(\rho)e^{i(\tau-\alpha)}$; then in the solution we go to Kelvin functions and determine the real and imaginary parts. Each of them is the solution of initial equation (we use the imaginary part). The solution equal to zero at infinity has the form

$$u(\rho,\tau)=C\left(\mathrm{kei}(\rho)\cos(\tau-\alpha)+\mathrm{ker}(\rho)\sin(\tau-\alpha)\right). \tag{16}$$

Together with the particular solution we obtain the function

$$p(\rho,\tau)=\sin\tau+C\left(\mathrm{kei}(\rho)\cos(\tau-\alpha)+\mathrm{ker}(\rho)\sin(\tau-\alpha)\right). \tag{17}$$

C and α are determine from the first boundary condition in (13) (the second boundary condition is fulfilled for any C and α). The final result is following. Dimensionless pressure in the well is

$$p_w(\tau)=p(\rho_0,\tau)=A_w\sin(\tau-\gamma), \tag{18}$$

Where

$$A_w=\frac{\chi B}{G},\quad \cos\gamma=\frac{B^2\chi-\rho_0 F}{BG},\quad G=\sqrt{B^2\chi^2-2\rho_0 F\chi+\rho_0^2 S},\quad B=\sqrt{\mathrm{kei}_1^2(\rho_0)+\mathrm{ker}_1^2(\rho_0)},$$

$$S=\mathrm{kei}^2(\rho_0)+\mathrm{ker}^2(\rho_0),\quad F=\mathrm{ker}(\rho_0)\frac{d\,\mathrm{kei}(\rho_0)}{d\rho_0}-\mathrm{kei}(\rho_0)\frac{d\,\mathrm{ker}(\rho_0)}{d\rho_0}.$$

Parameter γ is the phase difference between tidal strains and level variations. To return to dimensional values we use expressions (12) and level variation is $l(t)=P_w(t)/\sigma g$. Such simple solution enables easy construction of the synthetic function of tidal variations in level by summing variations at various frequencies. Let's make the short analysis of the solution (18). At $\rho_0\to 0$ we obtain the physically obvious result, i.e. $p_w\to\sin\tau$. Therefore the level amplitude is limited. Calculations also show that the phase difference (γ) depending on frequency has the maximum but the decrease in γ after it appears very weak. To calculate the amplitude value of level variation we assume: A = 10^{-7} (amplitude of the volume strain of the Earth tide), q = 7 $r_0=0.5$ m, $\kappa=25/864$ m²/s, $\chi=3\times10^{-5}$, $\beta=4\times10^{-1}$. (In general κ and χ should be defined from empirical data.) Then the level variation amplitude is equal to 1.2 mm and the amplitude limit is 175 mm.

PRECURSORY VARIATIONS OF LEVEL

The problem of calculation of precursory level variations is more complicated as compared to the previous one. The process is not oscillatory, the natural time scale is absent, initial condition is essentially zero and the boundary condition at infinity is also zero. The right-hand side of the differential equation depends on three variables (time, radius and angle) and involves two spatial scales: the local coordinate system where the filtration process occurs in the well vicinities and the global system of deformation of the Earth's crust due to an earthquake. We represent the solution of system (11) as the Fourier row by ϕ.

$$P(r,\varphi,t)=W(r,t)+\sum_{n\neq 0}W_n(r,t)e^{in\varphi}\;. \tag{19}$$

After averaging expression (19) on angle ϕ we receive $W(r,t)=\bar{P}$ and $P_w=W(r_0,t)$. As a result system (11) takes the form

$$\frac{\partial W}{\partial t}-\kappa\left(\frac{\partial^2 W}{\partial r^2}+\frac{1}{r}\frac{\partial W}{\partial r}\right)=-\frac{q}{\beta}\frac{\partial \bar{e}}{\partial t},\quad \left.\left(W-\frac{\kappa\chi}{r}\int_0^t\frac{\partial W}{\partial r}dt'\right)\right|_{r=r_0}=0,$$

$$W(\infty,t)=0,\quad W(r,0)=0,\quad P_w=W(r_0,t). \tag{20}$$

Let the medium get the volume strain ε at the end of the stage of long-term precursors. Assuming this moment as initial moment, we suppose that portion m of strain ε drops at the phase of short-term precursors after initial moment. We will assume this drop is instantaneous. Then we substitute function $e=-m\varepsilon E(t)$ into the right-hand side of differential equation from (11) or from (20). Further we introduce dimensionless variables according to expressions

$$w=\frac{\beta}{qmQ\sin 2\psi}W,\quad \rho=\frac{r}{\sqrt{\kappa T}},\quad \tau=\frac{t}{T},\quad n=\frac{\sqrt{\kappa T}}{H},\quad Q=\frac{\alpha s}{3G}. \tag{21}$$

If T = 86,400 s then the unit of dimensionless time corresponds to one day. At κ = 25 864 m^2 /s the unit of dimensionless distance will be 50 m; the diffusing particle travels this distance in a day and 500 m in 100 days. The meaning of parameters Q , n , ψ and the specific form of the right-hand side of the differential equation are given in Appendix A. As a result we obtain the following system

$$\frac{\partial w}{\partial \tau}-\left(\frac{\partial^2 w}{\partial \rho^2}+\frac{1}{\rho}\frac{\partial w}{\partial \rho}\right)=\delta(\tau)\theta(\tilde{\rho},n,\rho),\quad \left.\left(w-\frac{\chi}{\rho}\int_0^\tau\frac{\partial w}{\partial \rho}d\tau'\right)\right|_{\rho=\rho_0}=0,$$

$$w(\infty,\tau)=0,\quad w(\rho,0)=0,\quad p_w(\tau)=w(\rho_0,\tau). \tag{22}$$

The solution of system (22) is calculated as follows. Let u (ρ τ,) be the solution of nonuniform equation and it satisfies two zero conditions: initial condition and condition at infinity. Then, this solution is expressed via the Duhamel integral which in this case takes the form

$$u(\tilde{\rho},\rho,\tau)=\frac{1}{2\tau}e^{-\frac{\rho^2}{4\tau}}\int_0^\infty z\theta(\tilde{\rho},n,z)e^{-\frac{z^2}{4\tau}}I_0\left(\frac{\rho z}{2\tau}\right)dz, \tag{23}$$

where I_0 is the modified Bessel function. Due to presence of delta-function, formula (23) actually expresses the solution of the uniform equation with nonzero initial condition but it is unimportant. We take the Laplace transform to system (22) with respect to τ with parameter p . Designate the Laplace transform operator by L . We introduce an auxiliary function f satisfying the uniform equation and zero conditions at the initial moment and at infinity. The Laplace transformation of f is

$$Lf=C(p)K_0\left(\rho\sqrt{p}\right), \tag{24}$$

where K_0 is Macdonald function and C(p) is an arbitrary function. Now, the function v = f+u is constructing by choosing C(p) such that v satisfies the zero condition in the well, i.e. $v(\rho_0,\tau)=0$. Its Laplace transformation is

$$Lv(\rho,p)=Lu(\rho,p)-\frac{Lu(\rho_0,p)}{K_0\left(\rho_0\sqrt{p}\right)}K_0\left(\rho\sqrt{p}\right). \tag{25}$$

Finally, we take the solution of system (22) in the form $w=f+v$. This solution should satisfy only the boundary condition in the well of system (22) since other conditions are already fulfilled. As a result, the Laplace transform of pressure in the well is expressed by formula

$$Lw(\rho_0,p)=\frac{\chi\sqrt{p}K_1(\rho_0\sqrt{p})Lu(\rho_0,p)}{\rho_0 pK_0(\rho_0\sqrt{p})+\chi\sqrt{p}K_1(\rho_0\sqrt{p})}+\frac{\chi\sqrt{p}K_0(\rho_0\sqrt{p})Lu_\rho(\rho_0,p)}{\rho_0 pK_0(\rho_0\sqrt{p})+\chi\sqrt{p}K_1(\rho_0\sqrt{p})} \tag{26}$$

where $Lu_\rho(\rho_0,\tau)=L\left.\frac{\partial u(\rho,\tau)}{\partial\rho}\right|_{\rho=\rho_0}$.

Direct calculations and analysis conducted in the next section show that in calculations of precursory variations, we can neglect the second addend and accept

$$Lw(\rho_0,p)\cong\frac{\chi\sqrt{p}K_1(\rho_0\sqrt{p})}{\rho_0 pK_0(\rho_0\sqrt{p})+\chi\sqrt{p}K_1(\rho_0\sqrt{p})}Lu(\rho_0,p). \tag{27}$$

The inverse transformation of (27) is described in Appendix B.

RESULTS, DISCUSSION AND CONCLUSIONS

In this section we discuss only precursory level variations. The solution of this problem is given by expression (27). The original function is expressed as a convolution of two functions and it is reasonable first to analyze these functions.

Calculating function $u(\tilde{\rho},\rho,\tau)$ given by expression (23), we obtain the result unexpected, on the first glance. In the case of strong earthquakes (starting from magnitude 5 - 5.5) $u(\tilde{\rho},\rho,\tau)$, as a function of time, is almost constant during tens, hundreds, and thousands days (depending on the position of the well in the earthquake preparation zone) and is almost equal to its initial value. Namely $u(\tilde{\rho},\rho,\tau)\approx\theta(\tilde{\rho},n,\rho)$. It is caused that volumetric deformation is distributed in medium with small gradients which decrease with removal from an epicentre, decrease from increase in depth of focus and from growth magnitude. We remind that the value of precursory strain does not depend on magnitude and it does not exceed about 10^{-4}. At the same time, the size of area in a neighbourhood of the well that actually determines the level variation is not connected with earthquake parameters. Let this size equal to 0.5 km. Estimating the size of the near-field (epicentral) zone, where the volume strain distribution is most heterogeneous, as 4 average radius of heterogeneity, we obtain the values of 10, 25, and 65 km from magnitudes of 5, 6, and 7, respectively [1]. The conclusion is evident. If $u(\tilde{\rho},\rho,\tau)$ is constant then we obtain from (27)

$$Lw(\rho_0,p)=\frac{\chi K_1(\rho_0\sqrt{p})}{\sqrt{p}\left(\rho_0 pK_0(\rho_0\sqrt{p})+\chi\sqrt{p}K_1(\rho_0\sqrt{p})\right)}\theta(\tilde{\rho},n,\rho_0)=F(p)\theta(\tilde{\rho},n,\rho_0) \tag{28}$$

$$w(\rho_0,\tau)=F(\tau)\theta(\tilde{\rho},n,\rho_0). \tag{29}$$

here function F (τ) determines the time scale of the process of the short-term precursors. Function F (τ) presents in Figure 1. As seen from Figure 1, the level 3/4 reaches after 1 month and 9/10 after 2 months. Now we can write the approximate formula of precursory variations in level of liquid in the well applicable for strong earthquakes in the form

$$l\approx m\frac{q}{\beta\sigma g}\varepsilon(x,y,z)F(t). \tag{30}$$

There are no published examples of field observations that would make it possible to compare in detail theoretical results and empirical data. We will discuss only the following case. Work [3] give the plot of water level variations in the well at an epicentral distance of 30 km prior to the Przhevalsk earthquake of 1970 of magnitude 6.8 (Figure 2). Sudden transition to a short-term precursor was assumed in calculations. Actually so does not happen. Therefore we must introduce the conditional point of the beginning of a short-term precursor and corresponding coordinates (l t,) as it is showed on Figure 2. We see good conformity of calculated data (Figure 1) and empirical data (Figure 2) in coordinates (l t,).

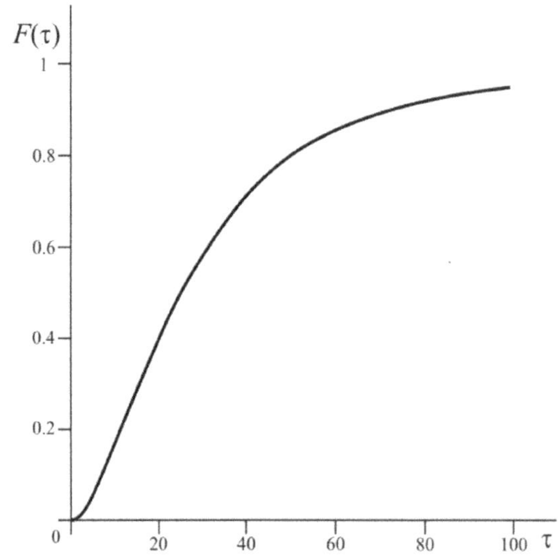

Figure 1: The plot of function F(τ). Unit of dimensionless time corresponds to one day.

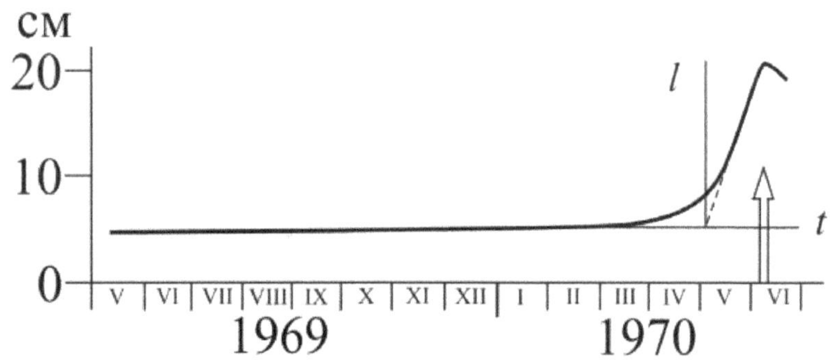

Figure 2: Plot of water level variations in the well before the Przevalsk earthquake of 1970. Figure is from [3]. The arrow is the moment of the earthquake. The origin of coordinates (l, t) is the point of the conditional beginning of short-term precursor.

On these data it is impossible to predict the moment of earthquake. In general, comparison of Figure 1 and Figure 2 enables us to conclude that the theory well determines the time scale of the phenomenon. The precise value of level variation cannot be estimated: there are no data on earthquake epicenter location and the point of precursor in the system of tectonic stresses acting in the region (do this data on tectonic stresses exist anywhere?), there are no precise enough data on hypocenter depth, whereas the decrease in effect with depth is proportional to the cube of relative depth, etc. Thus, we can give only the general estimate. Assuming the earthquake was shallow and we accept: $m = 5\%$, $q = 7$, $\beta = 4\times10^{10}$ Pa^{-1}, $s = 10^8$ Pa, $G = 2\times10^{10}$ Pa, $\alpha = 0.1$. We obtain that the maximum level varies within −50 - +50 cm at the epicentral circle with a radius of 30 km. It is necessary to note the definite uncertainty in description of properties of medium. Times of action of Earth tides and precursors differ by two orders of magnitude; maximum values of strains differ by 3 - 4 orders of value; the medium is strongly affected during an earthquake. How do the properties of medium change in these conditions? Despite of it we confirm that the existing theory of preparation of a tectonic earthquake and methods developed in the paper enables the effective calculation of variations in level of liquid in wells in the case of processes connected with tectonic earthquakes. This result is directly related with prediction problem of earthquake.

REFERENCES

1. Dobrovolsky, I.P. (2000) Theory of the Tectonic Earthquake Preparation. UIPE, Russian Academy of Sciences, Moscow, 134 p.
2. Rikitake, T. (1975) Earthquake Prediction. ESPC, 374 p.
3. Sadovsky, M.A., Nersesov, I.L., Nigmatullaev, S.K., et al. (1972) The Processes Preceding Strong Earthquakes in Some Regions of Middle Asia. Tectonophysics, 14, 295-307. http://dx.doi.org/10.1016/0040-1951(72)90078-9
4. Kissin, I.G., Belikov, V.M. and Ishankuliev, G.A. (1990) Extreme Groundwater Table Fluctuations in a Seismically Active Region. Doklady AN SSSR, 314, 1099-1103.
5. Dobrovolsky, I.P., Zubkov, S.I. and Miachkin, V.I. (1979) Estimation of the Size of Earthquake Preparation Zones. Pageoph, 117, 1025-1044. http://dx.doi.org/10.1007/BF00876083
6. Doetsch, G. (1956) Anletung Zum Praktiscen Gebrauch Der Laplace—Transformation. München, 208 p.
7. Watson, G. (1945) A Treatise on the Theory of Bessel Functions. Cambridge, 1, 243 p.

APPENDIX A. VOLUME STRAIN

The cycle of tectonic earthquake consists in beginning, development and disintegration of heterogeneity of properties in the Earth's crust [1] [5]. We introduce the Cartesian system of coordinates x , y , z in a half-space $z \geq 0$ with its origin at the heterogeneity epicenter and coordinate axes orient along tangential stresses s at infinity. To the beginning of acting of short-term precursors, the medium gets the volume strain [1]

$$\varepsilon = \frac{\alpha s R^3}{3G} 2xy \left(\frac{1}{r_1^5} + \frac{2}{r_2^5} - \frac{10H(z+H)}{r_2^7} + R^2 \left[\frac{14(z+H)^2}{r_2^9} - \frac{2}{r_2^7} \right] \right), \tag{A.1}$$

where $r_1 = \sqrt{x^2 + y^2 + (z-H)^2}$ and $r_2 = \sqrt{x^2 + y^2 + (z+H)^2}$.

Now we introduce dimensionless variables and polar coordinates $(\tilde{\rho}, \psi)$ by expressions

$$(\xi, \eta, \zeta) = \frac{(x, y, z)}{H}, \quad \tilde{\rho} = \sqrt{\xi^2 + \eta^2}, \quad \xi = \tilde{\rho}\cos\psi, \quad \eta = \tilde{\rho}\sin\psi. \tag{A.2}$$

Besides, we consider the near-surface layer and, thus, assume z = 0; this case is typical for wells and shallow boreholes. The earthquake considered is also shallow, so R =H, and the effect will be maximum. Then

$$\varepsilon = Q\theta_0 \sin 2\psi = Q\tilde{\rho}^2 \left(\frac{3}{Z^5} - \frac{12}{Z^7} + \frac{14}{Z^9} \right) \sin 2\psi, \tag{A.3}$$

where $Z = \sqrt{\tilde{\rho}^2 + 1}$, $Q = \frac{\alpha s}{3G}$ and notation θ_0 is obvious.

Evidently, ε is the product of two functions depending on angle and radius; the coordinate axes are zero lines. Further it is investigated the filtration problem in the well vicinity. Let the well axis is located at point x , y . We place the coordinate origin of the filtration problem at this point. Then x and y are parameters and variables will be r , ϕ . Then in (A.1) coordinates, considering (A.2) and (21), transform by the following scheme (for example, for x)

$$\frac{x}{H} \Rightarrow \frac{x + r\cos\varphi}{H} = \xi + \frac{\sqrt{\kappa T}}{H}\rho\cos\varphi = \xi + n\rho\cos\varphi. \tag{A.4}$$

here the scale ratio $n = \sqrt{\kappa T}/H$ is introduced automatically.

Now after average ε on angle ϕ we receive

$$\bar{\varepsilon} = Q\sin 2\psi\, \theta(\tilde{\rho}, n, \rho) \tag{A.5}$$

Function $\theta(\tilde{\rho}, n, \rho)$ is complicated. It involves K and E , which are the complete elliptic integrals of the first and second kinds accordingly, and consists 4 component

$$\theta(\tilde{\rho}, n, \rho) = S_1 + S_2 + S_3 + S_4, \tag{A.6}$$

$$S_1 = B\lambda^2(1-\lambda)\left(150(\lambda-1)^2 b^4 - 104\lambda(\lambda-1)(\lambda-2)b^2 + \lambda^2\left(24\lambda^2 - 71\lambda + 71\right)\right)\boldsymbol{K}\left(\sqrt{\lambda}\right) \tag{A.7}$$

$$S_2 = -2B\lambda\left(90(\lambda-1)^3 b^6 - 150\lambda(\lambda-2)(\lambda-1)^2 b^4 + 13\lambda^2(\lambda-1)\left(8\lambda^2 - 23\lambda + 23\right)b^2 - 4\lambda^3(\lambda-2)\left(6\lambda^2 - 11\lambda + 11\right)\right)\boldsymbol{E}\left(\sqrt{\lambda}\right), \tag{A.8}$$

$$S_3 = 12n^2\rho^2(\lambda-1)B\left(40(\lambda-1)^3 b^4 - 8\lambda\left(\lambda^2-1\right)(\lambda-2)(2\lambda-1)b^2 + \lambda^2\left(8\lambda^4 - 21\lambda^3 + 17\lambda^2 + 8\lambda - 4\right)\right)\boldsymbol{K} \tag{A.9}$$

$$S_4 = -24n^2\rho^2 B\left(10(\lambda-2)(\lambda-1)^3 b^4 - 2\lambda(\lambda-1)\left(8\lambda^4 - 19\lambda^3 + 15\lambda^2 + 8\lambda - 4\right)b^2 + \lambda^2(\lambda-2)\left(8\lambda^4 - 12\lambda^3 + 11\lambda^2 + 2\lambda - 1\right)\right)\boldsymbol{E}\left(\sqrt{\lambda}\right), \tag{A.10}$$

where $B=\dfrac{\sqrt{2\lambda}}{120\pi b^9(1-\lambda)^4}$, $\lambda=\dfrac{4\tilde{\rho}n\rho}{1+(\tilde{\rho}+n\rho)^4}$, $b=\sqrt{2\tilde{\rho}n\rho}$ and at $\rho\to\infty$ or $\tilde{\rho}\to\infty$ $\theta\to 0$.

The important remark is: obviously, it is possible to choose any other process of deformation of medium, but the technique will be the same.

Appendix B. Inverse Laplace Transformation

The inverse Laplace transformation is defined by the expression [6]

$$F(t)=\frac{1}{2\pi i}\int_{c-i\infty}^{c+i\infty} f(p)e^{pt}\mathrm{d}p, \tag{B.1}$$

where $c\geq c_0$ and c_0 is the convergence abscissa of the direct transformation. Now possible direct application of the formula (B.1), because existing computer software enables calculation of known, including special, functions in the whole complex area of their determination. Usually, such evaluations give numerical values in isolated points. It is convenient for check of a correctness of other methods. We will act usually and select the most convenient path of integration amongst the set of possible paths in the area, where the function under integral is analytical. Surely, this way does not reduce the attractiveness of analytical solution that is still preferable.

Apparently from (28) and (29), we should find the original from

$$F(p)=\frac{\sqrt{p}K_1\left(\sqrt{p}\right)}{pK_0\left(\sqrt{p}\right)+\chi\sqrt{p}K_1\left(\sqrt{p}\right)}. \tag{B.2}$$

Bessel functions are single-valued in complex plane with a branch cut along the negative semi-axis [7]. This property is also valid for function F(p). Then two paths of integration are really interesting: 1) along the imaginary axis and a positive semicircle centered at the origin of coordinates, 2) along two sides of the negative semi-axis. The second path is easier and the inversion expression for it after several transformations is

$$-\frac{2}{\pi}\int_0^{\infty} z\mathrm{Im}F(-z^2)e^{-z^2t}\mathrm{d}z, \tag{B.3}$$

where integral is taken from the imaginary part.

It is reasonable to represent function F(p) as the sum

$$F(p)=\frac{1}{\sqrt{p}}+\frac{\sqrt{p}K_1\left(\sqrt{p}\right)-\sqrt{p}K_0\left(\sqrt{p}\right)-\chi K_1\left(\sqrt{p}\right)}{pK_0\left(\sqrt{p}\right)+\chi\sqrt{p}K_1\left(\sqrt{p}\right)}. \tag{B.4}$$

here the first term has the original $1/\sqrt{\pi t}$ and the second term has no singularities.

CHAPTER 16 A WAVELET TRANSFORM METHOD TO DETECT P AND S-PHASES IN THREE COMPONENT SEISMIC DATA

Salam Al-Hashmi, Adrian Rawlins and Frank Vernon

Keywords: Discrete Time Wavelet Transform; P and S-phases; Automatic Detection; Rectilinearity Function

INTRODUCTION

Seismic events such as earthquakes cause a release of energy represented by seismic waves that can be recorded by seismic monitoring stations. Detection of seismic waves and estimation of their arrival times provides information about earthquake location and magnitude. Analysis and detection of seismic waves may be done by visual inspection by a trained analyst or automatic analysis software.

Rapid and accurate detection of seismic waves is of great importance in locating earthquakes [1-10]. Automatic detection techniques are of interest because they can be processed in near real-time; they apply a consistent set of metrics and are repeatable. However, it is often very difficult to determine consistent estimates of P and S waves if they have low signal-to-noise characteristics, particularly if arriving at seismic stations at regional distances.

The objectives of this research are:

- To design an algorithm for the detection and analysis of the two main types of seismic body waves (P and S-phases).
- To test the algorithm using seismic events recorded by various stations at local and regional distances, and then compares the results with the results obtained by other methods.

Citation: S. Al-Hashmi, A. Rawlins and F. Vernon, "A Wavelet Transform Method to Detect P and S-Phases in Three Component Seismic Data," Open Journal of Earthquake Research, Vol. 2 No. 1, 2013, pp. 1-20. doi: 10.4236/ojer.2013.21001.

Several methods have been tried recently these that have involved digital signal processing in both time and frequency domains [11-13]. The method adopted here is wavelet transform using an approach initially developed by K. Anant and F. Dowla [11].

The algorithms have been developed in Matlab 6.5™ (http://www.mathworks.com) including signal processing and WaveLab™ (http://www-stat.stanford.edu/~wavelab) toolboxes. WaveLab is a library of Matlab routines for wavelet analysis, wavelet-packet analysis, cosine-packet analysis and matching practice. The algorithms can be run on a laptop with 256 MB RAM or on a mainframe computer.

Seismic Data Used in Testing

Seismic data from Earthquake Monitoring Center in Oman has been used for testing the algorithms. The ANZA Broadband Seismic Network (http://eqinfo.ucsd.edu) also provided analyst reviewed data and STA/LTA automatic detected data used for comparison with the wavelet detector.

The paper structured as follows: An introduction about wavelet transform is given in Section 2. Section 2 also includes description of the discrete time wavelet transforms, its application on seismic signals to match specific features and how the selection of wavelet type has been done in this study. The multiresolution analysis technique and the perfect reconstruction filter banks are explained too. The developed software and methodology are explained in Section 3. The flow charts of the algorithms are illustrated in Appendix B and a detailed annotation is given in Section 3 too. The results are discussed in Section 4. Comments about the testing results and the comparison results with other methods are presented in Section 5.

WAVELET TRANSFORM ANALYSIS

Introduction

The wavelet transform is a very useful tool for analysing noisy and transient signals and has the ability to represent the signal in both its time and frequency domains. It is a very useful tool in the analysis of non-stationary signals such as seismic signals due to its ability to resolve features at various scales [14].

The wavelet theory arises in 1909 when Haar constructs the first orthonormal system of compactly supported functions called the Haar basis [15].

The term wavelet has been coined in the field of seismology in 1940 by Ricker, N. [15]. The wavelet theory has been applied on seismic signals by Grossmann and Morlet in 1984 [11]. In 1988, Daubechies developed an orthonormal, compactly supported wavelet basis that is smoother than Haar basis [14-16]. Application of wavelet transform on seismic signals has been done by Anant and Dowla [11], Oonincx, P. J. [12] as well as by Zhang et al. 2003 [13].

The wavelets form a family. The basic form is called the mother wavelet. All the daughter wavelets are derived from this wavelet according the following equation:

$$\Psi_{s,\tau}(t)=\frac{1}{\sqrt{s}}\Psi\left(\frac{t-\tau}{s}\right) \tag{2.1}$$

where, s and τ are the scale and translation of the daughter wavelet. The term $s^{-1/2}$ normalizes the energy for different scales, whereas the other terms define the width and translation of the wavelet.

Digital signal analysis using wavelet transforms begins with the construction of a single parent wavelet. The signal is then decomposed into a series of basis functions of finite length consisting of dilated (stretched) and translated (shifted) versions of this parent wavelet function, i.e., wavelets of different scales and positions in time or space. This process is similar to Fourier analysis, where the parent wavelet is analogous to the sine wave, and the basis functions in Fourier decomposition are sine waves of various amplitude, phase, and frequency variations of the parent sine wave [14,17].

Scaling a wavelet simply means stretching (or compressing) it. The smaller the scale factor, the more "compressed" the wavelet. The more stretched the wavelet, the longer the portion of the signal with which it is being compared, and thus the coarser the signal features being measured by the wavelet coefficients. Thus, there is a correspondence between wavelet scales and frequency as revealed by wavelet analysis:

- Low scale s $\Rightarrow$ compressed wavelet $\Rightarrow$ rapidly changing details $\Rightarrow$ High frequency ω.
- High scale s $\Rightarrow$ Stretched wavelet $\Rightarrow$ Slowly changing, coarse features $\Rightarrow$ Low frequency ω.

The decomposition advantage is very useful in dealing with signals contain features with various frequency characteristics. Another advantage of the wavelet transform is that its analysis can be chosen based on the application [11].

Figure 1 shows seismic signal at top and wavelet coefficients of six scales where "Daubechies 8" wavelet has been used. The smallest scale is the one that contains the highest frequency while the largest scale is the one that contain the lowest frequency.

THE DISCRETE TIME WAVELET TRANSFORM

A discrete type of wavelet transform exists, termed the discrete time wavelet transform (DTWT), where scales and shifts take on discrete values. It allows fast computation of the transform for the digitized signals and gives perfect signal reconstruction. It is a form of multiresolution analysis and is related to perfect reconstruction filter banks [14,18,19].

The wavelet transform can be applied to seismic signals in terms of the type of decomposition as well as in terms of pattern matching [11,12].

In wavelet analysis, we often speak of approximations and details. The approximations are the high-scale, lowfrequency components of the signal while the details are the low-scale, high-frequency components. Figure 2 shows that the input signal is filtered by a low-pass and high-pass filters in one stage wavelet transform. Then, the signal is down-sampled by a factor of 2 to produce DWT coefficients.

Subband Coding

The DTWT is implemented using the sub-band coding scheme represented in Figure 3" target="_self"> Figure 3. The signal is decomposed into 2 frequency bands: low (approximations) and high (details). For the approximation and detail coefficients to represent a discrete-time wavelet decomposition H and G filters must belong to a perfect reconstruction filter bank and required to be regular [14,18,20,21].

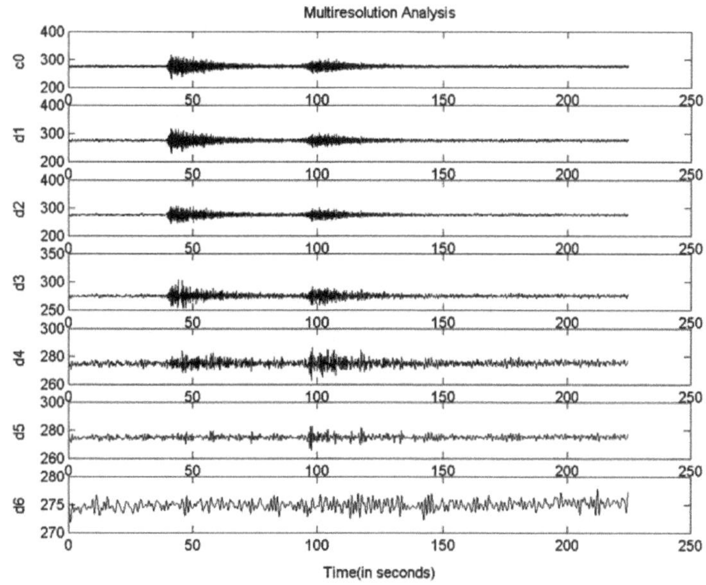

Figure 1: Seismic signal decomposed into several scales.

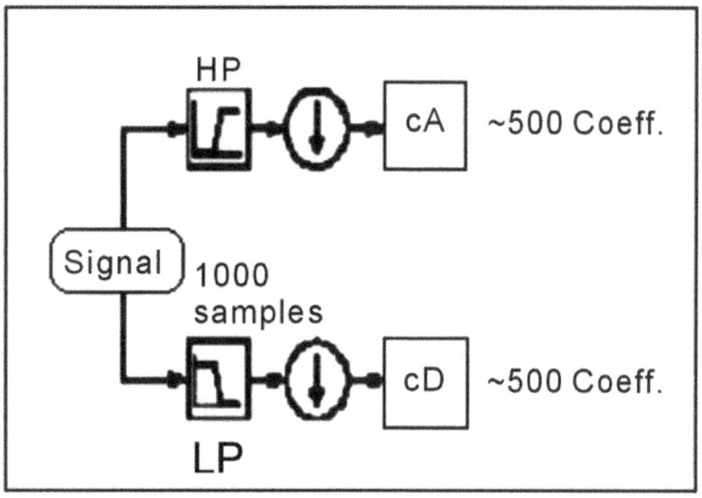

Figure 2: One stage wavelet transforms.

The wavelet analysis involves filtering and down sampling while the wavelet reconstruction process consists of up-sampling and filtering. Up-sampling is increasing the number of samples by inserting zeros between samples.

The following filters then perform an interpolation by filling in the zero points with the appropriate signal information.

The filtering part of the reconstruction process bears some discussion, because it is the choice of filters that is crucial in achieving perfect reconstruction of the original signal.

Figure 4 is a typical example of such filters magnitude responses.

The down-sampling of the signal components performed during the decomposition phase introduces a distortion called aliasing. It turns out that by carefully choosing filters for the decomposition and reconstruction phases that are closely related, we can "cancel out" the effects of aliasing.

In order for the lowand high-pass decomposition filters (H and G), together with their associated reconstruction filters (L^* and G^*) shown in Figure 3 to perform such decomposition and reconstruction a system of what is called Quadrature Mirror Filters is used [14,17,19].

Quadrature Mirror Filters

QMF banks are a set of finite impulse response (FIR) filters that enable a signal to be decomposed into subbands and allow reconstruction of the signal from the sub-bands without distortion. It is an example of decimation and interpolation and allows designing two-channel perfect reconstruction filter banks and therefore to generate wavelet bases.

The analysis filters have typically low pass and high pass frequency responses with a cutoff at $\pi/2$ (the "quadrature frequency") as illustrated in Figure 4. The philosophy of QMFs is to allow aliasing to be introduced by using overlapping filters for the analysis bank and design the synthesis filters in such a way that any aliasing is exactly cancelled out in the reconstruction process. Filters are also designed so that overall amplitude and phase distortion are minimized or eliminated.

Wavelet Type Selection

The selection of wavelet type is crucial in the processing. It can depend on seismic arrival shapes. Therefore, the wavelets used in the processing are chosen based on matching the seismic phase's arrival shapes. Thus, the selection will rely on the event as well as on the station that recorded this event. This is because the arrival shapes will vary from event to event and between the different stations as the seismic waves travel through different media and distances for each event and for each monitoring station.

Figure 5 shows some of the wavelets used in the wavelet transform decomposition.

In some cases where P-arrival shape clearly identified, it can be noticed that the wavelet that produce a composite function CT (the function that is used to locate S-arrival time) with the highest magnitude is closest in shape to P-arrival.

Figures 6 and 7 illustrate examples of P-arrival shapes of two different events. In the first example the P-arrival shape seems similar to Daubechies 8 while in the second example the P-arrival shape seems similar to Daubechies 6.

In our analysis the wavelet type selection was verified to be more important with respect to S-phase arrival, while the P-arrival is not affected significantly by the wavelet that is applied. This was also proved by Anant and Dowla [11].

SOFTWARE DEVELOPMENT

Introduction

The software is developed in Matlab 6.5™ including signal processing and Wavelab™ toolboxes. It consists of two algorithms: the P-phase detection algorithm and S-phase detection algorithm described in the sections below. The P-phase detection algorithm can be categorized into three main modules:

1) DTWT Processing that includes the following:
    - Multiresolution Analysis
        - Signal Decomposition (For east, north and vertical components).
        - Coefficients Reconstruction (For east, north and vertical components).
2) The Composite Rectilinearity Function Construction

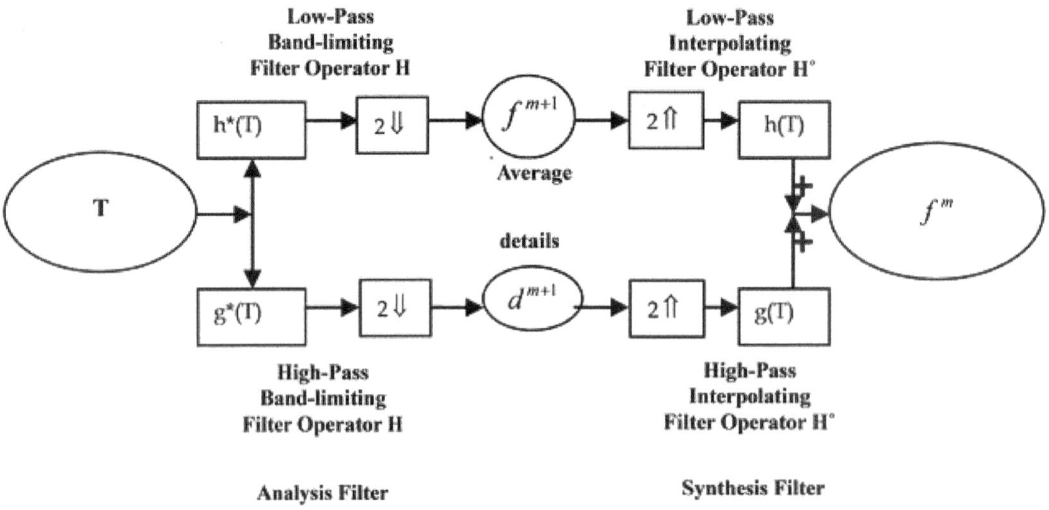

Figure 3: Sub-band coding scheme.

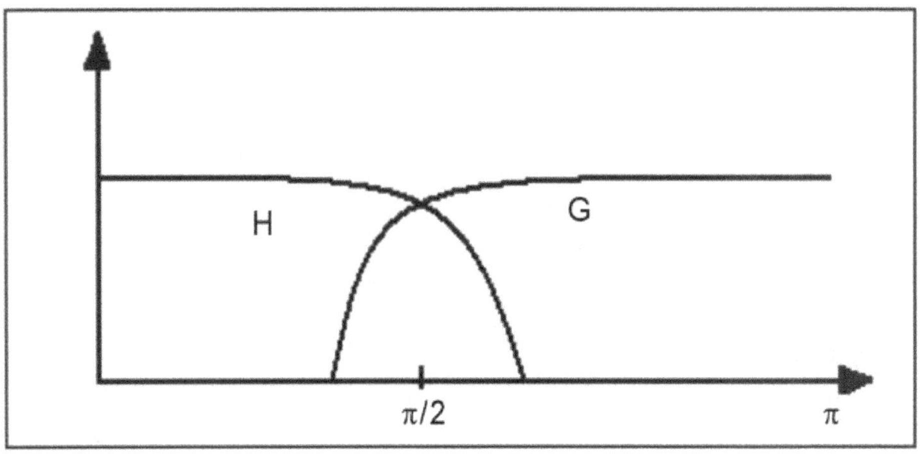

Figure 4: Magnitude responses of analysis filters.

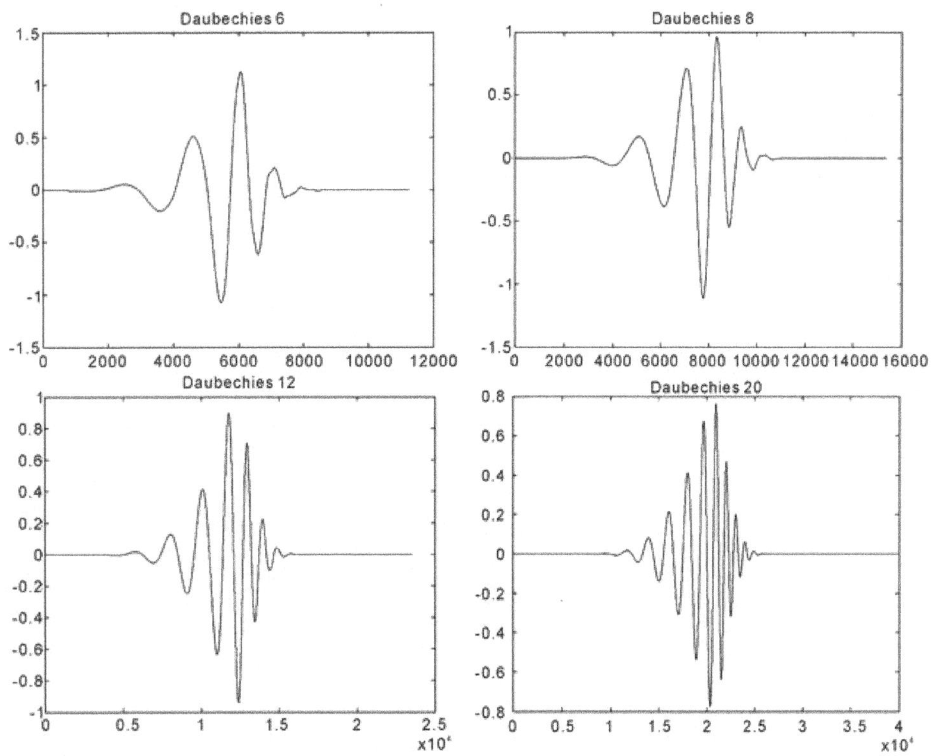

Figure 5: Some of the wavelets used in the research.

3) Backazimuth calculation

The S-phase detection algorithm can also be categorized into three main modules:

1. Components Rotation
2. DTWT Processing that includes the following:
    - Multiresolution Analysis
        - Signal Decomposition (For radial and transverse components).
        - Coefficients Reconstruction (For radial and transverse components).
3. Constructing the Composite Function of Transverse over Radial Amplitude Ratio.

P-Phase Detection Algorithm

It is known that the P-phase is linearly polarized with respect to the direction of propagation. Therefore, a metric that measures the degree of linear polarization is helpful to detect the P-phase arrival. Such a metric known as the rectilinearity function is defined by Kanasewich [22].

Its equation is:

$$\mathbf{F} = \mathbf{1} - (\lambda_2 / \lambda_1) \tag{3.1}$$

where λ_1 and λ_2 are the largest and the second largest eigenvalues of the covariance matrix respectively.

If the covariance matrix (Equation 3.2) is diagonalized, an estimate of the rectilinearity of particle motion trajectory over the specified time window can be obtained from the ratio of the principal axis of this matrix [22] i.e. the rectilinearity estimation can be obtained from the ratio of the largest and the second largest eigenvalues of the covariance matrix. The covariance matrix [23] is defined as:

$$\mathbf{M} = \begin{bmatrix} Cov(X,X) & Cov(X,Y) & Cov(X,Z) \\ Cov(Y,X) & Cov(Y,Y) & Cov(Y,Z) \\ Cov(Z,X) & Cov(Z,Y) & Cov(Z,Z) \end{bmatrix} \tag{3.2}$$

whereX: east component wavelet coefficients at scale j;

Y: north component wavelet coefficients at scale j;

Z: vertical component wavelet coefficients at scale j.

While the covariance of X and Y is defined as:

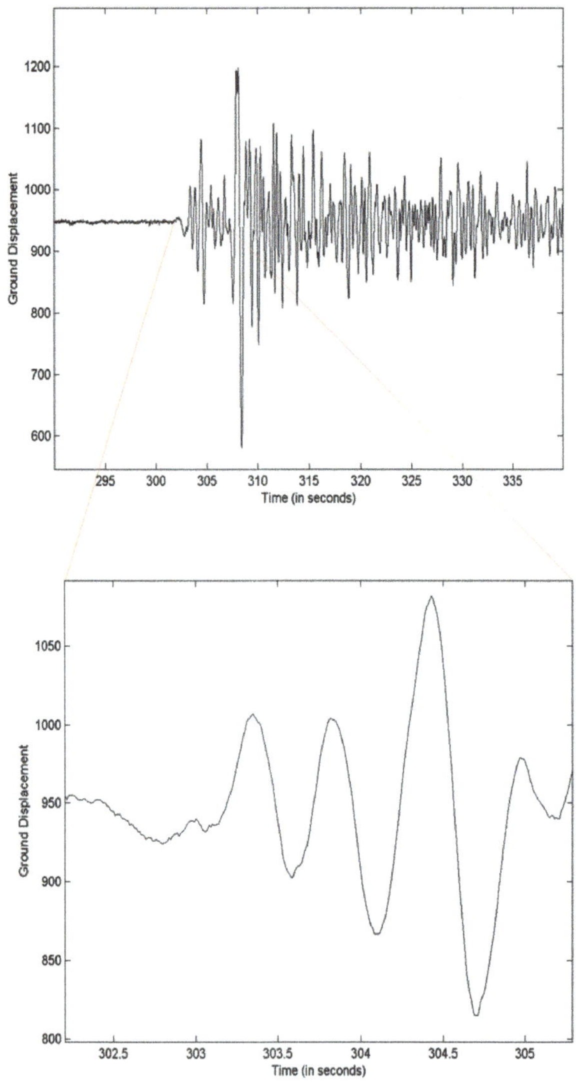

Figure 6: P-phase arrival of an earthquake record and the lower plot is a zoomed view of the arrival.

$$Cov(X,Y) = \frac{1}{N}\sum_{i=1}^{N}\left(X(i) - \mu_X\right)\left(Y(i) - \mu_Y\right) \tag{3.3}$$

where μ_X and μ_Y are the mean values of X and Y respectively.

The direction of polarization may be measured by considering the eigenvector of the largest principal axis [22]. If λ_1 is the largest eigenvalue and λ_2 is the next largest eigenvalue of the covariance matrix, then a function of the form as in (Equation 3.1) would be close to unity when rectilinearity is high ($\lambda_1 \gg \lambda_2$) and close to zero when two principal axes approach one another in magnitude (low rectilinearity).

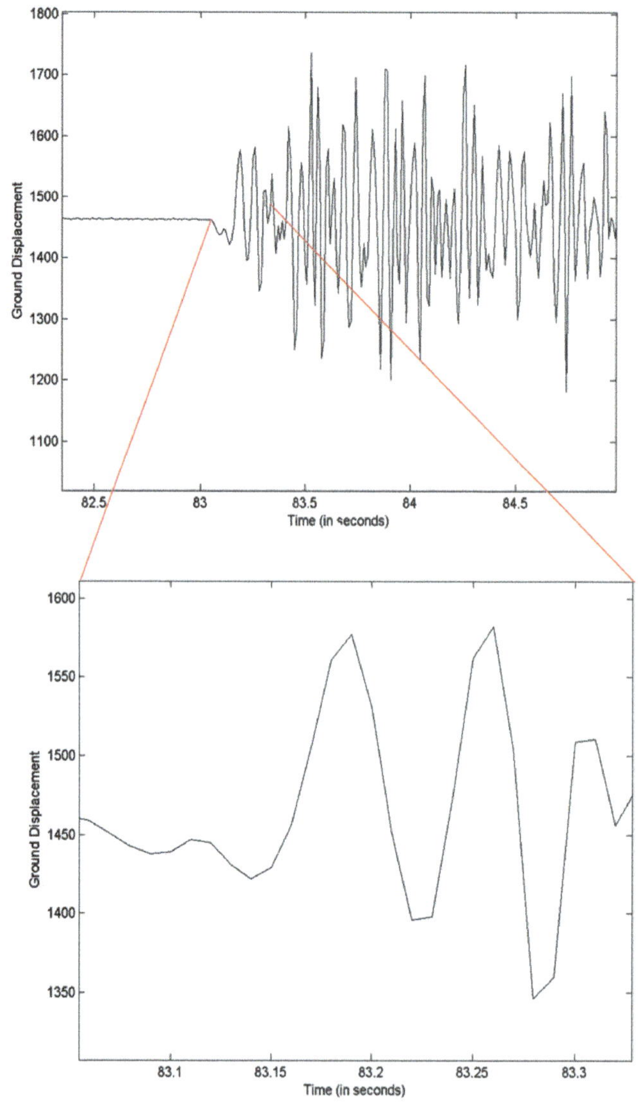

Figure 7: P-phase arrival of an earthquake record and a zoomed view of the arrival.

The direction of polarization can be determined by considering the components of the eigenvector associated with the largest eigenvalue with respect to the coordinate directions X, Y and Z.

The P-phase algorithm is developed as can be described in the following steps:

1) Processing the three component seismic signals using Discrete Time Wavelet Transform (DTWT)

A three component seismic signals represented by east (X), north (Y) and vertical (Z) are processed by the discrete time wavelet transform (DTWT) to calculate the wavelet coefficients wc1, wc2 and wc3 respectively.

"Daubechies" wavelets are used in this process because of their compact support basis functions and their shape that coincide with seismic wave arrivals shapes.

Multiresolution Analysis

Each component is decomposed into several scales. So, the results are the wavelet coefficients x_1^j, x_2^j and x_3^j for east, north and vertical components respectively (j is the number of scales).

2) Constructing the rectilinearity function

At each scale, a pointwise moving window is constructed over the three components as shown in Figure 8.

The window length is determined using a measure called the varimax norm that is described in Appendix A. At each window, a 3-by-3 covariance matrix is constructed by using Equation (3.2) with x_1^j substituted for X, x_2^j substituted for Y and x_3^j substituted for Z. Then, eigenvalues and their corresponding eigenvectors are calculated.

At each window the rectilinearity function is constructed so that the result is a rectilinearity function (Fj) for each scale. Then, a composite rectilinearity function (Cf) is constructed so that the rectilinearity function of each scale contributes in this function. This composite function is constructed by (Equation 3.4).

$$Cf = \prod_j Fj \tag{3.4}$$

where j is the scale number.

The location where this function gets its maximum value is taken as the P-arrival time.

3) Calculating the back azimuth angle

Back-azimuth is the angle measured from north to the direction from which the energy arrives at a given station [24,25]. It is used to determine the longitudinal and transverse directions for an incoming ray at a prescribed station. For an incident P wave the backazimuth is calculated by constructing the rectilinearity function and finding the eigenvector associated with the maximum eigenvalue for the detected P-wave. This eigenvector represents the direction of linear polarization that specifies the back azimuth angle.

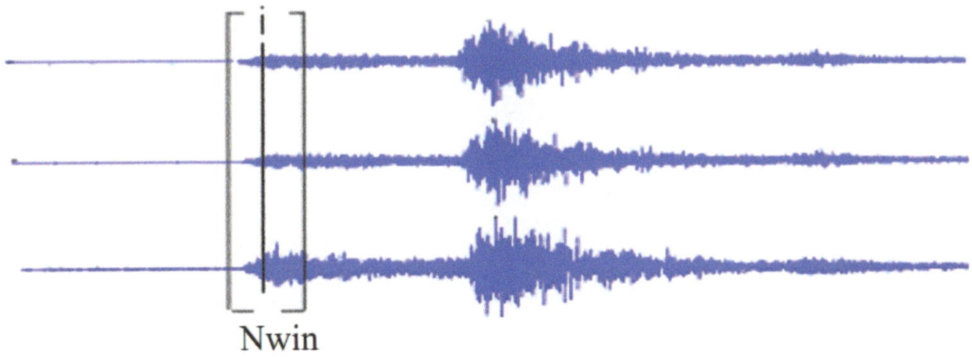

Figure 8: Illustration of windowing used to calculate covariance matrices.

This calculation can be conducted at each wavelet scale as well as on the original signals. But, since the original three component signals generally contain noise, the calculation is carried out at the third and higher scales while the first two scales that contain high frequency noise are excluded. It is found that the polarization is conserved along different scales i.e. the same backazimuth angle is obtained in the different scales. So, one value is considered in the processing.

S-Phase Detection Algorithm

The S-phase is a shear wave with particle motion in the transverse direction [22]. Accordingly, it has higher amplitude in the transverse component relative to its amplitude in the radial or vertical component. So, in order to locate the S-phase arrival, the amplitude ratio of the transverse to radial components of the earthquake record is analyzed.

The S-phase algorithm is developed and can be described as follows:

1) Rotating the east and north components

The backazimuth angle (θ) calculated in the previous part is used to rotate the east and north components to radial and transverse components respectively using the following equation:

$$\begin{bmatrix} dr \\ dt \end{bmatrix} = \begin{bmatrix} \sin\theta & \cos\theta \\ -\cos\theta & \sin\theta \end{bmatrix} \begin{bmatrix} X \\ Y \end{bmatrix} \tag{3.5}$$

where dr and dt are the radial and transverse component respectively.

2) Processing the radial and transverse components by the discrete time wavelet transform (DTWT)

The radial and transverse components processed by the DTWT. So, the result is several scales for each component. Referring to the algorithm the result is x^j and y^j that represents the different scales for radial and transverse components respectively (j is the number of scales).

3) Constructing the amplitude ratio

At each scale, a transverse over radial amplitude ratio is calculated as follows:

$$Atr^{j} = \frac{envt^{j}}{\left(envt^{j} + envr^{j}\right)} \quad (3.6)$$

where $envt^j$ and $envr^j$ are the envelope functions of transverse and radial components respectively.

The envelope function is used to avoid divide-by-zero problems. It is defined as

$$env(x) = \sqrt{x^2 + h^2} \quad (3.7)$$

where h is the Hilbert transform of x.

All scales are combined to construct a second composite function (CT) to be used to locate the S-arrival.

$$\mathbf{CT} = \prod_{j} Atr^{j} \quad (3.8)$$

The point after the P arrival time that has a value that is at least one half the maximum of CT is chosen for the S arrival time. The peak of CT itself is not used because it represents the time when S-wave attains its highest amplitude which is few seconds after its arrival time [11].

Several wavelets are used in the wavelet transform decomposition because the choice of wavelet that used in the processing is very important in how well CT locates the S arrival time.

RESULTS

Introduction

The algorithm has been tested by a real seismic data represented by several events recorded on various three components stations. The tested events apart from event 17 are distributed as shown in Figure 9. The Figure illustrates the tested events in orange circles as they located according to the stations in dark blue and black triangles for short period and broadband stations respectively. The purpose of this testing is to investigate if the project objectives listed in section 1 are achieved. To measure the performance of the algorithm, the algorithm results are compared with manual inspection results as well as with another automatic algorithm results (STA/ LTA method).

The performance of the system is measured by two ways: the first way is testing the ability of the software to detect P and S waves. However, the second way is comparing the results with another system results to see how accurate the system is.

P-Phase Detection Algorithm Results

The rectilinearity function proved to be an effective metric in locating the P-arrival time. Figure 10 shows the rectilinearity functions of six different scales obtained by testing the algorithm with event 21. It can be observed that the rectilinearity function is approximately equal to 1 when the waveform is linearly polarized and equal to zero when there is no linear polarization. This can be seen clearly in the first four scales.

Figure 11 shows the testing result of an earthquake (event 24). It illustrates the composite rectilinearity function Cf and how it is used to locate the P-arrival time.

Figure 12 shows the tested result of another earthquake (event 1) and the composite rectilinearity function Cf. It shows that the peak of this function is used to determine the P-arrival time. It can be observed from the figure that the position at this function is a maximum is where P-phase arrives.

Figure 13 illustrates the direction of Arabian Sea earthquake (event 4) as the result of testing the algorithm by three component data recorded by ABT station (one of the southern stations). Referring to Figure 9 it can be observed that ABT station is located in Southern Oman, which means that the algorithm indicates the event direction accurately as the event occurred in Arabian Sea.

In order to test the performance of the algorithm, it is tested by an event of magnitude 1.35 that occurs in California-Mexico border region (event 17).

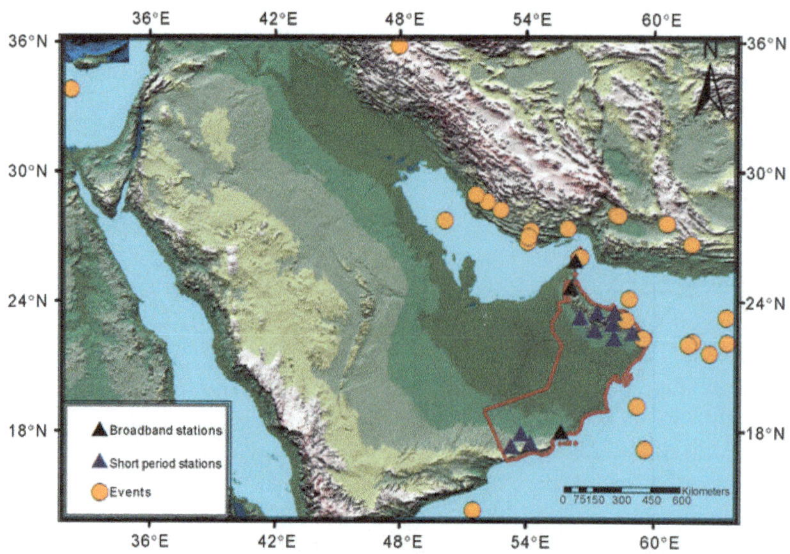

Figure 9: Distribution of tested events according to seismic stations.

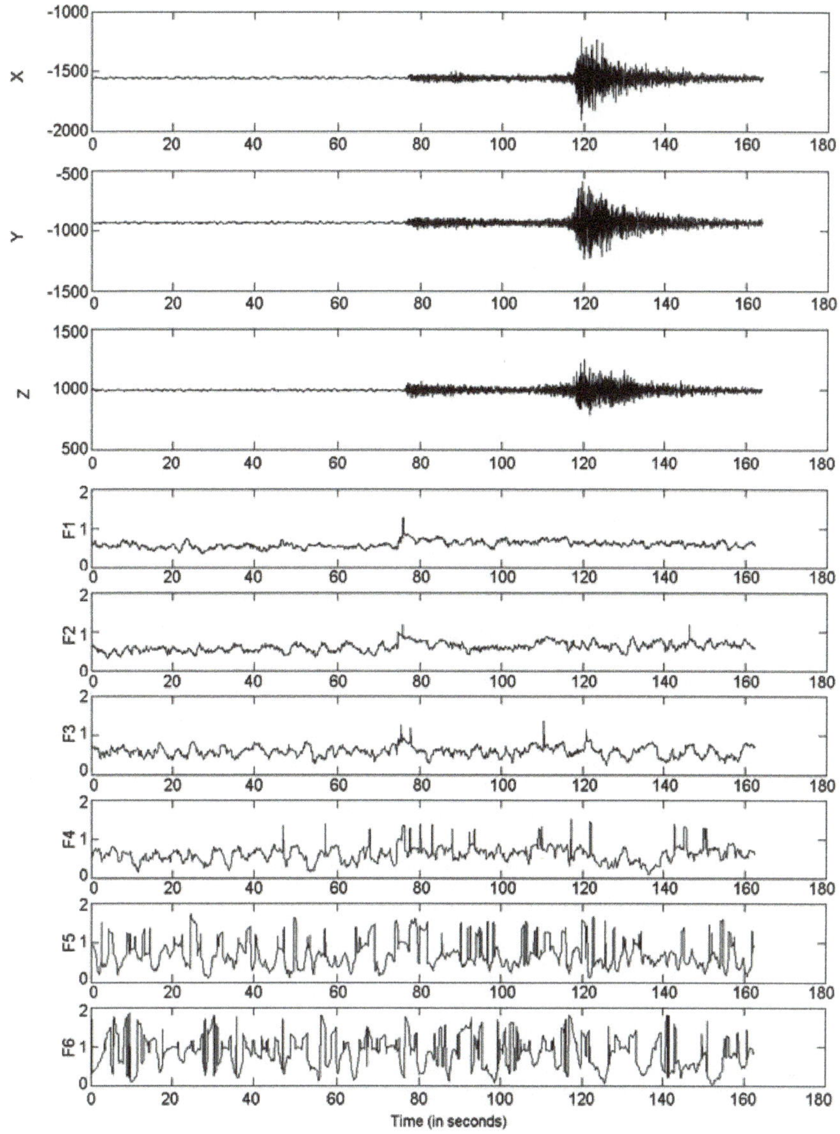

Figure 10: Three component waveforms of event 21 and its related rectilinearity functions of six different scales.

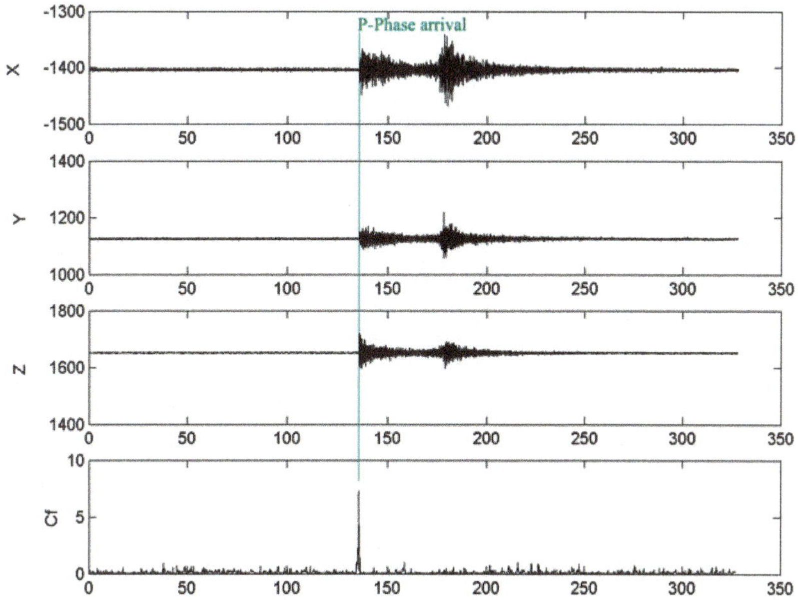

Figure 11: Three component data (east, north and vertical) of event 24 recorded by WBK station (one of the northern stations) with epicentral distance of 386.4 km and backazimuth of 106.9 degrees and the composite rectilinearity function Cf.

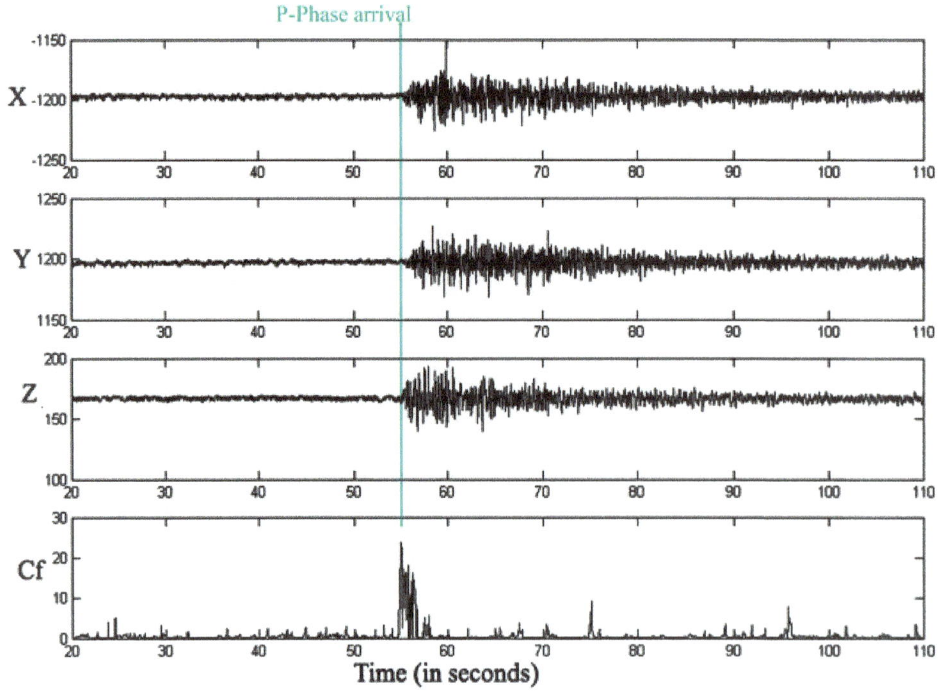

Figure 12: Three component data (east, north and vertical components) of event 1 recorded by JMD station with epicentral distance of 545.898 km and back azimuth of 96.03 degrees and the composite function Cf.

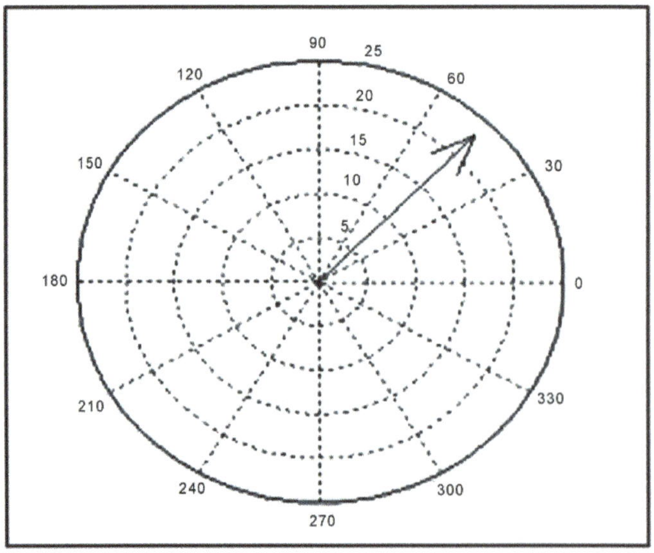

Figure 13: The direction of event 4 from ABT station.

The purpose of this testing is to compare the algorithm results with the results obtained by another automatic detector and manual inspection that has been done by an analyst from the Broadband Array data collection center at the University of California, San Diego. The comparison between the analyst result and the algorithm result is illustrated in Figure 14. It can be noticed that the results are consistent with each other.

The three components signals illustrated in Figure 14 are recorded by a station with noisy local conditions (BVDA2 of ANZA network). That's why; the STA/LTA (short term average/long term average) method couldn't give a result for P-arrival time for the data recorded by this station. However, the P-phase detection algorithm gives a good result though of this noise condition.

The algorithm result is compared with the analyst result as well as with STA/LTA results. The comparison shows that the three results are consistent with each other as illustrated in Figure 15.

To enhance the algorithm results, they compared with the results of STA/LTA and analyst results of all the events listed in Table 1. The time difference in the arrival pick of the algorithm and that of the analyst is shown in Figure 16. It can be observed that the mean magnitude of algorithm error is low (0.1952 seconds) as compared with STA/LTA mean magnitude error (0.3982 seconds). This proves that the performance of the algorithm is efficient.

S-Phase Detection Algorithm Results

As stated earlier in Section 3, various wavelets are used in the wavelet transform decomposition. The wavelet is chosen to match the pattern of S-phase to provide the best match in the wavelet scales and thus give the composite function CT with the greatest dynamic range. Therefore, the composite function with the highest amplitude peak is used to locate the S-arrival.

The S-phase detection algorithm is tested by the same events used to test the P-phase detection algorithm. Its testing results of event 17 are compared with the analyst results as shown in Figure 17 while STA/LTA method failed to give any result for S-arrival.

Figures 18 and 19 show the results of testing the algorithm by two different events (event 4 and event 18 respectively). They illustrate the radial and transverse components and the composite function that constructed to be used to detect the S-arrival time. They show that the composite function of transverse to radial amplitude ratio accurately succeeded to determine the S-phase arrival as can be noticed from the figures that the peak of this function indicates the position where S-phase attains its highest magnitude while the onset time of S-phase is few seconds before this peak.

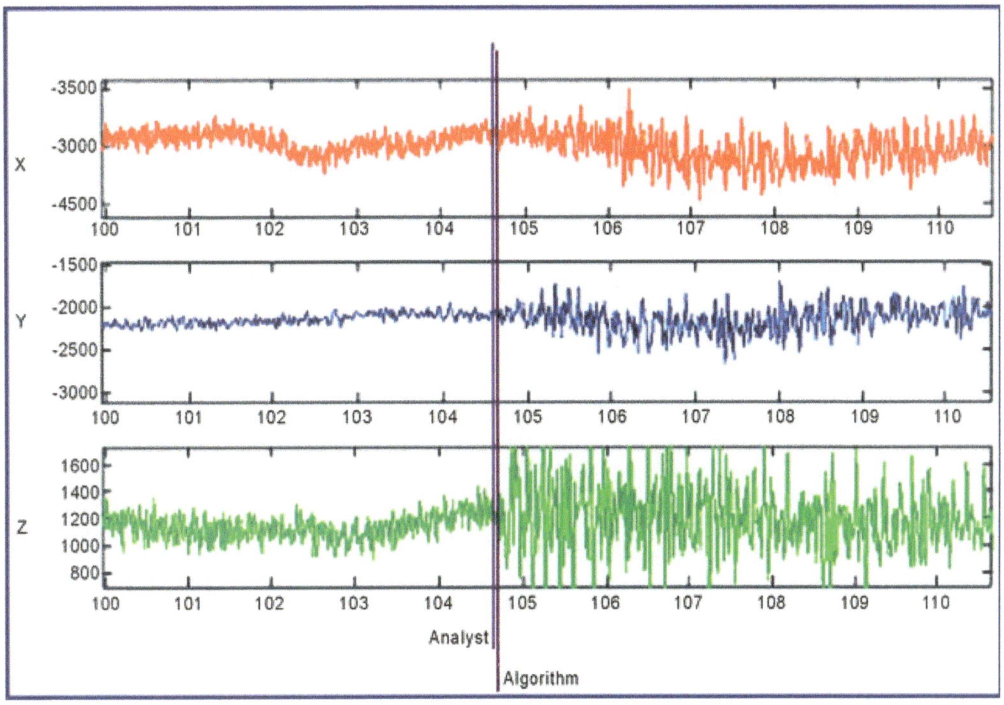

Figure 14: The three component signals of event 17 zoomed around P-arrival time to compare the analyst result with the algorithm result.

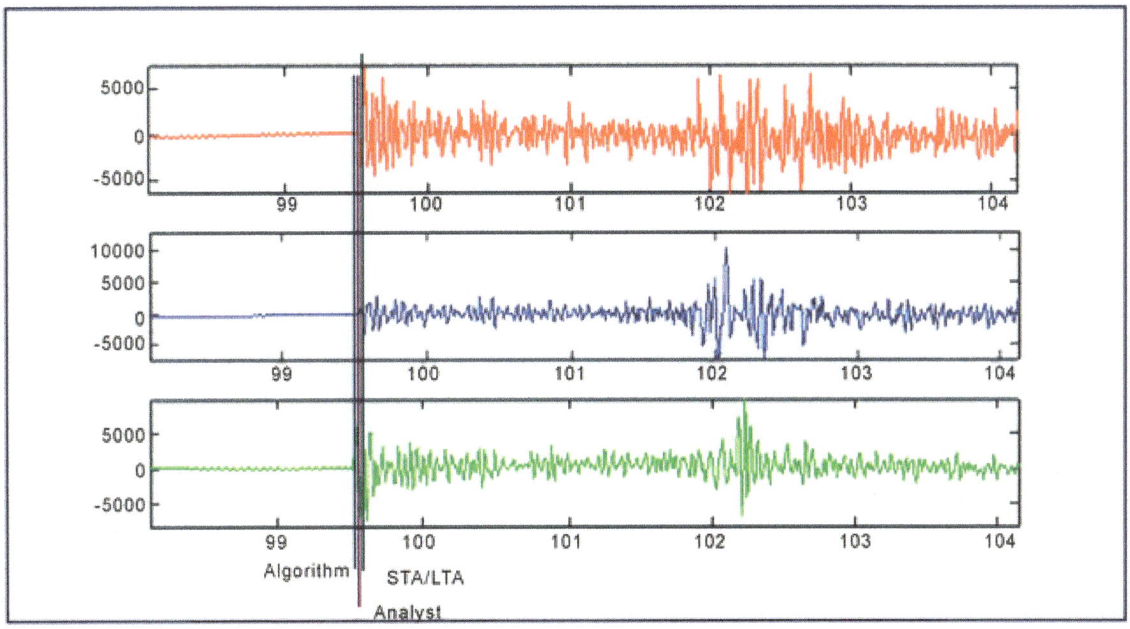

Figure 15: The result of comparing the algorithm with the analyst result and STA/LTA result when testing it by event 17 recorded by MONP station(Anza network) with epicentral distance of 10.656 km and backazimuth of 255.49 degrees.

Table 4: List of tested events

Event No.	Date of occurrence	Origin time	Latitude	Longitude	Depth	Location
1	01/12/2002	03:43:37	22.11	63.41	30	Arabian Sea
2	02/12/2002	05:00:28	33.8	32.22	60	Eastern Mediterranean
3	03/12/2002	21:42:48	28.26	52.77	50	Southern Iran
4	03/12/2002	23:08:47	17.21	59.56	50	Arabian Sea
5	09/05/2003	00:03:38	26.7817	54.047	50	Southern Iran
6	09/05/2003	00:47:03	27.2357	54.1686	20	Southern Iran
7	09/05/2003	05:54:43	27.4024	55.9109	20	Southern Iran
8	09/05/2003	10:34:07	27.2866	54.2007	35.4	Southern Iran
9	09/05/2003	12:01:30	27.0192	54.0923	30	Southern Iran
10	09/05/2003	16:42:17	28.0641	58.1985	50	Southern Iran
11	09/05/2003	19:17:03	28.6498	52.124	20	Southern Iran
12	10/05/2003	06:39:31	27.6593	60.5815	20	Southern Iran
13	10/05/2003	15:47:46	27.7768	50.1192	30	Persian Gulf
14	11/05/2003	10:17:28	28.9479	51.5855	20	Southern Iran
15	11/05/2003	10:56:14	24.1477	58.8162	36.7	Gulf of Oman
16	11/05/2003	18:01:53	19.1998	59.1707	20	Arabian Sea
17	26/07/2003	05:56:37	15.4463	116.3117	32.92	California-Mexico Border Region
18	24/06/2004	21:56:05	28.0082	58.2898	50	Southern Iran
19	24/07/2004	00:00:01	23.1694	58.662	30	Eastern Arabian Peninsula
20	27/07/2004	05:10:44	23.3008	63.3888	30	Near Coast of Pakistan
21	05/08/2004	08:38:24	14.3831	51.4685	20	Eastern Gulf of Aden
22	29/08/2004	00:43:18	22.1723	61.7808	30	Arabian Sea
23	30/08/2004	08:19:58	26.0658	56.5246	31.8	Southern Iran
24	13/12/2004	01:36:04	21.6386	62.568	35	Arabian Sea
25	13/03/2005	03:31:27	26.6853	61.7142	30	Southern Iran
26	23/03/2005	19:49:33	22.3243	59.4517	13.95	Eastern Arabian Peninsula
27	24/03/2005	00:37:12	22.0547	61.5609	38.6	Arabian Sea
28	03/05/2005	07:20:43	35.8229	47.9509	30	Western Iran

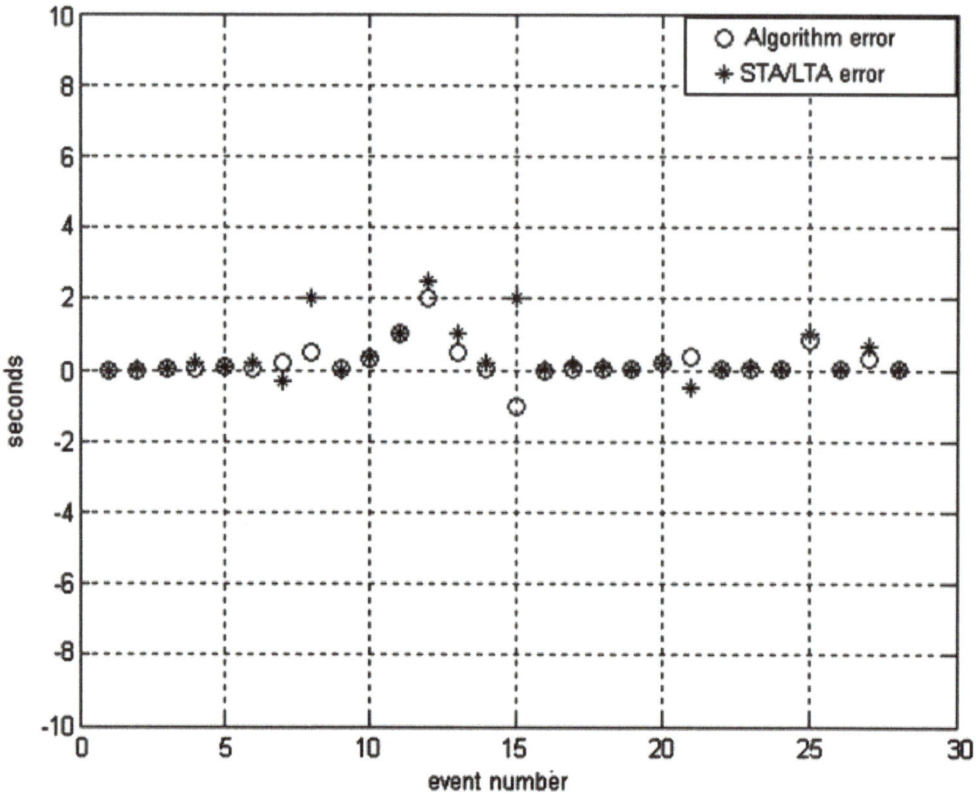

Figure 16: Residual plot of P-Phase arrival.

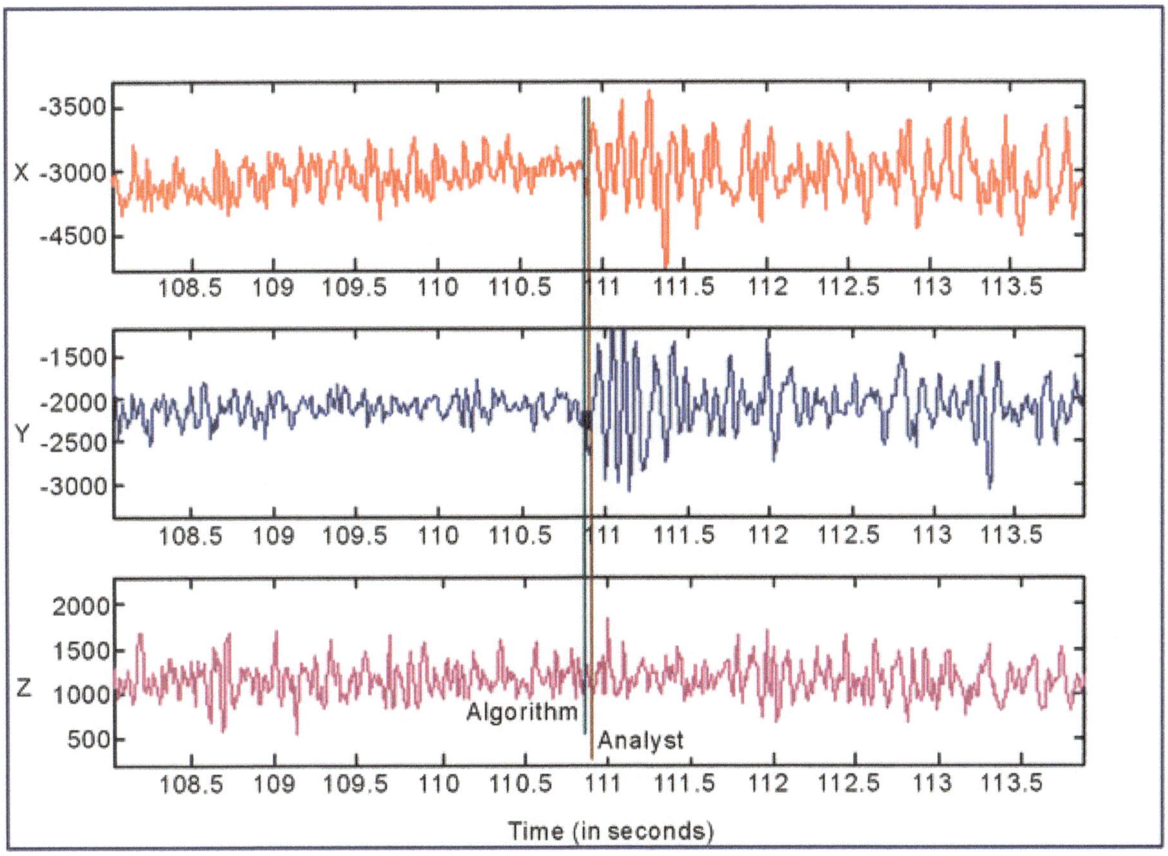

Figure 17: The three component signals of event 17 recorded by BVDA2 station with epicentral distance of 45.732 km and backazimuth of 353.73 degrees, zoomed around S-arrival time to compare the analyst result with the algorithm result.

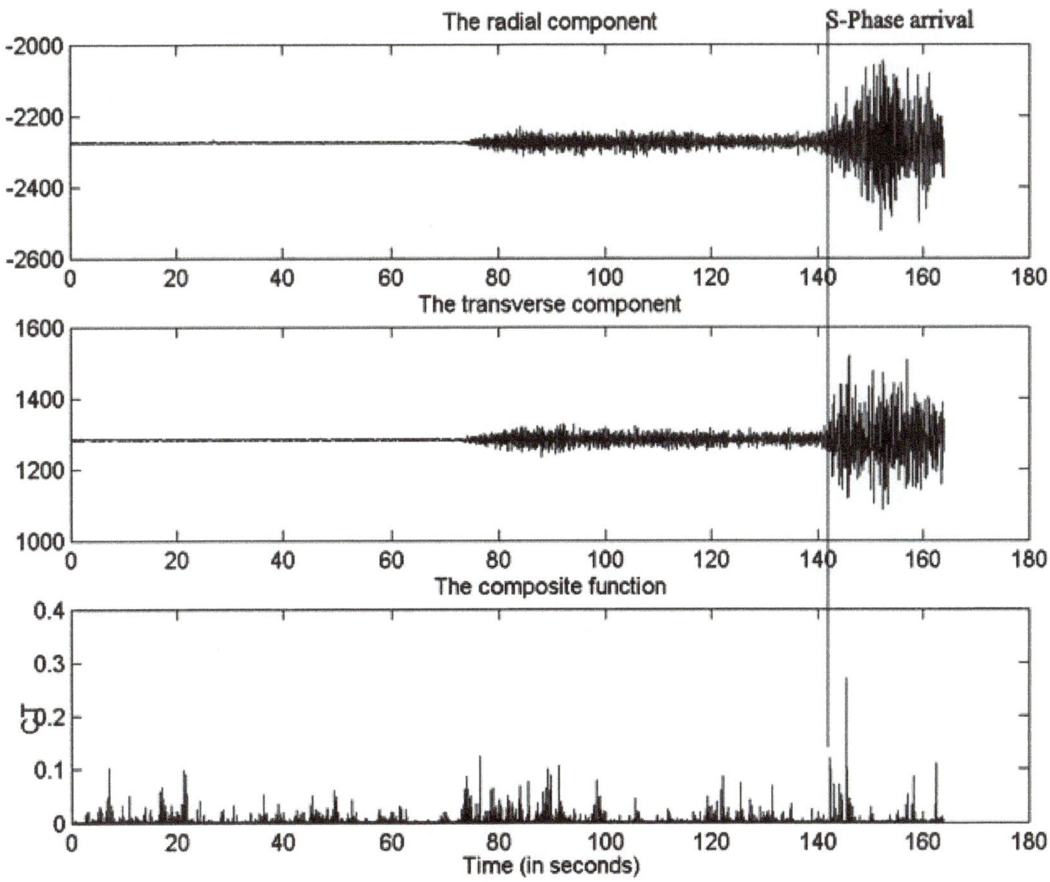

Figure 18: The radial and transverse components of event 4 recorded by ABT station with epicentral distance of 664.224 km and backazimuth of 43.9 degrees, and the composite function of the transverse to radial amplitude ratio.

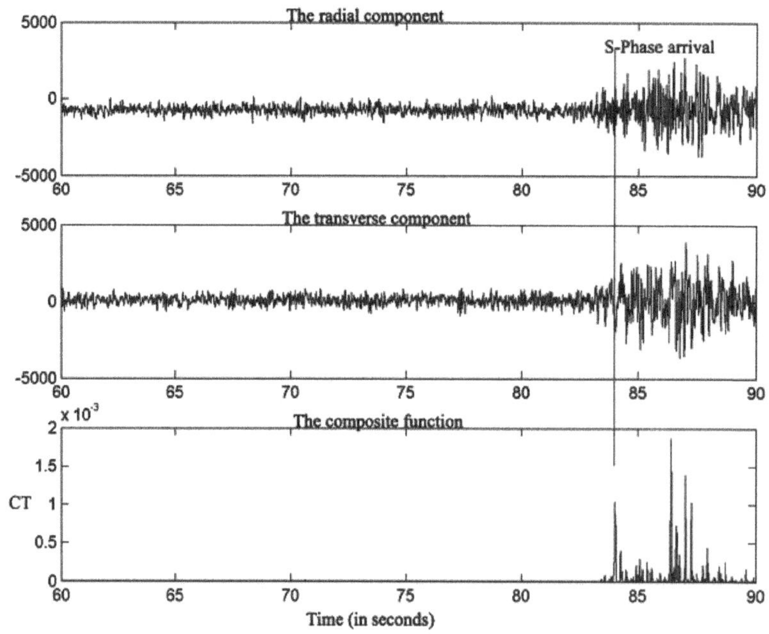

Figure 19: The radial and transverse components of event 18 recorded by BAN station (one of the broadband stations in northern Oman) with epicentral distance of 274.5 and backazimuth of 40.8 degrees and the composite function of transverse to radial amplitude ratio.

Figure 20 shows the residual plot of S-Phase arrival that illustrates the error between the algorithm and the analyst results (in circles) and the error between STA/ LTA results and the analyst results (in asterisks). It can be noticed that the mean magnitude error of the algorithm error is low (0.8197 seconds). In addition, it can be observed that there is no S-pick from STA/LTA for event 17 while the algorithm has good results and consistent with the analyst results.

As mentioned earlier in Section 4.3, several wavelets are independently used in the wavelet decomposition and the resulting CT of the highest dynamic range has been chosen to determine the S-arrival time. The results show that "Daubechies 8" wavelet is a good choice for most of the tested events. While the other wavelets such as "Daubechies 12" and "Daubechies 20" give good results for some events. This wavelet selection has been done according to the expected arrival shapes as described in details in Section 2.

CONCLUSIONS

Since the main objective of the project is to develop an algorithm that automatically detects particular classes of seismic wave, the software has been developed to detect P and S-phases in three component seismic data.

The algorithm has been tested by a real seismic data represented by multiple local and regional events recorded on various three components stations where it successfully detected P and S-phases.

Two important concepts have been presented in this research. The first concept is decomposing the signal into several resolutions or scales; important features in the seismic signals can be identified. Strong features in a signal can be maintained in several scales while weaker features will present in fewer scales or just one scale. The second concept is the wavelet type selection that depends on matching the wavelet to the arrival shapes.

The results of the software have been compared with a human analyst results as well as with the results obtained by software that uses STA/LTA (short term average/long term average) method. From the comparison results it can be concluded that the composite rectilinearity function and the composite function of transverse to radial amplitude ratio accurately determined P and S arrivals respectively. The results are consistent with the manual inspection results that done by an analyst as well as with other software based method in current professional use. The results showed that the STA/LTA failed to give any detection for S-phase of the tested events recorded by stations with low signal to noise ratio conditions.

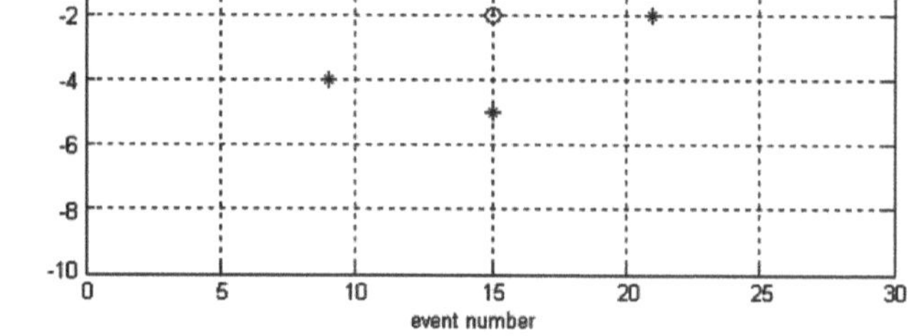

Figure 20: Residual plot of S-phase arrival.

In addition, the STA/LTA will fail to give a result for P-phase in low signal-to-noise conditions. However, the P and S-phase wavelet detection algorithms provided good results for P and S-arrivals and the results are consistent with the human analyst results.

To sum up, the developed wavelet algorithm can accurately determine the P and S arrivals in spite of the low signal-to-noise ratio. Furthermore, the software succeeded in determining back azimuths towards the earthquake sources.

The developed software can be used for a daily routine analysis in the seismological laboratories. In order to achieve this it should be tested by a huge amount of data set so that the parameters such as wavelet type that used in the DTWT analysis and the window size that used to construct the covariance matrices can be fixed.

ACKNOWLEDGEMENTS

Sincere thanks to Sultan Qaboos University for sponsoring this study. The first author would like to thank his colleagues at the earthquake monitoring center in Oman for sending the seismic data that are used in testing the software. In addition, many thanks to the staff at the Broadband Array data collection center, UCSD (http://eqinfo.ucsd.edu), for providing the seismic data that are used for results comparison. Partial support for this research was provided by NEHRP/USGS grants 03HQPA-0001 and 01HQAG0021. Furthermore, thanks to the University of Newcastle upon Tyne for providing computer facilities.

REFERENCES

1. H. Dai and C. MacBeth, "Application of Back-Propagation Neural Networks to Identification of Seismic Arrival Types," Physics of the Earth and Planetary Interiors, Vol. 101, No. 3-4, 1997, pp. 177-188. doi:10.1016/S0031-9201(97)00004-6
2. H. Dai and C. MacBeth, "The Application of Back-Propagation Neural Network to Automatic Picking Seismic Arrivals from Single-Component Recordings," Journal of Geophysical Research, Vol. 102, No. B7, 1997, pp. 15105- 15113. doi:10.1029/97JB00625
3. R. Allen, "Automatic Phase Pickers: Their Present Use and Future Prospects," Bulletin of the Seismological Society of America, Vol. 72, No. 68, 1982, pp. S225-S242.
4. M. Baer and U. Kradolfer, "An Automatic Phase Picker for Local and Teleseismic Events," Bulletin of the Seismological Society of America, Vol. 77, No. 4, 1987, pp. 1437-1445.
5. A. Cichowicz, "An Automatic S-Phase Picker," Bulletin of the Seismological Society of America, Vol. 83, No. 1, 1993, pp. 180-189.
6. P. S. Earle and P. M. Shearer, "Characterization of Global Seismograms Using an Automatic-Picking Algorithm," Bulletin of the Seismological Society of America, Vol. 84, No. 2, 1994, pp. 366-376.
7. D. Patane and F. Ferrari, "ASDP: A PC-Based Program Using a Multi-Algorithm Approach for Automatic Detection and Location of Local Earthquakes," Physics of the Earth and Planetary Interiors, Vol. 113, No. 1-4, 1999, pp. 57-74. doi:10.1016/S0031-9201(99)00030-8
8. R. Sleeman and T. Van Eck, "Robust Automatic P-Phase Picking: An On-Line Implementation in the Analysis of Broadband Seismogram Recordings," Physics of the Earth & Planetary Interiors, Vol. 113, No. 1-4, 1999, pp. 265-275.
9. P.-J. Chung, M. L. Jost and J. F. Böhme, "Estimation of Seismic-Wave Parameters and Signal Detection Using Maximum Likelihood Methods," Computers and Geosciences, Vol. 27, No. 2, 2001, pp. 147-156. doi:10.1016/S0098-3004(00)00088-1
10. M. Tarvainen, "Automatic Seismogram Analysis: Statistical Phase Picking and Locating Methods Using OneStation Three-Component Data," Bulletin of the Seismological Society of America, Vol. 82, No. 2, 1992, pp. 860- 869.
11. K. S. Anant and F. U. Dowla, "Wavelet Transform Methods for Phase Identification in Three-Component Seismograms," Bulletin of the Seismological Society of America, Vol. 87, No. 6, 1997, pp. 1598-1612. [Paper reference:6]
12. P. J. Oonincx, "A Wavelet Method for Detecting SWaves in Seismic Data," Computational Geosciences, Vol. 3, No. 2, 1999, pp. 111-134. doi:10.1023/A:1011527009040 [Paper reference:1]
13. H. Zhang, C. Thurber and C. Rowe, "Automatic P-Wave Arrival Detection and Picking with Multiscale Wavelet Analysis for Single-Component Recordings," Bulletin of the Seismological Society of America, Vol. 93, No. 5, 2003, pp. 1904-1912.doi:10.1785/0120020241 [Paper reference:1]
14. I. Daubechies, "Ten Lectures on Wavelets," The Society of Industrial and Applied Mathematics, 1992. [Paper reference:1]
15. E. J. Stollnitz, T. D. DeRose and D. H. Salesin, "Wavelets for Computer Graphics, Theory and Applications," Morgan Kaufmann, Burlington, 1996. [Paper reference:2]
16. G. Strang and T. Q. Nguyen, "Wavelets and Filter Banks," Revised Edition, Wellesley-Cambridge Press, Wellesley, 1998.
17. S. Mallat, "A Wavelet Tour of Signal Processing," Academic Press, Waltham, 1999.
18. S. Mallat, "A Theory for Multiresolution Signal Decomposition: The Wavelet Representation," IEEE Transactions on Pattern Analysis and Machine Intelligence, Vol. 11, No. 7, 1989, pp. 674-493. doi:10.1109/34.192463
19. B. Porat, "A Course in Digital Signal Processing," John Wiley & Sons, Inc., Canada, 1997.
20. M. Vetterli and C. Herley, "Wavelets and Filter Banks: Theory and Design," IEEE Transactions on Signal Processing, Vol. 40, No. 9. 1992, pp. 2207-2232. doi:10.1109/78.157221
21. M. Vetterli and D. Le Gall, "Perfect Reconstruction FIR Filter Banks: Some Properties and Factorizations," IEEE Transactions on Acoustics, Speech and Signal Processing, Vol. 37, No. 7. 1989, pp. 1057-1071. doi:10.1109/29.32283
22. E. R. Kanasewich, "Time Sequence Analysis in Geophysics," University of Alberta Press, Edmonton, 1981. [Paper reference:4]
23. A. Jurkevics, "Polarization Analysis of Three-Component Array Data," Bulletin of the Seismological

Society of America, Vol. 78, No. 5, 1988, pp. 1725-1743. [Paper reference:1]

24. Stein, S. and M. Wysession, "An Introduction to Seismology, Earthquakes and Earth Structure," Blackwell Publishing Ltd., Hoboken, 2003.
25. K. Aki and P. G. Richards, "Quantitative Seismology," University Science Books, Sausalito, 2002.
26. R. Wiggins, "Minimum Entropy Deconvolution," Geoexploration, Vol. 16, No. 1-2, 1978, pp. 21-35. doi:10.1016/0016-7142(78)90005-4 [Paper reference:1]

Appendix A

As mentioned in Section 3, a covariance matrix $M^j[i]$ is calculated at each point I of x_1^j, x_2^j and x_3^jover a T point window. The size of window affects the eventual output, Cf. The aim is to obtain a function Cf that has a one main spike with a minimum of secondary spikes. There is a measure called the varimax norm [26] can be used for this purpose. The higher the varimax norm of a signal, the fewer spikes the signal has.

The varimax norm of the composite function Cf is calculated as:

$$V_c = \left(\sum_i Cf^4[i]\right) \Big/ \left(\left(\sum_i Cf^2[i]\right)^2\right)$$

Thus, before selecting the size of the window (Nwin), several size windows are tested in order to maximize V_c. So, the window that gives a Cf with the highest varimax norm is selected as the window for a particular event.

Figure A1 and A2 show three component data of one of the tested events and a plot of the varimax norm versus the window size used to construct the covariance matrix. For this particular event, 6 seconds is chosen as the window length.

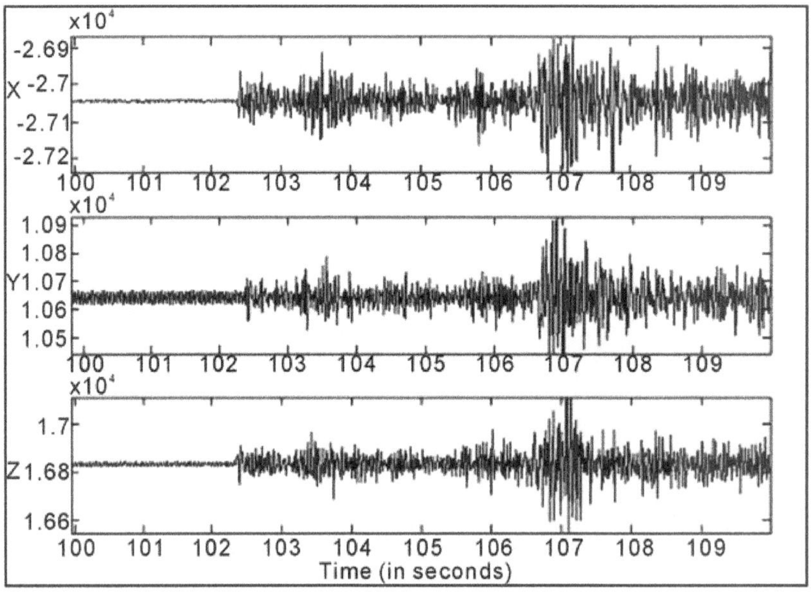

Figure A1: Three component data of an event.

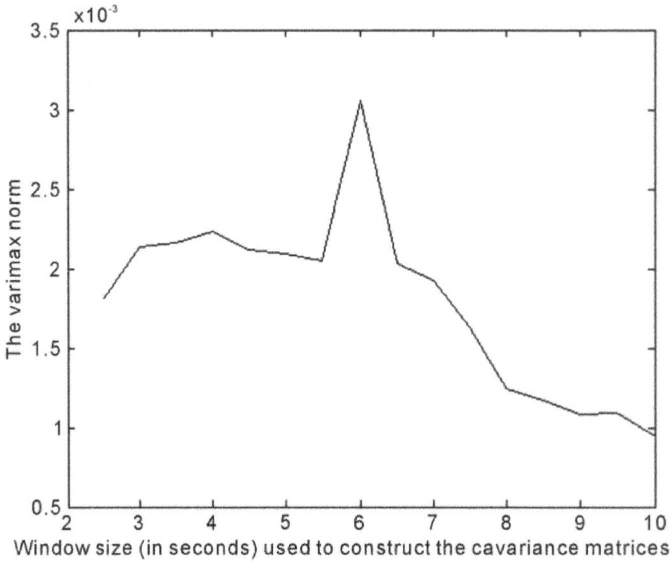

Figure A2: Demonstrating the effect of window size used to construct covariance matrices on the varimax norm of the composite rectilinearity function Cf.

Appendix B

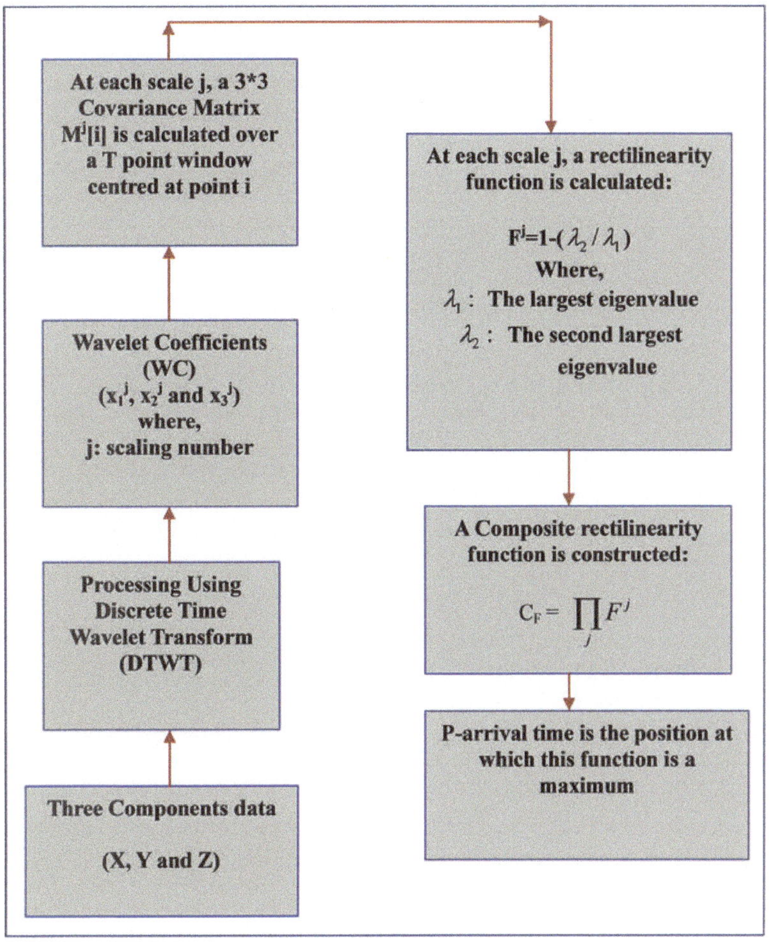

Figure B1: Flow Chart of P-wave detection algorithm.

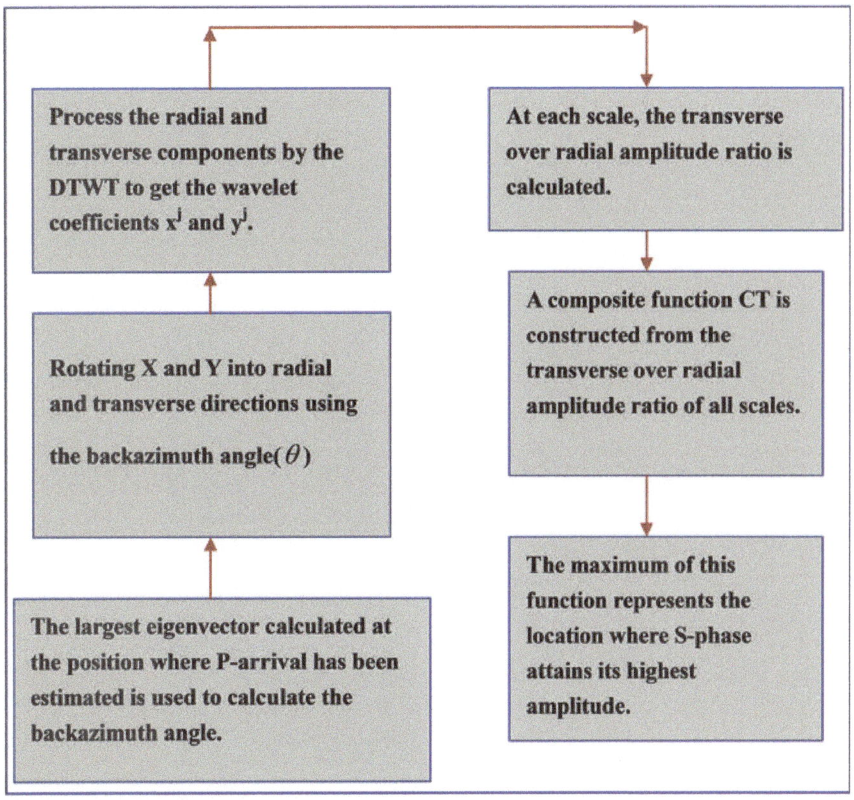

Figure B2: Flow Chart of S-wave detection algorithm.

CHAPTER 17 SEISMIC ACTIVITY OF THE EARTH, THE COSMOLOGICAL VECTORIAL POTENTIAL AND METHOD OF A SHORT-TERM EARTHQUAKES FORECASTING

Baurov Yury Alexeevich, Baurov Alexey Yur'evich, Baurov Alexandr Yur'evich (Jr.), Spitalnaya Alexandra Alfredovna, Abramyan Ara Arshavirovich, Solodovnikov Vladimir Alexandrovich

Keywords: Earthquake; New Anisotropic Interaction

INTRODUCTION

In References [1-3] theory of surrounding us world's creation and of physical space (vacuum) as a result of interaction of byuons-discrete objects, is given. The definition of the byuon contains the cosmological vector-potential Ag - a novel fundamental vector constant. This theory predicts a new anisotropic interaction of nature objects with physical vacuum. Peculiar "tap" to gain new energy (giving rise to an earthquake) are elementary particles because their masses are proportional to the modulus of some summary potential A_S that contains potentials of all known fields. The value of A_S cannot be larger than the modulus of A_g. In accordance with the experimental results shown in [1-9], this force ejects substance from the area of the weakened A_S along a conical formation with the opening of 100° ± 10° and the axis directed along the vector A_S. This vector has the following coor- dinates in the second equatorial coordinate system: right ascension a » 293° ± 10°, declination d » 36° ± 10° [3,4]. In the present paper, the hypothesis for connection of the Earth's seismic activity with fluctuations in the structure of physical space (vacuum) is considered. That is, our planet, the Earth, will be treated as a peculiar large-scale probe in the physical space the seismic activity of which allows to judge the fluctuations of physical space itself and in so doing to refine its structure and direction of the vector Ag. And vice versa, with a knowledge of space structure and of its fluctuations one would be able to predict more accurately the most dangerous day time for Earth's inhabitants at one or another place of our planet. The urgency of this problem has no need of much to talk because earthquakes lead to manythousand victims and catastrophic destructions.

In the work, basic models and standard mechanisms of earthquakes are briefly considered, results of processing of information on the earthquakes in the context of global spatial anisotropy caused by the existence of the vector A_g, are presented, and an analysis of them is given. A method of short-term earthquakes forecasting (1-2 days in advance) is advanced.

EARTHQUAKES AND THEIR MECHANISMS OF GENERATION

An earthquake is the sudden release of potential energy of bowels of the earth in the form of shock waves and elastic vibrations propagating every which way from the center (hypocenter), as well as in the form of shifts of the Earth's surface such as warpings of crust, displacements along fault lines, compactions of granular or not grouted precipitations, soil slips and mudflows, soil dilution, snow avalanches, new fracture formation in rocks [10]. The mechanism of

Citation: Alexeevich, B. , Yur'evich, B. , Yur'evich (Jr.), B. , Alfredovna, S. , Arshavirovich, A. and Alexandrovich, S. (2011) Seismic activity of the earth, the cosmological vectorial potential and method of a short-term earthquakes forecasting. Natural Science, 3, 109-119. doi: 10.4236/ns.2011.32016.

release of potential energy is not yet sufficiently studied for earthquakes at depths from 60 km to 720 km [11]. The Earth is never quiescent. Sensible seismographs constantly detect weak oscillations, microseisms, with periods from 4 to 6 seconds and variable amplitudes. Even "microseismic storms" take place, they are connected mainly with sharp changes in weather. During powerful earthquakes almost total seismic energy is released. An energy characteristic of earthquakes is the magnitude. A scale of magnitudes was elaborated by Karl Richter, a professor of Californian Technological Institute. The magnitude of an earthquake is a measure of the total energy quantity released in the form of elastic waves, and that is not equal to the total energy of earthquake. Part of elastic energy accumulated in the center of earthquake is turned into heat. The magnitude is characterized by the maximum amplitude of record by a seismograph of standard type at a fixed distance from the hypocenter of earthquake. The most potent disastrous earthquakes have amplitude M = 9. As an example of earthquakes with M = 9, that in Lissabon at November 1, 1755, may be considered. Any earthquake with M ³ 7 is the great disaster, particularly if that happens in the neighbourhood of populated area. The energy released during such earthquakes ranges from $2.1 \cdot 10^{22}$ to $1.6 \cdot 10^{25}$ erg. Very strong earthquakes lead to torsion and spheroidal oscillations of the Earth as a whole. For very weak shocks, negative values of the magnitude are used [10]. As to depth of occurrence (h) of earthquake centers, those are classified into three categories: crustal (0 £ h £ 70 km), middle focal (intermediate) (70 < h £ 300 km), and deep focal h > 300 km earthquakes.

The most deep focal shocks were recorded only in the southern hemisphere of the Earth:

1. in the vicinity of Kermadec runner (10047 m in depth, interval of latitudes 35°S £ j £ 25°S):
2. in the neighbourhood of Tonga runner (10882 m in depth, interval of latitudes 25°S £ j £ 15°S) (both runners are located to the north-west from New Zealand);
3. in the latitude j » 6° to the south from Sulawesi island (Indonesia). The centers of those earthquakes were up to 720 km deep from the surface.

In Table 1 are listed distributions of earthquakes of all types (from deep focal to crustal ones) registered at a period from 1904 till 19881 and having magnitudes M = 7 and more except for the deep focal earthquakes for which, to magnify the sample, we had to use those beginning with M ³ 5.9. The data were chosen from catalogues [10-15]. It follows from Table 1 that great majority of earthquakes of all type occurs at latitudes 35°-45° north as well as in southern latitudes of equatorial region and at 25°S. Therewith the crustal earthquakes are prevalent in the northern hemisphere, whereas deep focal earthquakes are in the southern one as evidenced by confidence intervals L obtained with the use of Student t-test for random values of the average with the level of significance of 0.01 and 0.001 (see Figure 1). The first fact was known as early as the beginning of 20-th century [16-19] and even earlier [20] in the context of the question of how and where the stresses accumulate in the Earth's crust. The question of a particular role of 35°-th parallel in the tectonic life of the Earth was repeatedly posed by geographers and geologists as in the Russian so in foreign literature. The first work along this line was probably that by A. A. Tillo [20] which has pointed out all the most high peaks and mountain chains as well as the most great depths in the northern hemisphere to gravitate towards the latitudes of 30°-40°. All researches of this sort persued prior 1962 (with voluminous list of literature) may be acquainted with in the articles of M. V. Stovas, an astronomer-geodesist [16,17], and V. A. Tsaregradsky, an astrogeologist [18]. One of important conclusions of those studies was that all processes connected with the most potent earthquakes have a common mechanism for the whole Globe. Following German geologists, the authors of Refs. [16-18] considered the periodic variation of the Earth's rotation velocity due to tide influence of Moon and Sun as one of the sources of energy for tectonic processes. The smaller is the angular velocity of rotation, the more globe-shaped becomes the form of the Earth.

Table 1: The dependence of number of earthquakes (n_i) from the latitude (or declination) of their epicenters and hypocenters.

$\varphi^\circ = \delta^\circ$	h > 300 км		70 < h ≤ 300 км		0 ≤ h ≤ 70 км		Все	
	n_i	$n_i/\bar{n}$	n_i	$n_i/\bar{n}$	n_i	$n_i/\bar{n}$	ni	$n_i/\bar{n}$
+85	0	0,00	0	0,00	0	0,00	0	0,00
+75	0	0,00	0	0,00	1	0,02	1	0,01
+65	0	0,00	2	0,12	14	0,25	16	0,15
+55	26	0,78	17	1,05	92	1,64	135	1,28
+45	69	2,06	20	1,23	120	2,14	209	1,98
+35	70	2,09	30	1,85	122	2,18	222	2,10
+25	45	1,34	16	0,99	75	1,34	136	1,29
+15	13	0,39	33	2,04	92	1,64	138	1,31
+5	23	0,69	19	1,17	85	1,52	127	1,20
−5	−99	2,96	42	2,59	137	2,45	278	2,63
−15	70	2,09	49	3,02	99	1,77	218	2,06
−25	168	5,01	48	2,96	68	1,22	284	2,69
−35	20	0,60	9	0,56	42	0,75	71	0,67
−45	0	0,00	1	0,06	25	0,45	26	0,25
−55	0	0,00	5	0,31	20	0,36	25	0,24
−65	0	0,00	0	0,00	15	0,27	15	0,14
−75	0	0,00	0	0,00	0	0,00	0	0,00
−85	0	0,00	0	0,00	0	0,00	0	0,00
Σ	603		291		1007		1901	
$\bar{n}$	33,5		16,2		55,9		105,6	
n_S	357	4,5 σ	154	σ	406	−6,1σ	917	1,5σ
n_N	246		137		601		984	

Σ– is the total number of earthquakes with M ≥ 7 over the period from 1904 till 1988; $\bar{n}$ –is the average number of earthquakes in 10°-interval of latitudes; σ–is the standard deviation characterizing the anisotropy of Northern and Southern hemispheres relative to the number of the most strong earthquakes; n_s–is the number of earthquakes in the southern hemisphere; n_N–is that in the Northern hemisphere.

And vice versa, with the growth of angular velocity the planet becomes more ellipsoid-shaped. The reconstruction of superficial form of the Earth will be the most considerably reflected in the regions of 35°-th parallels of northern and southern hemispheres. Those regions are named critical. Said parallels are also associated with compressive and tensile strains originated under the action of precession (slow motion of the Earth's axis of rotation over a circular cone around an axis orthogonal to the plane of ecliptic. Many epicenters of earthquakes concentrate near the critical parallels, too. To explain large number of meridionally extending structures, supporters of rotational hypothesis admitted the possibility of time variations of the Earth's axis of rotation. As the form of the Earth changes, all tectonic processes are to be symmetric about the equator. But from Figure 1, it is clearly seen that at southern latitudes of 30°-40°S the number of the strongest earthquakes is lesser than the average, and there is no symmetry in their distribution about the equator.

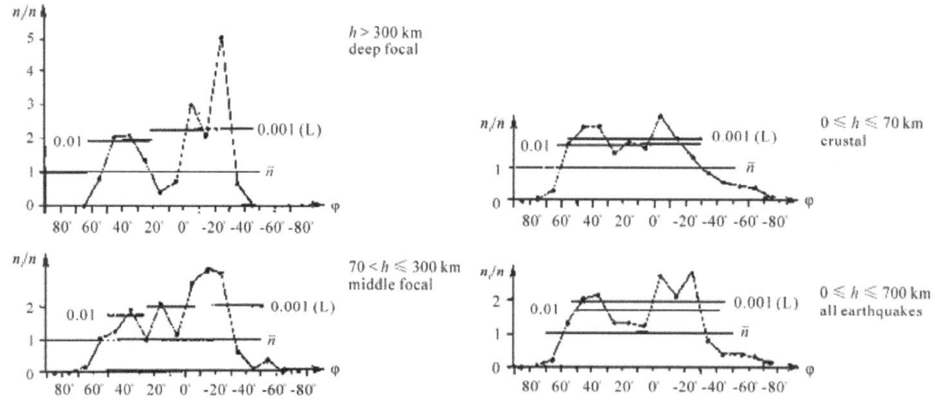

Figure 1: The distribution of earthquakes (in relative units $n/\bar{n}$) in dependence from latitude (φ) and depth (h) of earthquakes hypocenter. n –average numbers of earthquakes ($\bar{n} = 1$); L–confidence interval at significance level 0,01-0,001.

In the summary curve of latitude distribution of earthquakes, the maximums are the sames as in the curves of their depth-distribution, i.e. j = 25°S and j = 5° on the south, j = (45°-35°) N on the north. A substantial predominance of crustal earthquakes in the Northern hemisphere (6.1s) and deep focal earthquakes in the Southern one (4.5s) as well as their statistics equality (1s) for the middle focal earthquakes, are also important facts. In this regard it is interesting to note that the irregularity of distribution of oceans and continents over the Earth's surface has first drawn attention of Ch. Layel who has established an idea of evolution and principles of actualism in geology. In the first edition of his book "Basic principles of geology" published in the years 1830-1833, he gives a projection of Earth's hemispheres which is displaced in such a manner that one of them is mainly continental whereas the other is oceanic. Therewith the "northern pole" is represented by a district of London (j ~ 51°N) and the "southern pole" is located close to New Zealand. The plane of "equator" forms at that an angle of 39°with that of geographic equator. When investigating the planetary jointing of the Earth, it was elucidated [21,22] that the modern global relief presented mainly by lineaments (i.e. by extended straight lines manifested as contours of continental coasts, mountain chains, river valleys, and divides) is considerably a summary effect of manifestation of crustal jointing associated with epicenters of earthquakes. Hence the lineaments are very much to do with the most new tectonic displacements of the Earth's crust. The statistic distribution of lineamental azimuths reveals the existence of two quadrupole (cruciform) systems with dominant orientation along the "meridian-parallel" directions as well as diagonally to those [21,22]. The Sun's "zones of instability" (regions of predominant origin of spot groups and flare activity) manifest similar systems of preferable directions. The filaments arranged in parallel with such a zone, are more stable as compared with those of orthogonal orientation, i.e. the filaments themselves also form quadrupole systems of two types. The similarity of orientation of Sun's "instable zones" associated with the most potent non-stationary processes, and of lineaments connected with the present tectonic activeity on the Earth, has raised a question of the possibility of existence of some energy factors external to the Solar system and manifesting themselves in the form of the solar activity and tectonic actions on planets and their satellites [23,24]. The results of a number of works [23-30] dedicated to analyse of cosmic factors in the geotectonics and above all to elucidation of the relation between the solar activity (SA) and the seismicity of the Earth, have led to the conclusion that such relation does exist, that the earthquakes occur the most often in a phases of increase and decrease of solar cycle, and that the frequency of earthquake events depends more likely on SA-fluctuations than on its level itself. Several mechanisms for this relation were proposed:

- According to References [25-27], it can be accomplished through general atmospheric processes on planet. The intensification of corpuscular radiation of the Sun when an active region passes through the central meridian, increases the potential energy of the atmosphere, depresses zonal circulation, and leads to local accumulation of energy. Redistribution of atmospheric masses over the Globe disturbs the balance of the Earth's shape and results in displacements of crust blocks. The earthquakes are consequences of these displacements.
- Magneto-dynamic influences of the Sun on the Earth change its speed of rotation [27,31,32] which leads to the above mentioned rotational effect.
- In the Earth's crust, currents with daily variations are universally present [33]. Additional currents due to SA are meant to sufficiently act on mountain rocks and lead to detonation of earthquakes [27]. Thus the solar flares are considered as possible trigger mechanism of earthquakes in the Earth's areas being in instable state.
- In Reference [28] a question is raised whether or not the manifestations of Sun's and Earth's activity are simply events coincident in time, and the cause of them to be beyond as the Earth so the Sun. This question is posed on the basis of a number of common morphologic symptoms appearing in the course of development of chromospheric flares and earthquakes despite the fact that the process goes in plasma in the first case and in the solid body in the second case.

Let us analyze the seismic activity of the Earth in the context of action of the new force. Consider a mechanism of increase of this force due to Earth's currents. As was said above, there exists in the vicinity of the Earth some real summary potential A_S equal to the sum of Ag and potential from magnetic fields of Galaxy, Sun, Earth, etc. The new force is of complex nonlinear and nonlocal nature and can be represented as some series in DA_S [1-3,9]. As a first approximation we have

$$F \sim \Delta A_{\Sigma} \frac{\partial \Delta A_{\Sigma}}{\partial x} \qquad (1)$$

where x is the spatial coordinate. The quantity $|A_{\Sigma}|$ is always lesser than $|A_g|$ (see [1-3]). There are possible fluctuations of enormous magnitudes of vectorial potentials of magnetic sources remote from the Earth, for example, that of the Sun. Despite the large values of DA_S of those sources, magnitudes of $\frac{\partial \Delta A_{\Sigma}}{\partial x}$ from them are, as was said above, as a rule insignificantly small since variations of potentials take place through vast distances. Hence the values of the new force from remote sources are very small, too, and the values of $\frac{\partial \Delta A_{\Sigma}}{\partial x}$ due to currents in the Earth might be several orders greater than those from remote sources. Then the Earth might work as some amplifier of enormous fluctuations of DA_S from remote sources because the magnitude of the new force includes the product $\Delta A_{\Sigma} \frac{\partial \Delta A_{\Sigma}}{\partial x}$ where the first factor is created by remote cosmic sources and the second one is by the currents in the Earth. Give an example. In [1-3] the results of an uninterrupted experiment (from Feb. 24 to Mar. 22, 96) with a tide gravimeter developed in the Sternberg Astronomical Institute of Moscow University on base of a standard quartz gravimeter "Sodin" (Canadian production) were showed (Figures 2, 3). To measure the new interaction by the Sodin gravimeter, a constant magnet (60 mm in diameter, 15 mm in height, the field B in the center of 0.3T) was attached to it in such a way that the vector-potential lines of the magnet in the vicinity of a test platinum weight were directed perpendicular to the Earth's surface (i.e. towards the vertical component of the vector A r G). The major deflections of the gravimeter, being repeated every 24 hours, are associated with the Moon's gravity. Denote an average amplitude of Moon tide by L, an amplitude of accidental events, recorded by gravimeter and corresponding to an increase in Moon attraction, by K L+ , and that corresponding to a decrease by K L– , where K is a factor indicating the value of deflection in terms of Moon tide amplitudes. Three events were documented: on the 28th of February, at 10^{05}; 4th of March, at 10^{58}; 18th of March, at 20^{54}. Two last of them had a huge amplitude (13.6 L– and 15.2 L– , respectively) and Δt » 10 min. A time profile of the event on 18th of March 1996 is shown in Figure 3 at a larger time scale.

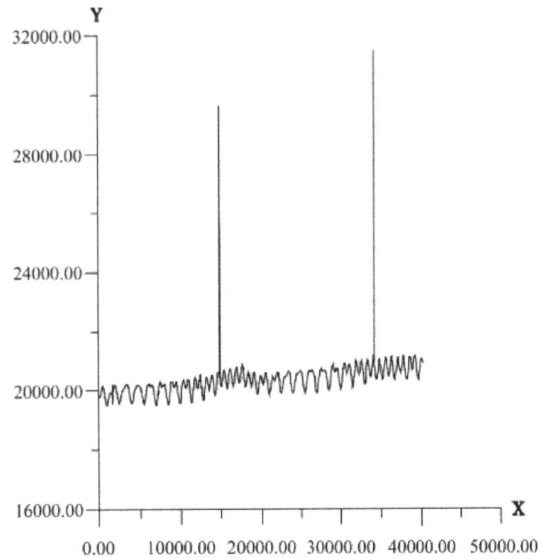

Figure 2: Readings of the gravimeter from Feb. 24, 1996, to March 22, 1996, inclusively. y is displacement of platinum weight. One division is 0.1 μm, or 0.2 μgal, x is time in minutes.

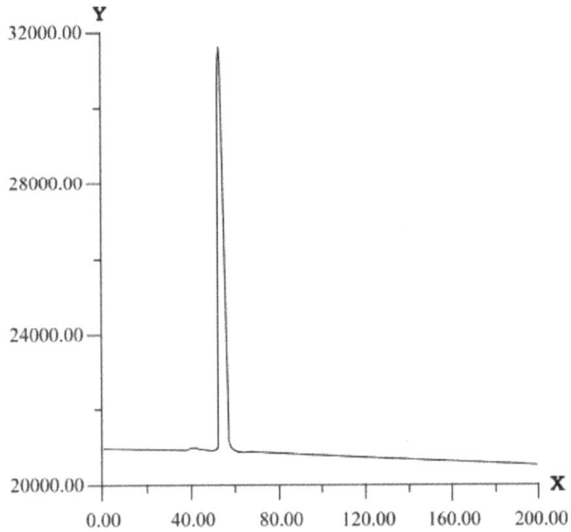

Figure 3: The event on March 18, 1996. y is displacement of platinum weight. One division is 0.1 μm, or 0.2 μgal, x is time in minutes.

Here DA was created by some natural sources, and $\frac{\partial \Delta A}{\partial x_1}$ was created by a constant magnet attached to the gravimeter, since DA_S from spatial sources varies at immensely long distances, with an infinitesimal value of $\frac{\partial \Delta A}{\partial x_1}$ at the point of the gravimeter location. The fact of local change in the gravitation field described in [1-3,34] (see Figures 2, 3) may act as trigger for the stresses accumulated in the Earth's crust, and cause earthquakes. At that the role of amplifier may be played, instead of a constant magnet, by the currents existing inside the Earth. Further, the maximum mass change in a local characteristic volume of the Earth takes place when this volume passes, during the Earth's rotation, through the region of maximum change in A_S by the vectorial potential A_E of the Earth. To estimate that, it will suffice to assume the existence of a circular current in the Earth (Figure 4) the magnetic field of which is observed by us as geomagnetism. Near the current the vectorial potential of its magnetic field is directed along the vector of current (Figure 4). Thus if the vector A_S in some volume of the Earth experiences the maximum change by AE, the masses of elementary particles in that same volume are also changed in accordance with the theory of byuons, and hence the total mass of the volume is changed, too. The latter leads to maximum change in the gravitation field of the Earth that causes earthquakes. Denote the zenith distance (the arc of the great circle of celestial sphere between the direction to zenith and

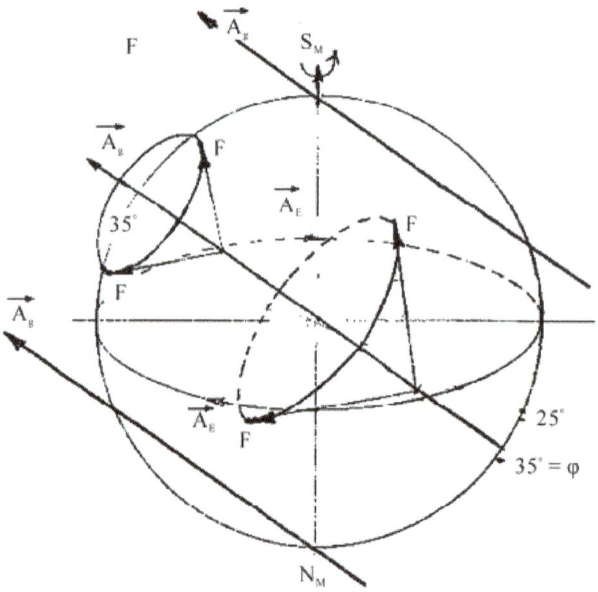

Figure 4: The direction of vector $\vec{A}_g$ and the cone of new force action (F) relatively to the globe. A_E–vectorial potential of Earth's magnetic field; S_M, N_M–South and North Pole of Earth's magnetic field.

that to the point of intersection of A_g with the sphere) by $Z_{\vec{A}_g}$. In Table 2 is given the dependence of distribution of $Z_{\vec{A}_g}$ arcs on the latitude j of earthquake epicenters: $\overline{n}$ is the number of earthquakes, n is their average, L – confidence interval, C_n – average number of earthquakes in column at the fixes $Z_{\vec{A}_g}$ calculated for two assumed values of the a-coordinate of A_g being equal to 270° and 290°; $\Sigma_{270°}$, $\Sigma_{290°}$ are the correspondent sums for columns of table. The coordinates φ°, $Z_{\vec{A}_g}$ for a cells of Table denote:

$$n_i(\alpha = 270^\circ) \quad n_i/\overline{n}(\alpha = 270^\circ)$$

$$n_i(\alpha = 290^\circ) \quad n_i/\overline{n}(\alpha = 290^\circ)$$

where n_i is the number of earthquake epicenters in a given interval of latitudes at the moment when apexes of A_g were in the interval of zenith distances $Z_{\vec{A}_{g\,av}} \pm 10°$. The average values $\overline{n}$ = n/9 are indicated for each interval of latitudes in 12-th column of table; $n_i/\overline{n}$ is the relative number of earthquakes normalized to the average $\overline{n}$. For example, at j = 55° and $Z_{\vec{A}_g} = 90°$ we have 57 ® 4.01 and 46 ® 3.24. That is in the interval of latitudes (50°-59°) there are happened 57 (a = 270°) and 46 (a = 290°) earthquakes when the apex of A_g was in the interval (80°-99°) of zenith distances $Z_{\vec{A}_g}$ i.e. in the region of sunset. At that $\overline{n}$ = 128/9 = 14,2; 4.01 = 57/14.2 and 3.24 = 46/14.2.

Table 2: The dependence of the distribution of arcs ZA_g on the latitude φ of earthquakes epicenter

φ°	ZA_g / n	10°		30°		50°		70°		90°		110°		130°		150°		170°		$\bar{n}$	L 0,1%	L 1%
55°	128	0	0,00	22	1,55	26	1,83	21	1,48	57	4,01	2	0,14	0	0,00	0	0,00	0	0,00	14,2	3,1	2,4
		0	0,00	22	1,55	26	1,83	31	2,18	46	3,24	3	0,21	0	0,00	0	0,00	0	0,00		2,9	2,2
45°	155	9	0,52	24	1,40	28	1,63	24	1,40	34	1,98	36	2,09	0	0,00	0	0,00	0	0,00	17,2	2,4	1,9
		4	0,23	26	1,51	27	1,57	30	1,74	38	2,21	30	1,74	0	0,00	0	0,00	0	0,00		2,4	2,0
35°	159	19	1,07	21	1,19	21	1,19	24	1,36	33	1,86	41	2,32	0	0,00	0	0,00	0	0,00	17,7	2,3	1,9
		19	1,07	25	1,44	20	1,13	23	1,30	38	2,75	34	1,92	0	0,00	0	0,00	0	0,00		2,3	1,9
25°	91	13	1,29	14	1,39	16	1,59	9	0,89	12	1,09	20	1,98	7	0,69	0	0,00	0	0,00	10,1	2,1	1,7
		18	1,78	12	1,19	10	0,99	9	0,89	15	1,49	13	1,29	14	0,13	0	0,00	0	0,00		2,0	1,6
15°	122	13	0,96	8	0,59	22	1,62	20	1,47	25	1,84	12	0,88	22	1,62	0	0,00	0	0,00	13,6	2,1	1,7
		5	0,37	23	1,69	14	1,03	21	1,54	17	1,54	19	1,40	23	1,69	0	0,00	0	0,00		2,1	1,8
5°	111	0	0,00	24	1,95	17	1,38	12	0,98	11	0,89	15	1,22	24	1,95	8	0,65	0	0,00	12,3	2,1	1,8
		0	0,00	20	1,63	15	1,22	16	1,30	12	0,98	15	1,22	21	1,71	12	0,98	0	0,00		2,0	1,7
−5°	194	0	0,00	20	0,33	42	1,94	20	0,93	18	0,83	33	1,53	35	1,62	26	1,20	0	0,00	21,6	2,1	1,7
		0	0,00	24	1,11	34	1,57	23	1,06	23	1,06	27	1,25	24	1,11	39	1,81	0	0,00		2,0	1,6
−10°	152	1	0,06	1	0,06	35	2,07	19	1,12	21	1,24	22	1,30	26	1,54	22	1,30	5	0,30	16,9	2,1	1,7
		0	0,00	2	0,13	36	2,13	17	1,01	16	0,95	26	1,54	23	1,36	25	1,48	7	0,41		2,2	1,8
−25°	141	0	0,00	1	0,06	23	1,46	29	1,85	21	1,34	16	1,02	22	1,40	17	1,08	12	0,76	15,7	1,60	1,3
		0	0,00	2	0,13	25	1,59	29	1,85	18	1,15	17	1,08	19	1,21	22	1,40	9	0,57		1,60	1,4
−35°	53	0	0,00	1	0,17	0	0,00	17	2,89	9	1,53	9	1,53	4	0,68	6	1,02	7	1,19	5,9	2,5	2,0
		0	0,00	0	0,00	2	0,34	14	2,37	9	1,53	9	1,53	9	1,53	6	1,02	4	0,68		2,3	1,8
−45°	26	0	0,00	0	0,00	1	0,34	5	1,72	4	1,38	3	1,03	3	1,03	7	2,41	3	1,03	2,9	2,3	1,8
		0	0,00	0	0,00	1	0,34	6	2,07	4	1,38	1	0,34	3	1,03	7	2,41	4	1,38		2,4	1,9
−55°	39	0	0,00	0	0,00	0	0,00	1	0,23	12	2,79	10	2,33	7	1,63	9	2,09	0	0,00	4,3	2,9	2,2
		0	0,00	0	0,00	0	0,00	2	0,47	13	3,02	9	2,09	4	0,93	11	2,56	0	0,00		2,9	2,3
c_p α = 270°	114,25	4,58		6,67		19,25		16,75		21,42		18,25		12,50		7,92		2,42				
c_p α = 290°	114,25	3,83		13,00		17,50		18,42		20,75		16,92		11,17		10,17		2,00				
$\Sigma_{270°}$	1371	55	0,36	80	0,53	231	1,52	201	1,32	257	169	219	1,44	150	0,99	95	0,62	29	0,19	152,3		
$\Sigma_{290°}$	1371	46	0,30	156	1,02	210	1,38	221	1,45	249	1,63	203	1,33	140	0,92	122	0,80	24	0,16	152,3		

It is seen from Table 2 that this theoretical statement is in total consistency with experiment. i.e. the maximum of earthquakes is observed at an angle of 90° between the direction to zenith and the vector Ag. As was above said, the local increase in the Earth's currents gives effects similar to those shown in Figures 2, 3 (i.e. currents in the Earth are playing part of magnet attached to the gravimeter), since in that case a considerable enhancement of $\frac{\partial \Delta A_v}{\partial x}$ may occur and, as the result, of the magnitude of new force functioning as trigger for earthquakes. As was shown in experiments with the plasma generator in gimbal suspension [3,4], the energy of the plasma jet increases up to 40% when the vectorial potential A of current of the plasma generator has an angle of ~ 135° relative to the vector Ag. On the basis of experiments with plasma generator and those on investigating the action of the new force on the b-decay rate of radioactive elements [1-3,5], in Figure 4 a cone of directions of action of the new force on the Earth's substance is shown (in the case that there are currents there flowing antiparallel to A_g).

As is seen from this figure, in the Northern hemisphere the maxi- mum of new force will fall into latitudes between 30°– 40° due to the current system of the Earth's magnetic field and the known direction of vector A_g. The substance at those latitudes should "bulge" from the Earth. As was said above, in those latitudes we have maximum number of earthquakes of substantionally superficial character. The mountain rocks themselves are of lesser hardness there than in the Southern hemisphere where the new force compresses the substance and rejects the epicenters of earthquakes deeper into the Earth. At that the epicenters become closer to the plane of equator as is shown in Figure 1. The maximum number of earthquakes occurs in southern latitudes, at 25°S and more close to the equator.

The long-term investigations of the influence of the new force on the rate of b-decay of radioactive elements [1-3,5] as well as experiments with plasma devices on energy generation with the aid of that force [1-4] allow to conclude that in the bowels of the earth similar manifestations of new force might take place which lead to local earth heating up, origination of various stresses in its deep layers, and hence to accumulation of potential energy liberated during earthquakes. The interaction of current systems with the physical space through the new force is possibly the main contribution of energy to maintain energy balance of celestial bodies: stars and planets. Such statement is naturally only a hypothesis requiring thorough and long verification.

Thus the following factors point to the fact of influence of new interaction and the global anisotropy caused by the vector Ag, on earth's tectonic processes and particularly on earthquakes:

- Predominance of crustal earthquakes in Northern hemisphere and of deep focal ones in the Southern hemisphere.
- Maximum probability of initiation of earthquakes when a local Earth's volume passes the direction drawn from the center of the Earth perpendicularly to the vector Ag (the region of maximum change in the gravitational field in the result of decreased $|A_\Sigma|$ under the action of vectorial potential of the Earth's magnetic field). A reverse estimation of direction of Ag gives therewith its coordinates: a » 270°-290°, d » 30° (from the distribution of earthquakes).
- Existence of an "oceanic" and a "continental" hemispheres of the Earth.
- Absence of symmetry in the number of the most potent earthquakes at 35° north and 35° south. The southern hemisphere is formed by stronger mountain rocks than the northern one.
- Stresses in the earth's crust experience daily variation along parallels of latitude and meridians (dependence of $Z_{\bar{A}_g}$ on latitude j and on hour angle t). That creates not only a system of parallel and meridional stresses but also a diagonal quadrupole system. The action of those stresses during long periods of time leads to structural rearrangements in the earth's crust. The influence of new interaction on the Earth has to have a nearyearly variation, too, which probably manifests itself in the Chandler's motion of poles and, precisely, in on-earth experiments as well [1,2].
- The higher is the latitude ±j, the lesser is the probability of potent earthquakes. In the Antarctic there are mountains and fractures but no very strong earthquake was on record in the entire period of observation on that territory.

METHOD OF A SHORT-TERM EARTHQUAKES FORECASTING

The analysis of collected data about strong earthquakes (see above) shows that practically all of them happened at the moments when a seismically dangerous region was in a certain manner oriented in space relative to the stars. To clarify the regularity observed, consider the diagram in Figure 5 that conventionally represents the positions of the Earth and the earthquake zone at the moment of the event (see Table 3). For simplicity, the diagram is brought to the plane of the ecliptic. The Earth 1 (globe) moves relative to the Sun 2 in an orbit 3 lying in the ecliptic plane.

Table 3: Greatest earthquakes in the world in the second part of the XX century and in XXI century

№	Place, coordinates	Date, UTC time (Greenwich time)	Magnitude by Richter scale	α	β
1	Kamchatka **52.76N 160.06E**	1952 November 04 16:58:26.0	9.0	1°	-
2	Andreanof Islands, Alaska **51.56N 175.39W**	1957 March 09	8.6	90°	-
3	Chile **38.24S 73.05W**	1960 May 22 19:11:14	9.5	12°	-
4	Prince William Sound, Alaska **61.02N 147.65E**	1964 March 28 03:36:14	9.2	14°	-
5	Armenia **41.0 N 44.2 E**	1988, December 7, 07:41	6,9	-	41°
6	West Coast of Northern Sumatra **3.295 N 95.982 E**	2004, December 26 00:58	9.0	90°	-
7	Sakhalin Island **52.63 N 142.83 E**	1995, May 27 13:03	6,7	-	40°
8	Pakistan **34.53 N 73.58E**	2005 October 08 19:46	7.6	90°	-
9	Haiti **19.0 N 72.0 W**	2010 January 13		81°	3°

α–Angle between projection of the vector Zenith and projection of the vector A_g on the plane of the ecliptic; β–Angle between projection of the vector V and projection of the vector A_g on the plane of the ecliptic.

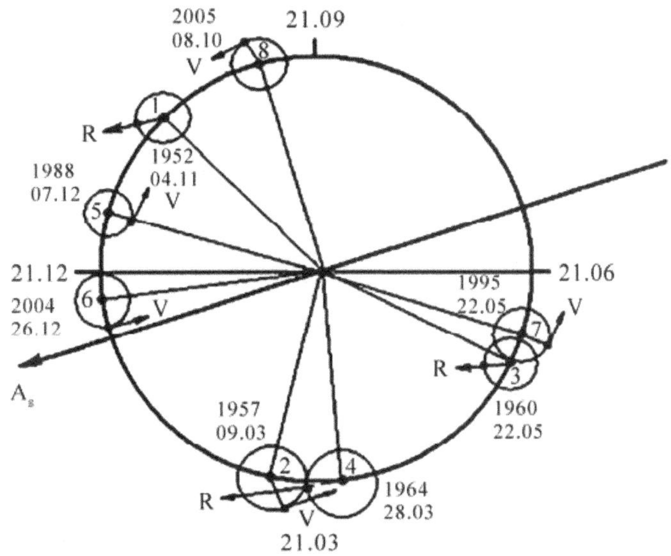

Figure 5: Directions of vectors: A_g, V and R (radius vector from Earth center or vector Zenith) for greatest earthquakes in the world in the second part of the XX century and in XXI century (Table 3).

The latter is inclined to the equatorial plane at an angle of e = 23°26' (in the diagram the plane of the ecliptic and the rotation of the Earth around its axis are shown as circles). Eight positions of the Earth are given at the moments of earthquakes (Positions: 1-8; (Table 3)). The Earth rotates around its axis, and the regions of earthquakes also move along circular trajectories 4 relative to this axis. At the moments of earthquakes their regions were in positions 1-8. The vectors V (Positions: 1-8) of linear velocities of daily rotation of earthquake's regions as well as the vector A_g take certain positions relative to each other. (Vectors V are antiparallel with the vector potential A from the magnetic field of the Earth). Namely, if one brings the vector A_g and vector V of linear velocities, by a parallel transfer, to the point of the earthquake epicenter, then in the moment of earthquake the vectors V of circular velocities of the earthquake regions fall into the limits of a zone restricted by two conic surfaces circumscribed about the vector A_g, and are directed along the generatrices of the family of conic surfaces inside the zone in consideration. The openings of the cones forming that zone are, respectively, 90° and 110°. The analysis of additional factors accompanying the earthquakes has shown that the seismic activity increased when the earthquakes were preceded by severe changes in the total activity of the Sun (i.e. changes in DA_S). From the above reasoning one can indicate, with an accuracy of ± 1 hour, the time points when a seismically dangerous region will go through the most hazardous spatial directions. Will an earthquake happen or not at a given timepoint, show special devices measuring changes in $¶A_S/\partial t$. The detected regularities are used as the basis for a method of short-term forecast of earthquakes with the probability close to 100%.

The invention [35] has received a recognition in the form of three gold medals in International Exhibitions in Brussels, Geneva, and Seoul.

CONCLUSIONS

In this paper we have considered our planet as a peculiar large-scale probe in the physical space the seismic activity of which allows to judge the fluctuations of physical space itself and so doing to refine its structure from scales (10-33 cm-10-17 cm) [1-3,5,36,] up to 1028 cm [1-3,37], ("probe in dark matter and dark energy" [38]) and as a peculiar large-scale probe and analogy of physics of seismic activity other astrophysical objects, for example, pulsar [39,40]. As distinct from all earlier models of earthquakes [16-33], the theory of byuon allows to establish the deep-seated nature of physics of earthquakes, and therefore to predict place and time of an earthquake with the probability close to 100%.

REFERENCES

1. Baurov, Y.A. (1998) Structure of physical space and new method of obtaining energy (theory, experiment, applications). Krechet, Moscow (in Russian).
2. Baurov, Y.A. (2000) On the structure of physical vacuum and a new interaction in Nature (theory, experiment and applications). Nova Science, New York.
3. Baurov, Y.A. (2004) Global anisotropy of physical space: Experimental and theoretical basis. Nova Science, New York.
4. Baurov, Y.A., Timofeev, I.B., Chernikov, V.A., Chalkin, S.F. and Konradov, A.A. (2003) Experimental investigation of the distribution of pulsed-plasma-generator at its various spatial orientation and global anisotropy of space. Physics Letters A, 311, 512. doi:10.1016/S0375-9601(03)00567-X
5. Baurov, Y.A., Konradov, A.A., Kuznetsov, E.A., Kushniruk, V.F., Ryabov, Y.B., Senkevich, A.P., Sobolev, Y.G. and Zadorozsny, S. (2001) Modern Physics Letters A, 16(32), 2089. doi:10.1142/S0217732301005187
6. Baurov, Yu.A., Klimenko, E.Yu., and Novikov, S.I., (1990), Doklady Akademii Nauk, 315, 1116, (in Russian).
7. Baurov, Y.A., Klimenko, E.Y. and Novikov, S.I. (1992) Physics Lettets A, 162, 32. doi:10.1016/0375-9601(92)90954-K

8. Baurov, Yu.A. and Ryabov P.M., (1992), Doklady Akademii Nauk, 326, 73 (in Russian).
9. Baurov, Y.A. (1993) Physics Letters A, 181, 283. doi:10.1016/0375-9601(93)90609-4
10. Aiby, J.A. (1978) Earthquakes, M, Progress, 1978 (in Russian).
11. Thomaas, C.N., Jr. (1974) Global summary of human response to natural hazards: Earthquakes. In: White, G.F., College Work: Natural Hazards Local, National, Global, University of Colorado, New York, London, Toronto, 274-284.
12. Golubeva, N.V. (1972) Catalogue of strong earthquakes of the Globe, 1953 to 1967 with M ³ 6. Academy of Science USSR, Institute of the Earth's Physics (in Russian).
13. Regional Catalague of Earthquakes, Berkshire, United Kingdom (1972-1981).
14. Catalogue of epicentres in the international seismological summary.
15. Seismic bulletin of the network of reference seismic stations of USSR (in Russian).
16. Stovas, M.V. (1963) On stressed state of crustal layer in region between 30° and 40°. In: Gosgeoltechizdat, M., College Work: Problems of Planetary Geology, 275-284 (in Russian).
17. Stovas, M.V. (1963) Several questions of tektogenesis. In: Gosgeoltechizdat, M., College Work: Problems of Planetary Geology, 222-274 (in Russian).
18. Tsaregradsky, V.A. (1963) To question on deformations of the earth's crust. In: Gosgeoltechizdat, M., College Work: Problems of Planetary Geology, 149-221 (in Russian).
19. Veronnet Al. La forme de la Terre et sa constitution interne. Paris, 1914.
20. Tillo, A.A. (1889) Average height of land and average depth of sea in Northern and Southern Hemispheres. Dependence of Average Depths of Seas on Geographic Latitude, Izvestiya Russkogo Geograficheskogo obshchestva, 25 (in Russian).
21. Schulz, S.S. (1965) In: Gosgeoltechizdat, M., College Work: Basic Problems of Study of Quaternary Period, 1965, 147-150. (in Russian).
22. Voronov, P.S. (1965) Essays on morphometry laws of global relief of earth. Nauka, (in Russian).
23. Shpitalnaya, A.A. (1978) On importance of choice of reference system during investigation of non-stationary cosmic processes. In: College Work: Problems of Investigation of Universe, 7 (in Russian).
24. Shpitalnaya, A.A. (1971) To the question of destruction of filaments by flares. Solar Data, 2, (in Russian).
25. Sytinsky, A.D. (1966) Geomagnetisdm and aeronomy. 3 (in Russian).
26. Sytinsky, A.D. (1973) Uspekhy Phyzicheskikh Nauk (UFN), 3, 367 (in Russian).
27. Simpson, J.F. (1968) Science Letters, 3, 418.
28. Shpitalnaya, A.A., Vasilieva, G.Y. and Pystina, N.S. (1975) On possibility of action of gravitation waves on activity of earth and sun. In: College Work: Problems of Investigation of Universe, 4, 129-137 (in Russian).
29. Yakovlev, B.A. (1960) Astronomical collective work of Lvov State University. 3, 152 (in Russian).
30. Afanassieva, V.N. (1963) Geomagnetism and aeronomy. 3, 561 (in Russian).
31. Danjon, A., (1962) CR Academy Science. Paris, 254, 3058.
32. Gribbin, J. and Plagemaun, S. (1973) Nature, 243, 26. doi:10.1038/243026a0
33. Hessler, V.P. and Wescott, E.M. (1959) Nature, 184 (Suppl. 9), 627. doi:10.1038/184627a0
34. Baurov, Y.A. and Kopaev, A.V. (2002) Experimental investigation of signals of a new nature with the aid of two high precision stationary quartz gravimeters. Hadronic Journal, 25, 697.
35. Baurov, Y.A., Baurov, A.Y., Baurov, A.Y., Shpitalnaya, A.A., Abramyan, A.A. and Solodovnikov, V.A. (2006) Method of short-term forecasting heightened seismic danger of seismically dangerous regions. Invention, 2006110310 RF.
36. Baurov, Y.A., et al. (2007) Physics of atomic nuclei. 70, 1825; Yadernaya fizika, 70, 1875.
37. Baurov, Y.A. (2009) La Trama Svelata, La Teoria Byuon: dall'origine dell' Universo alla tecnlogia del futuro, passando per I'Uomo. Edizioni Polistampa.
38. Baurov, Y.A. and Malov, I.F. (2010) Journal of Modern Physics, 1, 17. doi:10.4236/jmp.2010.11003
39. Baurov, Y.A., Shpitalnaya, A.A. and Malov, I.F. (2005) International Journal of Pure & applied Physics, 1, 71.
40. Malov, I.F. and Baurov Y.A. (2007) Astronomy Reports, 51, 830. doi:10.1134/S1063772907100071

CHAPTER 18

DOES THE SUN TRIGGER EARTHQUAKES?

Carlos A. Vargas and Emanuel D. Kästle

Keywords: Solar Cycles; Earthquakes; Geomagnetism; Earthquake Prediction

INTRODUCTION

The issue of a link between the solar activity and the occurrence of earthquakes is still not quite understood. There are several theories, e.g. Tavares and Azevedo [1] propose that a change in shape of the magnetosphere of the Earth could cause variable pressure in the crust, triggering earthquakes. Other theories are, that space weather could be responsible for changes in the angular velocity of the Earth [2], or that induced magnetic currents might cause heat and thus lower the friction in faults [2,3]. Solar activity seems to correlate quite well with atmospheric data, so that it's been theorized that air pressure, wind and water level are the link between earthquakes and solar events [4,5].

This work concentrates more on the actual relationship between earthquakes and solar activity and treats the effects causing the correlation only in the aspect of geomagnetic field strength variations. The approx. 11-year solar cycles, represented by the amount of Sunspots are correlated with the magnetic field strength on three different places (Colombia, Germany and Alaska), so to see, if the field shows the same long-term periodicity. Furthermore different types of earthquake data are used to check the results from Tavares and Azevedo [1].

Citation: Vargas, C. and Kastle, E. (2012) Does the sun trigger earthquakes?. Natural Science, 4, 595-600. doi: 10.4236/ns.2012.428079.

DATA

All data mentioned in this section are available as daily and yearly means of the respective values.

Solar Activity

The International Sunspot Number, provided by the SIDC in Belgium [6] with worldwide contributing observatories, represents a good indicator for the solar activity. The cycles shown in Figure 1 of approximately 11 years can easily be recognized. During the solar maxima, the number of CMEs (Coronal Mass Ejections) increases. Those CMEs consist mainly of Electrons and Protons, but also heavier elements, originating in solar spots [7,8].

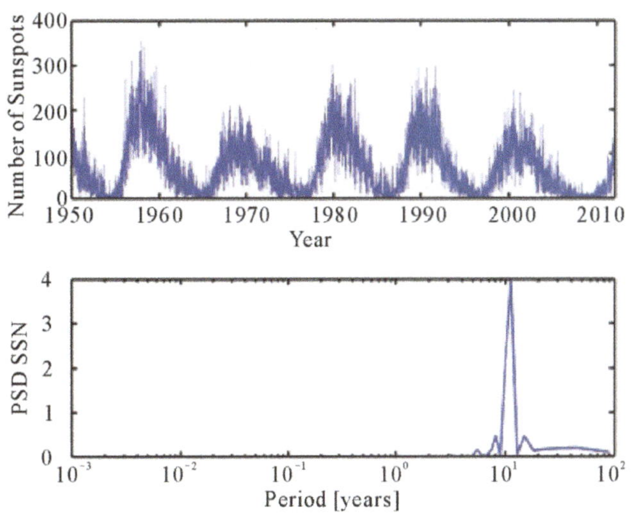

Figure 1: Above: Daily mean of the observed number of Sunspots showing the solar cycles between 1950 and 2010. Below: Spectral analysis shows a clear peak at an approx. 11 year period in the solar cycles.

Some of these ejections are directed towards the earth, called magnetic storms that cause e.g. the Northern Lights but also drastically compress the magnetosphere on the day-side of the earth and stretch it on the nightside [9].

Earthquake Occurrence

Regarding the seismic activity, the worldwide ANSS catalogue [10], the Centennial catalogue, as well as regional data are used to create a representative earthquake index. The data is restricted to events with a maximum depth of 40 km, assuming the postulated effect from the magnetic field to be limited to crustal depths. Furthermore only earthquakes of magnitude 5 and larger are used for the test statistics.

The worldwide ANSS catalogue is created by merging the master earthquake catalogues from contributing ANSS member institutions and then removing duplicate events, or non-unique solutions for the same event. For the study period the catalogue contains 39,343 events with the mentioned characteristics.

For comparison concerning large events also the Centennial earthquake catalogue, which is regarded complete down to magnitude 6.5 between the 1930's and today, is used [11]. The mentioned catalogue is a composition of different contributing catalogues giving a picture of worldwide significant earthquake events (Figure 2). There are 2081 earthquake registrations with magnitude greater 6.5 in the period of interest.

As Tavares and Azevedo [1] report regionally different effects and for better comparison with the magnetic field strength, the events (from the ANSS catalogue) around the 3 observatories (see following section) are separately taken into consideration. The regions can roughly be described as, Northern South America, Central Europe and Alaska.

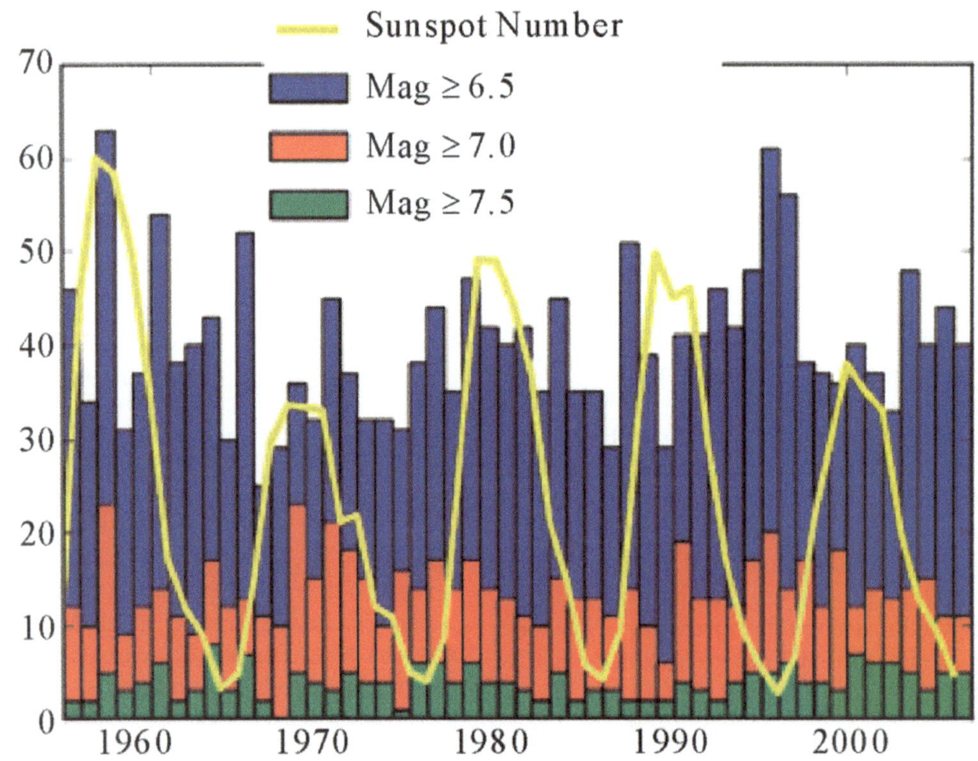

Figure 2: Comparison of the Centennial Earthquake Catalogue data with the sunspot number.

Magnetic Field Strength

Finally, in the attempt to prove Tavares and Azevedo [1] thesis, total field strength values obtained from the geomagnetic observatories in Fúquene (Colombia), Niemegk (Germany) and College (Fairbanks, Alaska, USA) are used. The Observatories represent not only different latitudes (Figure 3), but also different tectonic regimes. Alaska and Northern South America are mostly characterized by subduction earthquakes, whereas Central Europe represents in large parts stable continent with increasing activity towards the Mediterranean.

Especially the data from Fúquene at the beginning of the observation period (starting 1955) shows errors or missing values. In all three time series the data is revised and corrected for such flaws. However, as the goal of this study is to find the 11-year cycles of the solar activity, minor errors in the range of days in the appearance of the curves are not expected to influence the results.

METHODS

Cross-Correlation

The standard cross-correlation method is adopted to find patterns in the occurrence of Earthquakes that resemble the 11 year solar cycles. Thus, the results have to show significantly higher correlation coefficients than the null-hypothesis (cross-correlation of the Sunspot Index with a random time series). The standard numerical cross-correlation function of two vectors x and y of length N is given by:

Figure 3: Overview map of the geomagnetic observatories. The red dot indicates the position of the geomagnetic North Pole.

$$R_{xy}(m) = \sum_{n=0}^{N-m-1} x_{n+m} y_n \quad (1)$$

For comparison, the series x and y are normalized before using cross-correlation, so that their autocorrelations at zero lag (m = 0) is identically to 1.0. Moreover, possible trends were removed before applying the algorithm. The cross-correlation has also the characteristic of showing common periodicities in the correlated series, for example in the three magnetic field strength variations.

Cointegration

The cointegration method is a statistical method mostly used in economics to find dependencies in two or more time series. The basic concept can be expressed in the following manner (Definition from Sørensen [12]):

The two time series y t and x t are cointegrated if there exists a parameter α such that

$$u_t = y_t - \alpha x_t + \varepsilon \quad (2)$$

is a stationary process.

This definition applies to I(1) processes for x_t and y_t, which means they are integrated of the order 1, thus becoming stationary after differentiation. One also says the process has a unit root, i.e.

$$y_t = a\, y_{t-1} + \varepsilon_t \text{ for a = 1} \quad (3)$$

This is the equation of a random walk, so that at every time point the value of y_t depends on the value before and a random term. A standard unit root test would be e.g. testing the Null-Hypothesis of a = 1 (unit-root) versus a < 1 (stationary). For a = 1, the time series y_t is nonstationary and will be drifting from its initial value with increasing standard deviation. The series y_t and x_t are cointegrated if they are both following such a random walk, but having a common trend, i.e. they don't drift away too far from each other, such that there is always a relationship between the variables as in Equation (2).

The cointegration relationship is also valid for processes of higher order I(p) and there are methods to test cointegration for more than two variables, for example the Engle-Granger Cointegration Test, used in this article [13]. The basic concept of this test is a regression on the following relation (x_t^n being the n-th time series of same length as y_t):

$$y_t = \alpha x_t^1 + \beta x_t^2 + \cdots + \omega x_t^n + \varepsilon \quad (4)$$

After calculating the regression, the residuals are tested for a unit-root.

The big advantage of cointegration testing is that normal regression for uncorrelated traces can, in the case of random walks, yield results showing a correlation between the time series where in fact there is none. This phenomenon is called spurious regression.

RESULTS

Cross-Correlation

Figure 4 shows the results of cross-correlation between the Sunspot Number and the Earthquake and Magnetic Field Strength Data, respectively. A graphical comparison of the Centennial Data is also given in Figure 2. The maximum time lag in Figure 4(a) is restricted to ±2 years, assuming that the effects from high solar activity should be found in that time window. For comparison also the mean-cross-correlation-coefficient between 500 random series and the Sunspot Index is shown as a limiting line.

It becomes quite clear, that none of the series used shows a significant positive correlation. However, it seems like there is a negative correlation especially pronounced for the worldwide ANSS catalogue, which plots at exactly zero time lag with a coefficient of almost –0.4 (mean random + one standard deviation ≈ ±0.3).

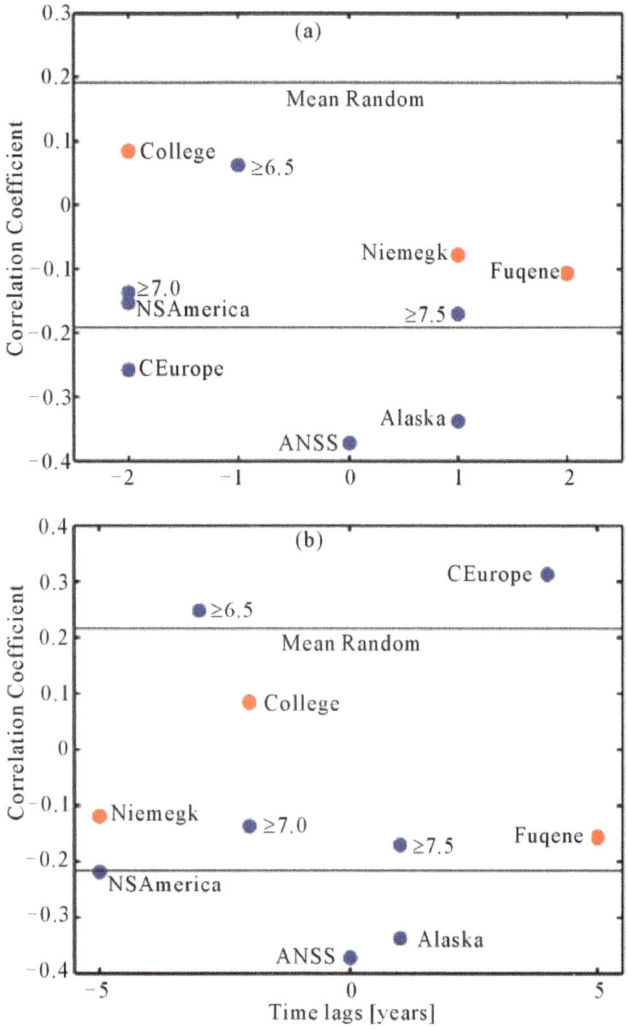

Figure 4: Maximum correlation coefficients for time lags ±2 (a) ±5 (b) years. Red: Sunspot Number and Magnetic Field Strength; Blue: Sunspot Number and Earthquake Data (≥x refers to Magnitudes from the Centennial Catalogue). Random Mean is the average correlation coefficient of 500 correlations between the Sunspot Number and a random series.

It is quite noticeable, that there seems to be stronger negative correlation. The picture changes only slightly when considering time lags of up to five years (Figure 4(b)). However, a time lag of five to six years means, as well as negative values, anti-cyclicality, i.e. more earthquakes in solar minimum phases. The maximum positive correlation coefficient can be found in the comparison of the Sunspot Number with the Earthquake Data from Central Europe, with a value slightly above 0.3 and a time lag of four years. A more detailed picture of this particular case is given in Figure 5(b).

Furthermore, common periodicity in the magnetic field strength data is found according to Figure 6. The most prominent feature is an approx. 20 year cycle that seems to be a stable feature of all three correlations.

Yet another approach is the comparison between the magnetic field strength and the regional earthquake data.

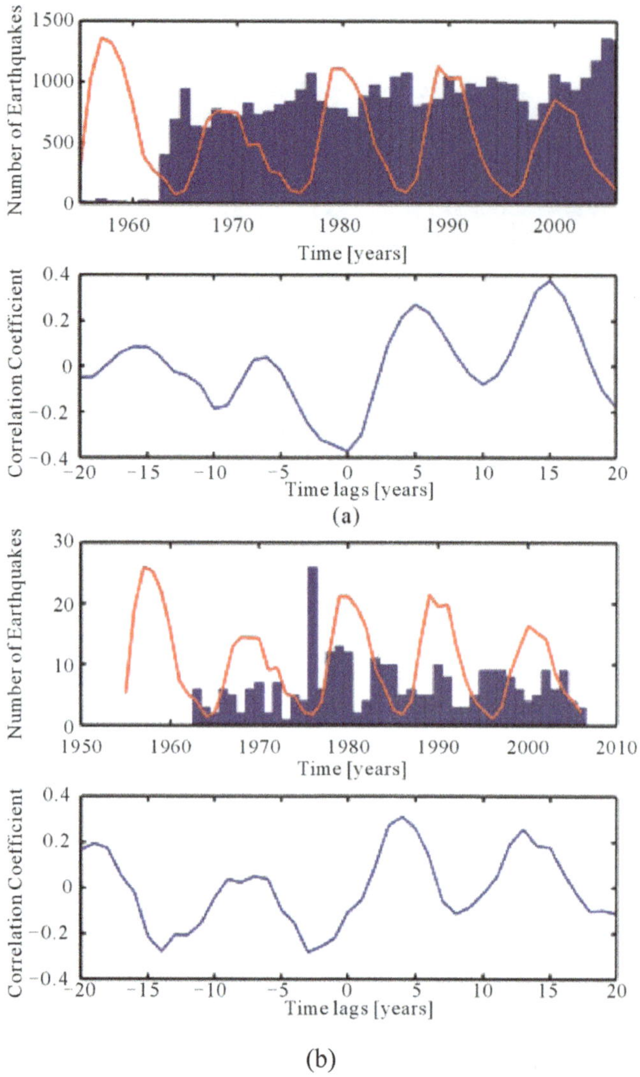

Figure 5: (a) Worldwide earthquake occurrence from ANSS catalogue ((b) for Central Europe) and Sunspot Number (red line). The correlation coefficients for up to 20 years time lag are given below.

It shows the highest correlation for the case of Niemegk and Central Europe. The correlation coefficient is about 0.4, whereas correlation with a random earthquake profile plus one standard deviation is about 0.3. However, Figure 7 shows that the correlation almost exactly resembles the input values of field strength at Niemegk. This also means that there is equally high correlation not only for zero, but also for about 18 years time lag, which is difficult to explain.

Cointegration

As mentioned in Section 3.2, in order to adopt the cointegration method it is important, that the analysed time series have a unit root. Intuitively one already suspects the geomagnetic records to meet the criteria for cointegrated time series.

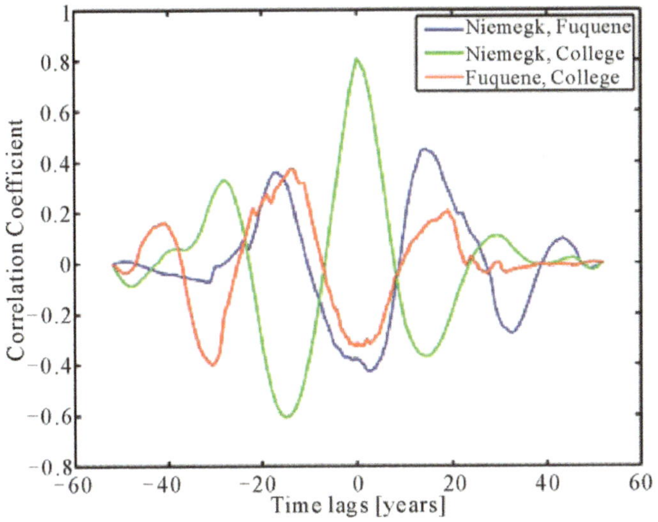

Figure 6: Correlation of the magnetic field strength recorded at Fúquene (Colombia), Niemegk (Germany) and the College Observatory (Alaska).

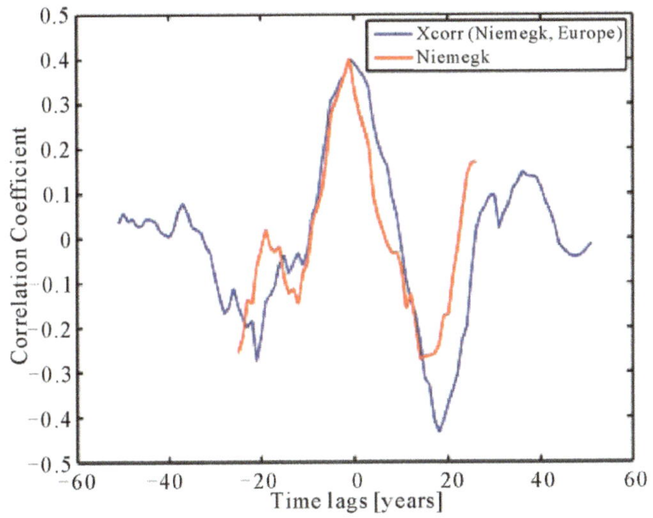

Figure 7: Correlation of the magnetic field strength recorded at Niemegk (Germany) and the Earthquake occurrence in Central Europe from the ANSS catalogue. The red line shows the curve form of the magnetic field strength in Niemegk (de-trended).

The series follow a random walk (secular variations) with some degree of individual, random freedom (magnetic storms, currents in the ionosphere...), but undoubtedly their general behaviour is linked to each other.

On the other hand, the Sunspot Number does not show those random walk characteristics, resembling a stationary process with a rather strict periodicity. Also the earthquake occurrence is stationary, following (in a basic model) a Poisson Distribution.

The test statistics show indeed a cointegration relation between the three magnetic field strength records. The graph for the Residuals u_t in **Figure 8** is given by

$$u_t = F_{College} - \alpha_1 F_{fúquene} - \alpha_2 F_{Niemegk} - c \qquad (5)$$

where F_x are the records from the observatories, α_i constant factors and c is a constant that shifts the curve to zero mean.

Applying the cointegration with the Sunspot data, although it contains no unit root, results either in an associated factor of α almost zero, or, in the case of SSN = y_t in Equation (4), in the residuals being identically to the Index. Although this is not exactly a valid test, this could mean that Sunspot Number and magnetic field strength (in a relation of year-long cycles) are independent. If there was a common trend or cyclicality the curve form of the Sunspot Index wouldn't appear in the Residuals.

As the earthquake data is linked to the same problem, further cointegration testing is omitted.

DISCUSSION

First of all it's notably that the geomagnetic field strength does not show a clear correlation with solar activity (Figure 4). This fact, however, is expected, given the data used in this study.

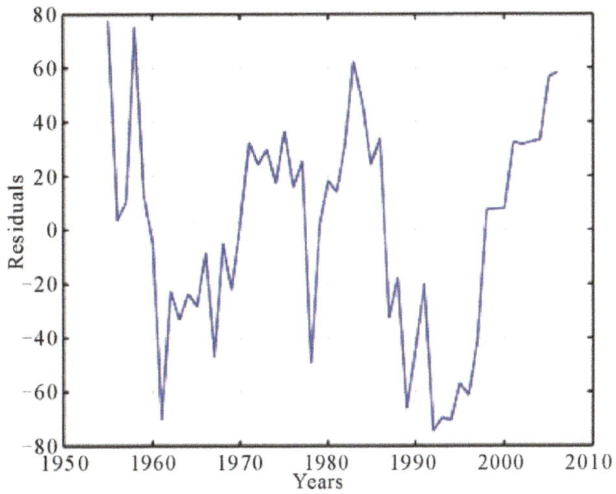

Figure 8: Residuals from the cointegration analysis according to Equation (5). $a_1 = 0.422$; $a_2 = 1.440$. The Residuals basically show the difference in the three magnetic field strength records, whilst common features are cancelled out.

During solar maxima the number of magnetic storms increases considerably, yet, those are effects, typically in the range of hours to few days. Hence, the long-term magnetic field strength is not expected to be influenced strongly by these disturbances. The correlation between field strength and regional earthquake events doesn't resolve this issue (Figure 7), but is rather inconclusive, although the correlation coefficients are slightly higher than random correlation.

The cointegration analysis including the Sunspot Number, yields the same result, as the residuals resemble the variations in solar activity. This could mean that there is no direct link between the two values, i.e. no cointegration, although the results cannot be seen as proof, as the basic prerequisite condition of a unit root has been ignored.

Although the three geomagnetic observations are cointegrated, the approx. 20 year periodicity in Figure 6 is possibly due to geomagnetic jerks. These are non-predictable impulses or inversion of the trend-direction of the declination of the magnetic field [14]. Those variations, that are still not very well understood, have roughly shown that periodicity in the last 50 years. However they include no direct information for this report.

Concerning the earthquake data, there also seems to be no strong relation between their occurrence and the solar activity (Figure 4). It's remarkably that more anti-cyclical correlation is found, although one has to keep in mind, that the regional catalogues (CEurope, Alaska, NSAmerica) make part of the ANSS catalogue and thus don't really show independent results. Taking greater time shifts between the series into account, there is data plotting above the mean random-correlation limit. But not only the time lags are difficult to explain, but it's also still more likely for those values to plot randomly in this field.

Tavares and Azevedo [1], on the other hand, state a positive result in their study regarding the connection between the solar cycles and the earthquake activity. Positive in the sense of a existing relationship and in the sense of higher earthquake activity during solar maxima. Firstly, they examine historic events, that coincide with the Maunder (1645-1720) and the Dalton (1790-1820) minima. They find an overall lower number of events during those two periods, though the activity in tectonic regions of mostly strike-slip faults is slightly increased. Tavares and Azevedo [1] propose that high solar activity could compress the magnetic field, exerting a pressure on certain tectonic settings, whereas during a minimum, the energy is liberated in zones characterized by strike-slip faulting or subduction zones.

Their observations are backed by observations during the last 50 years, similar to this study. The last maximum (around 2000), earthquakes seem to have occurred more frequently on different tectonic plates. This result, however, is not universal for all parts of the earth and not during all of the 5 maxima between 1950 and 2005. Their observations during the modern period seems exclusively based on graphical comparison and not one number is shown in this part. Also, the comparison of occurrence during the two historic minima vs. the whole time period are hardly statistically robust, as there are only small deviations from a normal pattern observed. Those effects could probably as well be explained by natural random behaviour.

The results of this work rather contradict the observations from Tavares and Azevedo [1], as the data presented in this article generally shows no or anti-cyclic correlation. Yet it is interesting that—though not predominant—they also describe in their work an inverted relationship between seismicity and solar activity. Comparing the two results, in the authors' opinion these effects seem rather arbitrary and it's highly questionable if not only the choice of the earthquake catalogue is the crucial factor in finding a positive or negative correlation.

CONCLUSIONS

The influence of the solar activity on earthquakes proves to be an elusive phenomenon. Magnetic storms caused by CMEs are supposed not only to affect modern technology such as GPS, but also the solid Earth's crust, triggering earthquakes. As such events happen considerably more frequently during solar Sunspot Maxima, it is of interest, whether earthquake occurrence resembles these cycles.

This study doubts the results presented by Tavares and Azevedo [1], as the only clear signs of correlation found, predict a negative correlation between earthquakes and solar activity. Yet, future studies could come to more significant conclusions, by, for example, correlating directly CME events with earthquakes. In any case, it is important to take the complexity of an earthquake process into consideration, as there are many factors influencing the behaviour of fault zones and if at all, solar activity represents only one part of the puzzle. Thus, effects are presumably small, so all the more there is need for robust statistical methods. Otherwise, this topic stays in its niche, widely ignored by the scientific community.

ACKNOWLEDGEMENTS

We want to thank the Colombian National University and the University of Potsdam, for making this project possible. For providing the data we express our thanks to the Geomagnetic Observatory in Fúquene (Instituto Geográfico Agustín Codazzi—IGAC), Niemegk (GFZ Potsdam) and College (USGS), as well as the NCEDC, the SIDC and the USGS.

REFERENCES

1. Tavares, M. and Azevedo, A. (2011) Influences of solar cycles on earthquakes. Natural Science, 3, 436-443. doi:10.4236/ns.2011.36060 [Paper reference:9]
2. Simpson, J.F. (1968) Solar activity as a triggering mechanism for earthquakes. Earth and Planetary Science Letters, 3, 417-425. doi:10.1016/0012-821X(67)90071-4 [Paper reference:1]
3. Han, Y. Guo, Z., Wu, J. and Ma, L. (2004) Possible triggering of solar activity to big earthquakes (Ms $\geq$ 8) in faults with near west-east strike in China. Science in China Series G: Physics, Mechanics and Astronomy, 472, 173-181.
4. Gousheva, M.N., Georgieva, K.Y., Kirov, B.B. and Antanasov, D. (2003) On the relation between solar activity and seismicity. IEEE Proceedings of International Conference on Recent Advances in Space Technologies, Istanbul, 11-13 June 2003, 236-240.
5. Odintsov, S., Boyarchuk, K., Georgieva, B., Kirov, B. and Antanasov, D. (2006) Long-period trends in global seismic and geomagnetic activity and their relation to solar activity. Physics and Chemistry of the Earth, Parts A/B/C 31.1-3, 88-93.
6. Solar Influences Data Center (SIDC) (2012) International sunspot number.http://sidc.oma.be/ [Paper reference:1]
7. Kunow, H. Crooker, N.U. and Linker, J.A. (2007) Coronal mass ejections. Springer, Berlin.
8. NASA (2012) Coronal mass ejections. http://helios.gsfc.nasa.gov/cme.html
9. Dandouras, I.S., Rème, H., Cao, J. and Escoubet, P. (2009) Magnetosphere response to the 2005 and 2006 extreme solar events as observed by the Cluster and Double Star spacecraft. Advances in Space Research, 43, 618-623. doi:10.1016/j.asr.2008.10.015 [Paper reference:1]
10. Northern California Earthquake Data Center (NCEDC) (2012) ANSS worldwide earthquake catalogue. http://www.ncedc.org/anss/ [Paper reference:1]
11. Engdahl, E.R. and Villasenor, A. (2002) 41 global seismicity: 1900-1999. International Geophysics, 81, 665-690. doi:10.1016/S0074-6142(02)80244-3 [Paper reference:1]

12. Sørensen, B.E. (2005) Cointegration. http://www.uh.edu/~bsorense/coint.pdf [Paper reference:1]

13. Engle, R.F. and Granger, C.W.J. (1987) Cointegration and error correction: Representation, estimation and testing. Econometrica, 55, 251-276. doi:10.2307/1913236 [Paper reference:1]

14. British Geological Survey (BGS) (2012) The Earth's magnetic field: An overview.http://www.geomag.bgs.ac.uk/education/earthmag.html

CHAPTER 19

STUDY ON CORRELATION OF TIDAL FORCES WITH GLOBAL GREAT EARTHQUAKES

Youjin Su, Hong Fu and Hui Hu

Keywords: Tidal Force; Earthquake; Run Test; Seismic Belt

INTRODUCTION

The occurrence of a moonquake is closely affected by tidal forces on the moon generated by the Earth and the Sun. This has already been studied and confirmed [1,2]. As for the correlation between the celestial tidal forces and earthquakes, it remains although the research history is very long and many papers have been published it remains controversial problem [3]. The studies of this subject carried out both in China and elsewhere can be roughly divided into two categories:

- The time sequence of certain tidal component is calculated. The statistics of phases corresponding to the earthquake-generating times or the earthquake-generating frequencies are evaluated by different methods [4,5], alternatively, the correlation between the earthquakes and tidal-component is directly found values from the statistics of the tidal periodicity of the earthquake-generating times [6].
- The tidal-force indeed shear stress in the sliding direction of a causative fault is calculated, mainly based on the angle of stress. The relation between the earthquakes and tidal effects is consequently discussed from the result in different aspects [7,8].

Different and even contradicting conclusions have been reached by adopting different methods in these papers. In this paper the method of run test is used. It is different from the methods is previous studies. The relation between the great earthquakes over the world and the celestial tidal forces is thus studied in the perspective of entire time sequence, leading to the conclusion that they do relate to each other. Another difference is that not the earthquakes in a single region but all earthquakes in the world are studied, according to the seismic belts or regions. The phenomenon that earthquakes generally happen with a period of 18.6 years in the main seismic regions of the world suggests that the lunar node tide (i.e. the tidal effect from the Moon at lunar node) should be an important astronomical factor which influences the seismic belts or regions hosting potential new earthquakes.

CELESTIAL TIDAL FORCE HAS CLOSE RELATION TO THE OCCURRENCE OF AN EARTHQUAKE

Tidal Force

The resultant of forces (vector difference) of the lunisolar gravitation exerted on the unit mass of the Earth and the inertial centrifugal force produced from the revolution of the Earth round the common center of mass of the Moon and Earth is called the tide generating force, that is to say, the difference between the gravitation exerted on some part of the attracted celestial object and that exerted on the part with the same mass at the center of the object is regarded as the tide generating force (Fang, 1984; Fu 1985; Melchior, 1978). From this the calculation formula of the tide generating force exerted on a particle with a unit mass can be obtained as follows [9]:

Citation: Y. Su, H. Fu and H. Hu, "Study on Correlation of Tidal Forces with Global Great Earthquakes," International Journal of Geosciences, Vol. 3 No. 2, 2012, pp. 373-378. doi: 10.4236/ijg.2012.32041.

$$F_r = \frac{KMr}{D^3}(3\cos^2 Z - 1)$$

and

$$F_s = \frac{3}{2}\frac{KMr}{D^3}\sin 2Z \tag{1}$$

here F_r and F_s are respectively the vertical and horizontal components of the tide generating force, with F_r being taken as negative when pointing to the geocentre and F_s being positive when pointing to the celestial object, K is the gravitational constant, r is the earth's radius, and M is the mass of the celestial object. D is the distance from the object to the geocentre. Z is the zenith distance of the celestial object. From the above-mentioned formula it can be calculated that though the solar mass is 2.7 million as much as the lunar mass, the distance between the Moon and Earth is only 1/390 of the distance between the Sun and Earth, and therefore the tide raising force of the Moon is 2.25 times as much as that of the Sun, with both belonging to the same order of magnitude while the tidal force on a point of the earth's surface exerted by each of the other planets is several orders of magnitude less than that exerted by the Moon, and therefore the tidal forces of the other planets can be neglected.

Run Test

One of the most strong points of the run test is that the distribution of the stochastic variable need not be taken into account and it is a nonparametric method. A sequence is considered which consists of n observed values of the stochastic variable x. The variable x is divided into two kinds which repel each other, and can simply be expressed with the positive sign (+) and negative sign (–). The simplest example is the sequence, which is obtained by throwing a coin, with each observed value being either positive (+) or negative (–). The second example, assuming that a sequential measured value is x_i (where i = 1, 2, 3, n) and their mean value is x_0, and then it is (+) when $x_i \geq x_0$ and it is (–) when $x_i < x_0$. The third example, a double sequence consists of two serial observations xi and yi (where i = 1, 2, 3, n), where each pair of the observations are (+) when $x_i \geq y_i$ or (–) when $x_i < y_i$. In any case the observed sequence obtained by the positive sign (+) and negative sign (–) can be written as:

+ + –   + + –   + + + – +   – – +   – –   +   – – –

1 2 3   4 5   6 7   8 9   10   11   12

A run is defined as a segment of the sequence to include the same category of outcomes as many as possible. There are N = 20 of runs observations in the above illustration, with the number of runs being r = 12. The number of an observational sequence shows whether the outcomes of the stochastic variable are independent or correlated. The run test can be directly used for analyzing data and a test for the existence of systematic pattern of a single observational sequence may be given as follows. Let the assumption be that there is no clear pattern in a sequence which consists of N independent observations of the same stochastic variable. Under the assumption that the observations with + are equal to those with −, the sampling distribution of the number of runs in the sequence is determined by the run-distribution theory. From the comparison of the observed number of runs to the opened interval ($r_{n,\,1-\alpha/2}$, $r_{n,\,\alpha/2}$), the assumption can be tested under any required level of significance α, where n = N/2. If the number of runs lies inside this opened interval the assumption is accepted within the confidence level. Otherwise, the assumption is rejected as not having sufficiently confident evidence [10].

Test of the Influences on Great Earthquakes from the Celestial Tidal Forces

Engdahl and Villasenor [11] collected the statistics of seismic records of all countries in the world and found, as a result, that the records of earthquakes of M ≥ 7.0 have been complete and reliable since 1900. In order to determine whether the occurrence of an earthquake is correlated with the celestial tide generating force or not, and to let the conclusions obtained from analysis be reliable and reveal the true characteristics of the matter, all the earthquakes events analyzed in this article are selected from the "Global seismicity: 1900-1999" with Engdahl and Villasenor (2002). There occurred 1553 earthquakes of M ≥ 7.0 in the whole world in the time interval from 1900.0 to 2000.0, and there were 67 earthquakes of M ≥ 8.0 during that same interval. The analysis in the present paper is carried out for all the earthquakes with M ≥ 7.0. It is common knowledge that the energy E released by an earthquake and its magnitude M have the following relation [12]:

$$E = 10^{11.8+1.5M} \tag{2}$$

here M correspond the magnitude determined by surface wave. Unit of E is erg. Form which it is not difficult to know that if earthquake is one order greater than another in magnitude the energy released by the former will be 30 times greater than the latter. Thus as far as the energy released by earthquakes is concerned, it is mainly concentrated on great earthquakes. The destruction caused by earthquakes mainly also comes from the great earthquakes. It is considered that all over the word the earthquakes of M ≥ 8.0 are too few while those of M < 7.0 are too many. In addition, there are some earthquakes of M < 7.0 caused by artificially created factors (such as the nuclear explosion, construction of large dams and reservoirs, etc.), and the reliability of the catalogue of the earthquakes of M < 7.0 is not good as that of M ≥ 7.0. Based on these reasons all the earthquakes of M ≥ 7.0 are selected by us as the object for analyses. As for analyses of every year we take the frequencies of the earthquakes of M ≥ 7.0 in that year and do not compute their energies. Based on the time of occurrence of an earthquake and the longitude and latitude of the epicenter, the celestial tidal force exerted on the epicenter when the earthquake occurs is calculated and the run tests are respectively given to three sequences of the tidal force (i.e., the resultant, vertical component and horizontal component of the tide generating force), or the runs at which the earthquake occurs near the extreme and non-extreme values are calculated. It is found from the results that for the resultant, vertical component and horizontal component, the runs at which the earthquake occurs near the extreme and non-extreme values are respectively 1030, 1046 and 510. The null hypothesis is taken as that the occurrence of the earthquake has no relation to the celestial tidal force, then for the level of significance $\alpha = 0.05$, the acceptance region of the assumption $n = N/2 = 776$ should lie in the opened interval (656, 906). Evidently, all the runs of the above mentioned three sequences lie outside this opened interval, and therefore the null hypothesis is rejected. In other words, the occurrence of an earthquake and the tidal force have obvious potential tendency, i.e. the occurrence of an earthquake has some relation to the celestial tidal force.

SEISMIC ACTIVITIES OF PRINCIPAL SEIMIC BELTS IN THE WHOLE WORLD AND LUNAR NODE TIDE

Existence of Period of 18.6 Years in Seismic Activities in Principal Seismic Belts in the World

As already pointed out, among various celestial tidal forces the lunar-origin tidal force is the greatest, and the next important is the solar-origin tide fore. Therefore the tidal force of the lunar origin becomes the prior subject in our study. During the studies of the characteristics of seismic activities, many researchers have found that the earthquakes in a certain region or belt have periodicity in activities [13,14]. There were 5 active periods of strong earthquakes and 5 calm periods in China in the 20th century [15] with the period being 18.6 years [16]. There may be similar periodicity of earthquakes over the entire world as well. If the earthquakes of $M \geq 7.0$ in the entire in the 20th century are collectively analyzed, there is no evidence for such periodicity. However, the period of 18.6 years apparently exists for the earthquakes in a limited region associated with boundaries between tectonic plates. The phases of the cycles of earthquake frequentcies in different seismic belts or regions are different. This causes the period of 18.6 years not seen in the data covering the entire world. The differences in phases are understandable. Because the occurrences of earthquakes are related to certain geological structures, only in the areas with the same or similar seismotectonics, their seismic activities can have the same or similar laws. Therefore, the earthquakes in different regions need to be studied separately, otherwise little pattern of the seismic activities can be found. It is the very reason why the regularity of the seismic activities has not been revealed from the analyses of the combined data of all earthquakes of $M \geq 7.0$ over the world. As pointed out by Chen Yong, regularity and randomness coexist in seismic activities, of which above 90% are distributed near the boundaries between tectonic plates [17]. The theory of tectonic plates holds that the crust of the Earth consists of some rigid blocks or plates, the interior of a plate is relatively stable, many phenomena, such as submerging, colliding and shearing, occur at the boundaries between the plates and the mutual movements between the plates cause the formations and occurrences of earthquakes. A plate drifts under the joined influence of the centripetal force from the Earth rotation and the mantle plume heat convection. The Celestial tidal forces periodically trigger the release of the stress energy accumulated along the boundaries between plates, in the forms of the seismic energy and volcanic energy [18]. Since the major earthquakes are mainly concentrated near the boundaries between the tectonic plates [19,20], we divided the seismic belts of the major earthquakes over the world along the plate boundaries into 11 regions (Figure 1). We find for separate analysis that the seismic activities of each region have a period of 18.6 years. There are 1416 earthquakes of $M \geq 7.0$ in the 11 regions in total making up 91.2% of all the earthquakes of $M \geq 7.0$ of the world. The remaining 137 earthquakes of $M \geq 7.0$ did not appear to be in the 11 regions.

Lunar Node Tide Is an Important Earthquake-Pregnant Astronomical Factor in the Principal Seismic Belts and Regions in the World

The period of 18.6 years equals to the period of the Earth main nutation term, and also of the motion of the lunar node. The Moon revolves around the Earth in its orbit, while the lunar orbit shifts under the action of the solar gravitation. Therefore, the node of the lunar orbit on the ecliptic (i.e., the plane of the Earth's orbit) moves westward, causing the ecliptic longitude of the lunar ascending node to steadily decrease. The ecliptic longitude of the lunar ascending node is expressed by the equation [21]:

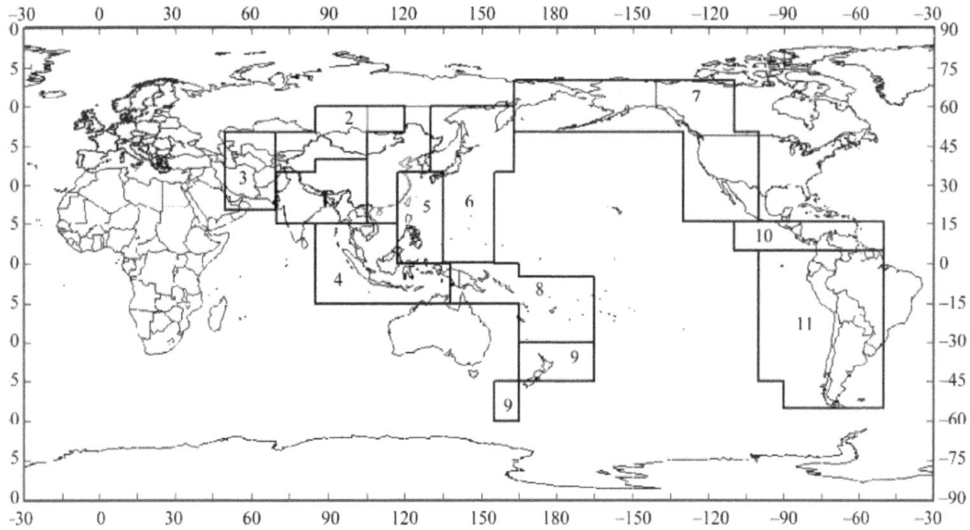

Figure 1. The studied region of 18.6 years recurrence of the great earthquakes in the world.

$$\Omega = -125°02'40.280'' - 1934°08'10''.539T + 7''.455T^2 + 0''.008T^3 \quad (3)$$

where T is the number of Julian centuries elapsed from J2000.0. This equation can be used to calculate the longitude of the lunar ascending node for the moment of any earthquake. The distribution of the ecliptic longitudes of the lunar ascending node at the time of earthquakes in the region under study can thus be obtained. This allows us to find potential concentration patterns of the ecliptic longitude of the lunar ascending node at the time of occurrence of the earthquakes in each studied region, i.e. including the periods of the seismic activities there.

By taking the uniform distribution as the null hypothesis we carry out the $\chi 2$ test for each region. A sampling point is taken in the active period or in the calm period, and the degree of freedom evidently equals to one. Under degree of freedom is equal to 1, if $\chi 2 > 3.841$, the null hypothesis can be rejected with the 95 percent this confidence level, and the periodicity of earthquakes to concentrate within the mentioned active period is highly likely. The test of results show that the values of $\chi 2$ are all greater than 3.841 as listed in Table 1. Therefore, the seismic activities in each region studied have significantly shown a period of 18.6 years. The interval of strong seismic activities N is associated with an opened interval of the ecliptic longitude of the lunar ascending node in the Table 1.

Table 1: Statistic results of seismic recurrence of 18.6 years of the principal seismic regions in the whole world

No. Studies region	Total sample	Seismic active interval		
		Sample	N	(°)
1) Himalayas and its region	48	45	(260 180)	9.000
2) Mt.Tianshan and Baikal	45	37	(0 270)	4.900
3) Eastern Alps plate	19	15	(0 200)	5.221
4) Nalay Penisaula-Sund Islands	110	86	(70 330)	3.958
5) Weslern Philipping sea plate	187	162	(150 80)	8.480
6) Philippine sea-Northeast Japan	236	168	(130 10)	16.041
7) Aleut-North Mexico	132	73	(0 250)	6.303
8) New Guinea-Cumtuok	371	355	(350 330)	7.851
9) South Cumtuok	30	26	(330 200)	8.244
10) Central America Caribbean	71	59	(60 330)	4.187
11) West of southern America	167	126	(350 230)	9.674
Total	1416	1152		

Total sample is the time of occurrence of an earthquake in the studies region from 1900.0 to 2000.0; N is an opened interval of strong seismic activities with the ecliptic longitude of the lunar ascending node; Sample is the time of occurrence of an earthquake in the opened interval N.

The tidal forces of solar and lunar origins not only cause the sea tides on the Earth surface but also deform the elastic Earth mantle, consequently redistribute the matter and change the gravitational field of the Earth [22,23]. The tidal force of the lunar origin is 2.25 times as much as that of the solar origin. During the steady westward motion of the lunar ascending node, the obliquity between the lunar orbit and to the ecliptic varies in the range 4°57' to 5°19', thereby causing the obliquity between the lunar orbit and the Earth equator to vary, in the range18.4° to 28.8°, with a period of 18.6 years. The variation in the obliquity between the lunar orbit and the Earth equator translates into the variation in the range of the declination of the lunar motion in a tropical month and ultimately the latitude range of strong tidal effects of lunar origin on the Earth surface, thereby controlling the range of motion of the Moon in the equatorial zone and evidently controlling the latitude zone of the tidal force on the Earth's surface affected by the Moon. Therefore, the influence of the inclination of the Moon's path with the equator results in the tidal force exerted by the Moon on the Earth's surface.

It is well known that the pattern of seasons in the southern hemisphere of the Earth is opposite to that of the northern hemisphere due to the difference between the amounts of solar radiation received (which in turn depend upon the incident angles of the sunlight beams), e.g., when there is a hot summer in the southern hemisphere, there is a severe winter in the northern hemisphere (Pen & Lu, 1983). The most fundamental reason for this phenomenon is that there exists the obliquity of the ecliptic from the equatorial plane of the Earth. Simi larly the obliquity of the Lunar orbit from the terrestrial equatorial plane is probably the astronomical factor causing the latitude dependence of the seismic activities.

DISCUSSION

- A new method, which is to study earthquakes in separate regions and employs run tests, is adopted in this paper. If the earthquakes of the entire world are considered collectively, there is no evidence of the period of 18.6 years. The situation changes little if the earthquakes in apparently ill-defined regions are studied. The key to the question is to divide earthquake regions reasonably. Through the run tests of the occurrences of earthquakes near the extreme values and non-extreme values of the celestial tidal forces we reach the conclusion that the occurrences of earthquakes are related to the celestial tidal forces. We thus emphasize the importance of adopting proper regions of seismic activities and using the run tests in probing the influences of tidal forces on earthquakes.
- The contribution of astronomical factors on the earth is a global effect. However, at different time one of the astronomical factors plays a role in different region on the earth. Response regions are confined to the area

where potential movement in the atmosphere and crust is highly unstable. Therefore natural calamities induced by astronomical factors are regional effects [24].

- The variation of the obliquity of the lunar orbit causes variation in the tidal force of the lunar origin. The amplitude of the tidal force of the lunar node is smaller than that of the tidal force at the semidiurnal diurnal, or leading to semi-monthly tides (Gao, 1997). Studies have already verified that the tidal forces of the semidiurnal, diurnal, and semi-monthly tides play triggering roles in the earthquakes in some regions [25-30]. According to the theory of dissipative structures, when the earthquakeformation and energy accumulation have already been in an extreme unstable state, which is far away from the equilibrium, any small fluctuation can be enlarged to cause the giant fluctuation of the geological system, thereby inducing an earthquake [31]. The so-called trigger means that an earthquake may be triggered at any moment when the earthquake-pregnency has arrived at the moment of occurrence of the earthquake. So that the lunar node tide may not play a trigger role in the occurrence of an earthquake or may and probably it participates in the earthquake-pregnant activities, or the longterm action of the lunar node tide joins the earthquakepregnant process in the seismic belt and the accumulative effect of its action is added to the accumulation of the regional energy and to deviate the area from slowly stable [32,33]. The details of such modulation mechanism need to be studied in future.
- The period of 18.6 years in the seismic activities is only statistical. From Table 1 it can be clearly seen that except for the phase differences, the patterns of cycles of different regions are also different: some have longer active times and some have considerably short active times. An earthquake is an extremely complicated process, as it is affected by many factors internal and external to the Earth's (the internal factors are more important, since the external factors affect through the internal factors), and therefore, earthquakes cannot have exact periodically.

ACKNOWLEDGEMENTS

We heartily thank for the referee's valuable comments. In response to our request, Professor E. R. Engdahl of the University of Colorado at Boulder (USA) sent us twice their paper and computer readable catalogue of global earthquakes, which is adopted in this paper. He also gave us tremendous help in the process of this work. Dr. Chen I-wan, British (ancestor Chinese), Advisor of Committee of Natural Hazard Prediction of China Geophysics Society gave us valuable discussion and help. We express our heartfelt thanks to these colleagues.

REFERENCES

1. N. R. Goutly, "Tidal Triggering of Deep Moonquakes," Physics of the Earth and Planetary Interiors, Vol. 9, No. 1, 1979, pp. 52-58.
2. M. N. Toksöz, "Lunar and Planetary Seismology," Reviews of Geophysics and Space Physics, Vol. 13, No. 3, 1975, pp. 306-311. doi:10.1029/RG013i003p00306
3. X. P. Wu, et al., "Statistical Analysis of Tidal Stress Effect on Seismic Faults," Chinese Journal of Geophysics, Vol. 42, 1999, pp. 65-74.
4. S. Shlien, "Earthquake-Tide Correlation," Geophysical Journal. Royal Astronomical Society, Vol. 28, No. 1, 1972, pp. 27-34. doi:10.1111/j.1365-246X.1972.tb06108.x
5. H. Tsuruoda, M. ohtake and H. Sato, "Statistical Test of the Tidal Triggering of Earthquakes: Contribution of the Ocean Tide Loading Effect," Geophysical Journal International, Vol. 122, No. 1, 1995, pp. 183-194. doi:10.1111/j.1365-246X.1995.tb03546.x
6. P. R. Du, "18.6 Years Seismic Cycle and the Preliminary Exploration for Its Cause," Chinese Journal of Geophysics, Vol. 37, No. 3, 1994, pp. 362-369.
7. T. H. Heatön, "Tidal Triggering of Earthquakes," Bulletin of the Seismological Society of America, Vol. 72, No. 6, 1982, pp. 2181-2200.
8. M. Souriau, A. Souriau and J. Gagnepain, "Modeling and Detecting Interactions between Earth Tides and Earthquakes with Application to an Aftershock Sequence in the Pyrenees," Bulletin of the Seismological Society of America, Vol. 72, No. 1, 1982, pp. 165-180.
9. G. B. Peng and W. Lu, "The 4th Sort of Natural Factors of the Climate," Science Press, Beijing, 1983.
10. J. S. Bendat and A. G. Piersol, "Random Data: Analysis and Measurement Procedures," 1971.
11. E. R. Engdahl and A. Villasenor, "Global Seismicity: 1900-1990," In: H. K. Lee, H. Kanamori, P. C. Jennings, et al., Eds., International Handbook of Earth- quake and Engineering Seismology, Part A, Academic Press, Amsterdam, 2002. doi:10.1016/S0074-6142(02)80244-3
12. Z. L. Shi, "The World's Earthquake Catalogue," Seismolog Press, Beijing, 1981.
13. S. Z. Hong, "Application of the Optimum Section to Seismic Instalments," Northewest Seismology Journal, Vol. 6, No. 1, 1984, pp. 49-57.
14. G. M. Zhang and L. M. Gen, "Computer Simulation Study of the Recurrent Activities of the Strong Earthquakes in Chinese Mainland," Earthquake Research in China, Vol. 9, No. 1, 1993, pp. 20-22.
15. B. X. Gao, "Principle of Astro-Geodynamics," Science Press, Beijing, 1997, pp. 31-50.
16. H. Hu and X. M. Li, "Research on Correlation of Positions of Celestial Objects with Earthquakes," Natural Hazards, Vol. 23, No. 2-3, 2003, pp. 339-348.
17. Y. Chen and P. J. Shi, "Natural Disaster," Beijing Normal University Press, Beijing, 1996, pp. 70-71.
18. X. C. Jin, "Theory of Structure of the Plates," Shanghai Science and Technology Press, Shanghai, 1981, p. 310.
19. GINSB, "Seismic-Tectonic Map of Europ-Asia," Map Press, Beijing, 1981.
20. G. W. Moore, "Plate-Tectonic Map of the Circum-Pacific Region," The America Association of Petroleum Geologists, 1982.

21. D. D. McCarthy Ed., "IERS Standards," IERS Technical Note 3, Observatoire de Paris, 1989, pp. 1-77.

22. P. Melchior, "The Tides of the Planet Earth," Pergamon Press, Oxford, 1978.

23. J. Fang, "Solid Earth Tides," Science Press, Beijing, 1984.

24. Z. J. Guo, et al., "Earth—Atmosphere Coupling and Natural Disasters Forecast," Seismological Press, Beijing, 1996.

25. F. W. Klein, "Earthquakes Warms and the Semidiurnal Solid Tide," Geophysical Journal. Royal Astronomical Society, Vol. 45, No. 2, 1979, pp. 245-295.

26. D. Young and W. Zurn, "Tidal Triggering of Earthquakes in the Swabian Jura?" Journal of Geophysics, Vol. 45, No. 1, 1976, pp. 171-182.

27. A. Palumbo, "Lunar and Solar Tidal Components in the Occurrence of Earthquake in Italy," Geophysical Journal. Royal Astronomical Society, Vol. 84, No. 1, 1986, pp. 93-99. doi:10.1111/j.1365-246X.1986.tb04346.x

28. P. A. Rydelek, I. S. Sacks and R. Scarpa, "On Tidal Triggering of Earthquakes at Campi Flegrei, Italy," Geophysical Journal International, Vol. 109, No. 1, 1992, pp. 125-147.

29. X. M. Gao, et al., "Triggering of Earthquakes by the Tidal Stress Tensor," Acta Seismologica Sinica, Vol. 3, No. 3, 1981, pp. 264-275.

30. J. Zhao, Y. B. Han and Z. A. Li, "Variation in the Solar and Lunar Tide Raising Forces and Earthquakes in Taiwan," Journal of Natural Disasters, Vol. 10, No. 4, 2001, pp. 137-141.

31. L. D. Chen, "Cerebration of the Scientific Guide Lines, Methodology and Difficulties in Earthquake Prediction," Seismological Research, Vol. 15, No. 2, 1996, pp. 67-74.

32. P. R. Du and D. Y. Xu, "Astroseismologic Introduction," Seismological Press, Beijing, 1979.

33. J. Q. Luan, "Motion of Celestial Object and Long-Term Prediction of Weather and Earthquakes," Beijing Normal University Press, Beijing, 1988.

CHAPTER 20

FORECASTING THE LOCATIONS OF FUTURE LARGE EARTHQUAKES: AN ANALYSIS AND VERIFICATION

Robert Shcherbakov, Donald L. Turcotte, John B. Rundle, Kristy F. Tiampo and James R. Holliday

Keywords: Earthquake forecasting, ROC diagram, error diagram, forecast verification

INTRODUCTION

Forecasts of the locations of future major earthquakes play an important role in earthquake preparedness and determining earthquake insurance costs. Many such forecasts have been carried out, one example is the National Seismic Hazard Map for the United States (FRANKEL *ET AL.*, 1996). This is a probabilistic estimate of the ground-shaking risk. A major input into this assessment is the observed rate of occurrence of past earthquakes. KOSSOBOKOV *ET AL.* (2000) utilized the rates of occurrence of $m \geq 4.0$ earthquakes globally to forecast the expected rates of occurrence of larger earthquakes. Recently, the Regional Earthquake Likelihood Models (RELM) forecasts were published (FIELD, 2007). This represents the first-generation of forecasts based on a variety of approaches and methods. It also introduces a testing methodology to verify and compare these forecasts (SCHORLEMMER *ET AL.*, 2007).

In this paper we have presented a method to construct a binary forecast of where future large earthquakes are likely to occur globally. We divide the Earth's surface into $l^\circ \times l^\circ$ cells in order to construct the forecast and evaluate its success. The input for our forecast is a relative intensity (RI) map of $m \geq mc$ earthquakes that occur in the cells during a specified training period, ΔT_{tr}. As a measure of RI we have utilized the cumulative Benioff strain in each cell during that period. The cumulative Benioff strain has been computed using the data from the CMT (Harvard) catalog (http://www.globalcmt.org). This catalog is complete from 1/1/1976 to 1/1/2008 to magnitude $mc = 5.5$. The RI map is then converted to a binary forecast by introducing a threshold cumulative Benioff strain. Cells with Benioff strains greater than this threshold constitute *alarm cells* where future earthquakes are forecast to occur. A high threshold reduces the forecast areas but leads to more events that are not predicted. A low threshold reduces the failures to predict but increases the false alarms. We have introduced a standard optimization procedure for binary forecasts in order to select the optimal threshold.

In order to test our method we have performed a retrospective analysis. To define training and forecasting time periods we subdivide the catalog using a base year Tb, a forecasting year Tf, and an end year Te. We have utilized seismicity during the training period from Tb = 1/1/1976 to Tf = 1/1/2004 ($\Delta T_{tr} \equiv Tf - Tb$ = 28 years) to make a forecast for the locations of future target earthquakes of a specified magnitude, $mT \geq 7.2$. The locations of earthquakes during the forecasting period Tf = 1/1/2004 to Te = 1/1/2008 ($\Delta T_{for} \equiv Te - Tf$ = 4 years) are then used to test the validity of the method. The success of the binary forecast has been evaluated using techniques developed in the atmospheric sciences (MASON, 2003). The evaluation is in terms of a hit rate (fraction of events that are successfully forecast) versus a false alarm rate (the fraction of forecast cells in which earthquakes do not occur). Both hit rate and false alarm rate are functions of the tuning parameter (in our case the threshold Benioff strain). Their parametric plot produces a receiver operating characteristic (ROC) diagram. This approach has been used by HOLLIDAY *ET AL.* (2005) to test the success of forecasting algorithms in California. Similar analysis was also applied to the verification of the intermediate-term earthquake prediction algorithm M 8 by MINSTER AND WILLIAMS (1999).

Citation: Robert Shcherbakov, Donald L. Turcotte, John B. Rundle, Kristy F. Tiampo, James R. Holliday, Forecasting the Locations of Future Large Earthquakes: An Analysis and Verification, doi:10.1007/s00024-010-0069-1

We have also quantified our approach using the variation of the error diagram introduced by MOLCHAN (1991), MOLCHAN AND KEILIS-BOROK (2008). The comparison of the error diagram approach and the maximum likelihood method for evaluation of earthquake forecasting methods was given by KAGAN (2007). The connection between the entropy score and log-likelihood estimates with applications to several seismicity models is given in HARTE AND VERE-JONES (2005). The analysis of the error diagram approach applied to California seismicity was performed by ZECHAR AND JORDAN (2008). KOSSOBOKOV AND SHEBALIN (2003) and ZALIAPIN*ET AL*. (2003) have applied the error diagram to test a number of earthquake predictions algorithms.

Relative Intensity of Global Seismicity

The proposed analysis is based on the RI of past seismicity where large target earthquakes with magnitudes greater than $mT \geq 7.2$ are forecasted. The moment magnitude m of earthquakes has been calculated from the scalar moment according to the formula $m = \frac{2}{3}\log_{10}(M_0) - 10.7$, where M_0 is a scalar moment reported in the CMT catalog. The grid size plays an important role in this analysis. In this study the Earth's surface has been subdivided into 2° × 2° cells

As a measure of the RI the cumulative Benioff strain has been computed in each cell for a specified training time period ΔT_{tr}. The cumulative Benioff strain B up to time t is defined as

$$B_{xy}(t) = \sum_{i=1}^{N_{xy}(t)} \sqrt{E_{xy}^{(i)}}, \tag{1}$$

Where $E_{xy}^{(i)}$ is the seismic energy release in the ith earthquake and $N_{xy}(t)$ is the number of earthquakes considered up to time t, and (xy) is the coordinates of a given cell. The seismic energy E is related to the seismic moment M_0 as E ~ M_0.

The normalized RI in a cell is given by B_{xy}/B_{max} where B_{xy} is the cumulative Benioff strain in the active cell under consideration and B_{max} is the Benioff strain for the most active cell. The cumulative Benioff strain is also normalized by the actual area of the corresponding cell measured in square kilometers.

In order to construct a binary map a decision threshold of RI is applied. All cells with intensities greater than the specified threshold are alarm cells and constitute the binary forecast. To illustrate our approach we consider a specific example. The analysis is based on the seismicity during a training period from T_b = 1/1/1976 to T_f = 1/1/2004 (ΔT_{tr} = 28 years). The threshold for the normalized Benioff strain B/B_{max} = 0.0417 is used. The reason for selecting this particular threshold will be discussed in the next section. The alarm cells in which earthquakes are forecast to occur are illustrated in Fig. 1. During the entire time period (T_b = 1/1/1976 toT_e = 1/1/2008) there were 1,503 active cells (cells with at least one earthquake with magnitudesm ≥ 5.5) out of total of 16,200 cells. Note that the area of the active cells depends on the chosen cell size and their geographic location. Also shown in Fig. 1 are the 34 earthquakes with$m_T \geq 7.2$ that occurred during the forecast period T_f = 1/1/2004 to T_e = 1/1/2008 (ΔT_{for}= 4 years). The next step is to quantify the success of our binary forecast of the locations of these earthquakes.

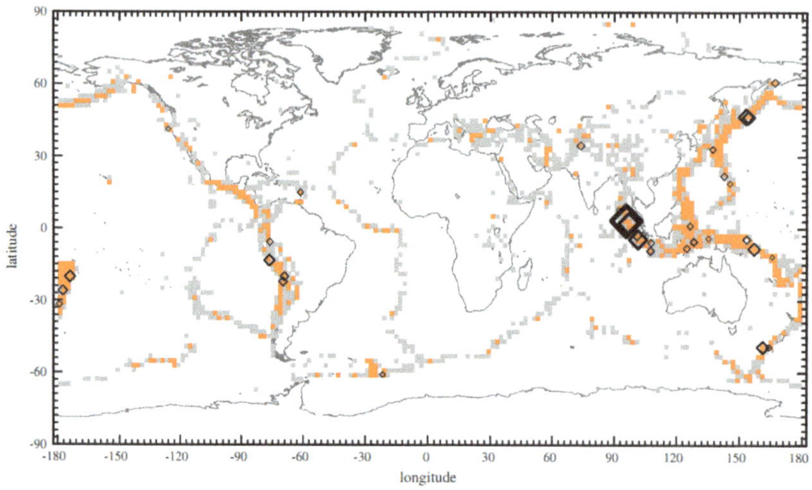

Figure 1: Spatial distribution of 2° × 2° alarm cells. Earthquakes with magnitudes greater thanm ≥ 5.5 are considered during the time period 1/1/1976 until 1/1/2004 (ΔT_{tr} = 28 years). Also shown as squares are the locations of earthquakes with $m_T \geq 7.2$ that occurred during the forecast period T_f = 1/1/2004 to T_e = 1/1/2008 (ΔT_{for} = 4 years). The threshold value corresponds to the maximum Pierce's skill score given in Fig. 3a

FORECAST VERIFICATION

The success of a forecast is based on maximizing the fraction of the earthquakes that occur in alarm cells and minimizing the fraction of the alarm cells in which no earthquake occurs. This analysis is carried out by utilizing a contingency table approach. Each cell in which an earthquake is forecast to occur constitutes an alarm cell. Earthquakes occur in some of these alarm cells but also in cells where alarms have not been issued. As a result of this test each cell falls in one of four categories: a, alarm and earthquake (successful forecast); b, alarm but not earthquake (false alarm); c, no alarm but an earthquake (failure to predict); d, no alarm and no earthquake (successful forecast of non-occurrence). In addition a is the number of target earthquakes that occurred in alarm cells, b is the number of alarm cells with no earthquakes in them, c is the number of target earthquakes that occurred outside alarm cells, and d is the number of cells with no target earthquakes and which are not alarm cells.

As a specific example we consider the forecast given in Fig. 1. We retrospectively test this forecast in terms of the locations of $m_T \geq 7.2$ target earthquakes that occurred during the forecast period T_f = 1/1/2004 to T_e = 1/1/2008. The total number of 2° × 2° cells in which am ≥ 5.5 occurred during the entire time period (active cells) was 1,503. During the forecast period a + c = 34 earthquakes with $m_T \geq 7.2$ occurred. Of these a = 23 occurred in alarm cells (were forecast) and c = 11 occurred in the other cells (were not forecast).

The success of a forecast can be quantified using a hit rate, H, and a false alarm rate, F:

$$H = \frac{a}{a+c}, \qquad F = \frac{b}{b+d}. \tag{2}$$

The hit rate H is the fraction of earthquakes that are successfully forecast. For this example we have H = 23/34 = 0.676. The false alarm rate F is the fraction of cells with no earthquake for which an alarm was issued. The hit rate H and the false alarm rate F define conditional probabilities H = P(A|E) and F = P(A|not E) respectively, where A denotes an event of having an alarm in given cells and E denotes an event of having a target earthquake in a given cell. For this example we have F = 371/1469 = 0.253. The result of a forecast is given in a contingency table (Table 1).

Table 1: Contingency table for earthquakes with $m_T \geq 7.2$ that occurred during the forecast periodT_f = 1/1/2004 to T_e = 1/1/2008 (ΔT_{for} = 4 years)

Forecast	**Observed**		
	Yes	**No**	**Total**
Yes	a = 23	b = 371	a + b = 394
No	c = 11	d = 1098	c + d = 1109
Total	a + c = 34	b + d = 1469	a + b + c + d = 1503

The above example has given the values of H and F for the specific value of the tuning parameter B/B_{max} = 0.0417. The plot of H and F for a range of B/B_{max} values is known as a ROC diagram. An ROC diagram for our forecast is shown in Fig. 2a. Results are given for large earthquakes with magnitudes $m_T \geq 7.2$. The diagonal line from (0.0) to (1.1) would be the long-run expectation for alarms that are declared randomly. A perfect forecast would have F = 0 (no false alarms) and H = 1 (all earthquakes are forecast).

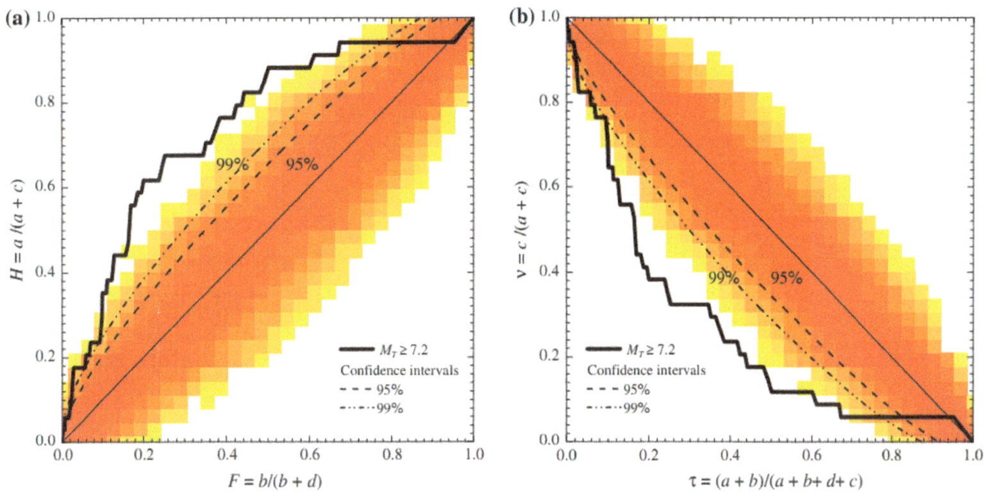

Figure 2: a The ROC diagram for our forecast of global seismicity. **b** The error diagram for our forecast. Also included the results of 150,300 Monte-Carlo simulations (shaded area) in which alarm cells have been redistributed randomly and the corresponding confidence intervals

The choice of the optimal forecast (preferred value of B/B_{max}) is the subject of extensive discussions in the atmospheric science literature (Mason, 2003). Several quantitative measures find wide use. One of these is Pierce's skill score:

$$\mathrm{PSS_{ROC}} = H - F = \frac{ad - bc}{(a+c)(b+d)}, \tag{3}$$

which is the difference between the observed hit rate H and the value on the diagonal line corresponding to the random forecast. This score is also known as Hansen-Kuiper's score orKuiper's performance index (Mason, 2003; Harte and Vere-Jones, 2005). The history and applications of this score along with its properties can be found in Mason (2003). A perfect forecast would be PSS_{ROC} = 1.0. The values of PSS_{ROC} corresponding to the results in Fig. 2a are given in Fig. 3a. They are plotted as a function of the tuning parameter values B/B_{max} (Fig. 2a).

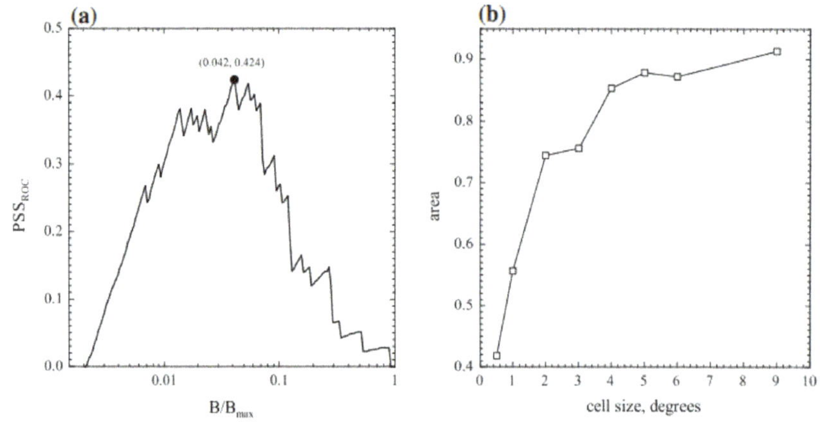

Figure 3: a Pierce's skill scores PSS_{ROC} from the ROC diagram given in Fig. 2a as a function of the threshold values of the normalized cumulative Benioff strain B/B_{max}. The maximum value of the Pierce's skill score is $PSS_{ROC}^{max} = 0.424$ corresponding to the threshold value $B/B_{max} = 0.042$. **b** Dependence of the area under ROC diagram on different cell sizes

The maximum value of $PSS_{ROC}^{max} = PSS_{ROC}$ from an ROC diagram can be used to find an optimizing value of the tuning parameter. For large earthquakes $m_T \geq 7.2$ the maximum value of Pierce's skill score is $PSS_{ROC}^{max} = 0.424$ corresponding to the threshold value $B/B_{max} = 0.0417$. This corresponds to H = 0.676 and F = 0.253 and was the basis for selecting this threshold value for the example given in Fig. 1. The value of the hit rate H = 0.676 (67.6%) gives the probability of having a large earthquake in the alarm cells during the forecast period. The threshold values obtained from retrospective forecasts can be applied a priori in subsequent prospective forecasts. This can be done by analyzing a particular region of study where the magnitude of completeness of a catalog is obtained and the optimal threshold and cell size values were determined from the retrospective analysis.

Another method of measuring the success of a forecast is the error or Molchan's diagramapproach (Molchan, 1991; Molchan and Kagan, 1992; Molchan and Keilis-Borok, 2008). In order to construct the error diagram the miss and alarm rates

$$\nu = \frac{c}{a+c} = 1 - H, \qquad \tau = \frac{a+b}{a+b+c+d} \tag{4}$$

are introduced. The sum of the hit rate and miss rate is unity, i.e. H + N = 1. The alarm rate τ is the fraction of all cells that are alarm cells and defines the probability of a forecast of occurrence in weather forecast verification (Mason, 2003). The parametric plot of ν and τ for a range of threshold parameters B/B_{max} is known as an error diagram. The error diagram for the above example is given in Fig. 2b. An analog of the Pierce's skill score for the error diagram is

$$SS_{ED} = 1 - \nu - \tau = \frac{ad - bc}{(a+c)(a+b+c+d)}. \tag{5}$$

From the analysis of the error diagram curve the maximum value of this skill score is $SS_{ED}^{max} = 1 - 0.324 - 0.262 = 0.414$ corresponding to the threshold value $B/B_{max} = 0.0417$. It is worthwhile to note that the threshold values corresponding to the maximum values of the skill scores for both ROC and error diagrams are the same. This is due to the fact that the values of F and τ are very close to each other for corresponding threshold values as they are both dominated by large b + d values compared to small a + c. As a result the error diagram curve is almost a mirror image of the ROC curve due to the fact that ν = 1 − H. In general this is not always the case.

In the above analysis we have used a uniform prior distribution for the seismically active cells and assumed a null hypothesis where alarm cells are redistributed randomly. To reject this null hypothesis in favor of the method presented here we have also included in Fig. 2a and b the results of 150,300 Monte-Carlo simulations in which alarm cells have been redistributed randomly and the same target earthquakes were used to construct ROC and error diagrams. The confidence intervals were evaluated using the exact analytical result for the binomial distribution given in Zechar and Jordan (2008).

DISCUSSION

The primary purpose of this paper is to quantify the extent to which locations of past smaller earthquakes can be used to forecast the locations of future large earthquakes. Our study has been carried out globally. The surface of the Earth has been divided into a grid of $l° \times l°$ cells with $l = 2°$. The objective is to forecast the locations of future large target earthquakes with magnitudes greater than mT based on the past seismicity. Another important aspect of this work is to introduce different optimization procedures to find the best forecast and optimal values for model parameters. It is also important to note that in the presented analysis the measure of seismicity based on cumulative Benioff strain does not perform better than the predictor based on a long-run rate of occurrence of earthquakes.

The sensitivity of the method to the model parameters is an important issue. The cell size $l = 2°$ has been chosen based on the analysis of the areas under the ROC curves constructed for different cell sizes and different forecasting magnitudes. The analysis shows that the areas tend to increase (better forecast) with increasing cell sizes (Fig. 3b). Bigger cell sizes cover a larger Earth's surface and increase the total success rate. However, bigger cell sizes are less practical in forecasting as they include larger geographical regions and decrease the value of the forecast. In this analysis we assume all earthquakes are point events. Cell size can be an important factor if forecast earthquakes are defined through their rupture areas which can span several cell sizes.

Other important parameters are the durations of the training and forecasting intervals and the lower magnitude cutoff. The analysis of the areas under the ROC curves shows that increasing the training time interval (ΔT_{tr}) usually improves the forecast. As to the forecasting time interval (ΔT_{for}) changes in the duration of the interval result in different numbers for forecasting earthquakes which, as was shown by ZECHAR AND JORDAN (2008), affects the ROC and error diagrams. To investigate the performance of the method for different forecasting time intervals a contour plot of the values of the areas between the ROC curve and 95% confidence interval for different forecasting years, T_f, and base years, T_b, is given in Fig. 4. Darker areas indicate a better performance with respect to the 95% confidence interval of the random forecast. It shows a prominent increase in the values for intervals less than 4 years and another increase for 15-year interval. This was a reason to use a 4-year forecasting interval in our analysis.

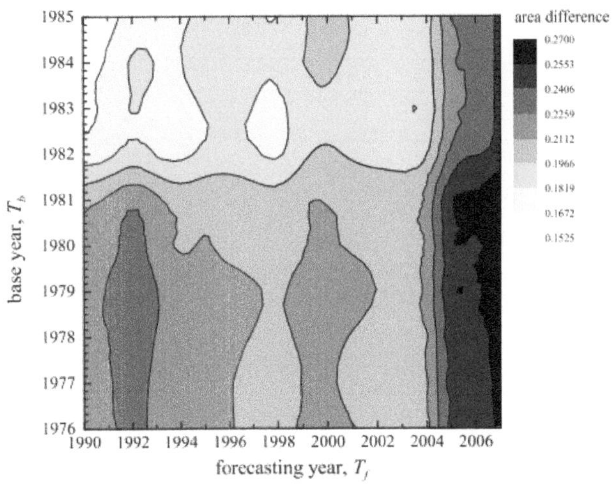

Figure 4: A contour plot of the values of the areas between the ROC curve and the 95% confidence interval for the random forecast for different forecasting (T_f) and base (T_b) years

Finally, the dependence of the method on the lower magnitude cutoff, m_c, shows a strong trend in the performance increase of the method with decreasing m_c. However, the completeness of the catalog prevents the use of events smaller than $m < 5.5$. The full analysis of the dependence of the method on the above parameters will be presented elsewhere. By varying the thresholdB/B_{max} the method generates a spectrum of forecasts. The use of the Pierce's skill score is probably the most accepted optimization tool, but several others have been proposed (Mason,2003; Harte and Vere-Jones, 2005). An actual optimization should also be based on other factors such as available resources, population densities, quality of construction, and other risk factors.

ACKNOWLEDGMENTS

This work has been supported by NSERC Discovery grant (RS), by NSF grant ATM-0327558 (DLT), by NASA grant NGT5 (JRH), by a US Department of Energy grant to UC Davis DEFG03-14499 (JRH, JBR), and by the NSERC and Benfield/ICLR Chair in Earthquake Hazard Assessment (KFT).

REFERENCES

1. Field, E.H. (2007), Overview of the working group for the development of Regional Earthquake Likelihood Models (RELM), Seismol. Res. Lett. 78, 7–16.
2. Frankel, A., Mueller, C., Barnhard, T., Perkins, D., Leyendecker, E., Dickman, N., Hanson, S., and Hopper, M. (1996), National seismic-hazard maps, Tech. Rep. Open-File Report 96–532, USGS.
3. Harte, D. and Vere-Jones, D. (2005), The entropy score and its uses in earthquake forecasting, Pure Appl. Geophys. 162, 1229–1253.
4. Holliday, J.R., Nanjo, K.Z., Tiampo, K.F., Rundle, J.B., and Turcotte, D.L. (2005), Earthquake forecasting and its verification, Nonlinear Proc. Geophys. 12, 965–977.
5. Kagan, Y.Y. (2007), On earthquake predictability measurement: Information score and error diagram, Pure Appl. Geophys. 164, 1947–1962.
6. Kossobokov, V.G., Keilis-Borok, V.I., Turcotte, D.L., and Malamud, B.D. (2000), Implications of a statistical physics approach for earthquake hazard assessment and forecasting, Pure Appl. Geophys. 157, 2323–2349.
7. Kossobokov, V.G. and Shebalin, P., Earthquake prediction. In Nonlinear Dynamics of the Lithosphere and Earthquake Prediction (eds. Keilis-Borok, V.I., and Soloviev, A.A.) (Springer-Verlag, Berlin 2003), pp. 141–207.
8. Mason, I.B., Binary events. In Forecast Verification (eds. Jolliffe, I.T. and Stephenson, D.B.) (Wiley, Chichester 2003), pp. 37–76.
9. Minster, J.-B. and Williams, N., Systematic global testing of intermediate-term earthquake prediction algorithm. In 1-st ACES Workshop Proceedings (ed. Mora, P.) (GOPRINT, Brisbane 1999).
10. Molchan, G. and Keilis-Borok, V. (2008), Earthquake prediction: Probabilistic aspect, Geophys. J. Int. 173, 1012–1017.
11. Molchan, G.M. (1991), Structure of optimal strategies in earthquake prediction, Tectonophysics 193, 267–276.
12. Molchan, G.M. and Kagan, Y.Y. (1992), Earthquake prediction and its optimization, J. Geophys. Res. 97, 4823–4838.
13. Schorlemmer, D., Gerstenberger, M.C., Wiemer, S., Jackson, D.D., and Rhoades, D.A. (2007), Earthquake likelihood model testing, Seismol. Res. Lett. 78, 17–29.
14. Zaliapin, I., Keilis-Borok, V., and Ghil, M. (2003), A Boolean delay equation model of colliding cascades. Part II: Prediction of critical transitions, J. Stat. Phys. 111, 839–861.
15. Zechar, J.D. and Jordan, T.H. (2008), Testing alarm-based earthquake predictions, Geophys. J. Int. 172, 715–724.

INDEX